Studies on Entrepreneurship, Structural Change and Industrial Dynamics

The ‘Studies on Entrepreneurship, Structural Change and Industrial Dynamics’ series showcases exceptional scholarly work being developed on the still unexplored complex relationship between entrepreneurship, structural change and industrial dynamics, by addressing structural and technological determinants of the evolutionary pathway of innovative and entrepreneurial activity.

The series invites proposals based on sound research methodologies and approaches to the above topics. Volumes in the series may include research monographs and edited/contributed works.

This is a SCOPUS-indexed series.

Conceição Rego • Maria Raquel Lucas
María Isabel Sánchez-Hernández
Luísa Cagica Carvalho • Adriana Backx Noronha
Editors

Entrepreneurship, Technological Change and Circular Economy for a Green Transition

Research Contributions for a More Productive Environment

Springer

Editors
Conceição Rego
Department of Economics
University of Évora
Évora, Portugal

Maria Raquel Lucas
Department of Management
University of Évora
Évora, Portugal

María Isabel Sánchez-Hernández
Faculty of Economics and Business
University of Extremadura
Badajoz, Spain

Luísa Cagica Carvalho
Instituto Politécnico de Setúbal
Setúbal, Portugal

Adriana Backx Noronha
School of Economics, Business and Accounting
University of São Paulo
São Paulo, Brazil

ISSN 2511-2023 ISSN 2511-2031 (electronic)
Studies on Entrepreneurship, Structural Change and Industrial Dynamics
ISBN 978-3-031-48078-2 ISBN 978-3-031-48079-9 (eBook)
https://doi.org/10.1007/978-3-031-48079-9

This Springer imprint is published by the registered company Springer Nature Switzerland AG
The registered company address is: Gewerbestrasse 11, 6330 Cham, Switzerland

Paper in this product is recyclable.

Acknowledgments

The editors are grateful for the support of FCT—Fundação para a Ciência e a Tecnologia, I.P., within the scope of the project UIDB 04007/2020.

Contents

Introduction

Maria Raquel Lucas, María Isabel Sánchez-Hernández, Conceição Rego, Luísa Cagica Carvalho, and Adriana Backx Noronha

Abstract The challenges that societies face today show the need to evolve, among other challenges, to the green transition. The evolution of the linear economy, in which societies are still, mostly, towards a circular economy, where the 5 R' predominate—Rethink, Refuse, Reduce, Reuse and Recycle—it is fundamental. Alongside this evolution, the decarbonization of the predominant production model as well as the way of life that dominates the most developed societies is also essential. It is well known that this transition will not be immediate and will not be cost-free. Therefore, it is important to understand and know more about how to proceed in this transition paying attention to technological change and knowing the advantages and risks of this process. Only in this way is it possible to change in a sustainable way and seek to minimize the losses associated with this transition. For example, knowing the characteristics of businesses models based on circular economy and aimed at the green transition, as well as knowing the necessary technological changes in this domain or the training needs associated with technological change and the green transition, are essential.

M. R. Lucas (✉)
Department of Management and CEFAGE – Center for Advanced Studies in Management and Economics, Universidade de Évora, Évora, Portugal
e-mail: mrlucas@uevora.pt

M. I. Sánchez-Hernández
Department of Business Management and Sociology – School of Economics and Business Sciences, Universidad de Extremadura, Badajoz, Spain
e-mail: isanchez@unex.es

C. Rego
Department of Economics and CEFAGE – Center for Advanced Studies in Management and Economics, Universidade de Évora, Évora, Portugal
e-mail: mcpr@uevora.pt

C. Rego et al. (eds.), *Entrepreneurship, Technological Change and Circular Economy for a Green Transition*, Studies on Entrepreneurship, Structural Change and Industrial Dynamics, https://doi.org/10.1007/978-3-031-48079-9_1

Keywords Entrepreneurship · Technological change · Circular economy · Green transition · Productive environment

The relationship between entrepreneurship, technological change, and the advancement of the circular economy (CE) is instrumental in driving the green transition. Entrepreneurs, through their innovative pursuits, bring forth green technologies and circular business models that optimize resource use and minimize waste. Technological advancements, in turn, empower these entrepreneurs to implement sustainable practices effectively. Together, they create a transformative force that propels society towards a more sustainable and environmentally friendly future.

First, this book deals with the concept of entrepreneurship referring the process of identifying, creating, and pursuing opportunities to develop new products, services, or business models. Entrepreneurs are innovative risk-takers who seek to solve problems and meet market demands. In the context of a green transition, entrepreneurs play a crucial role in driving sustainable solutions and green business practices. They are at the forefront of developing and implementing environmentally friendly technologies and CE strategies. Second, the book considers the current technological changes as the advancements and innovations in technology that can significantly impact the way of living, work, and conduct businesses. In the context of a green transition, technological change encompasses the development and adoption of clean technologies, renewable energy sources, resource-efficient processes, new solution for by-products and eco-friendly products. These technological advancements are vital for reducing environmental impacts and achieving a more sustainable economy. Finally, in this book the CE is presented as a relatively new economic model designed to minimize waste and promote the continual use and regeneration of resources. Unlike the traditional linear economy (take-make-dispose), the CE focuses on closing the loop, where products and materials are reused, repaired, remanufactured, and recycled, thus extending their life cycles. This approach aims to create a more sustainable and restorative system that maximizes resource efficiency and reduces environmental harm.

In fact, CE has emerged as a new paradigm to address the green transition in different sectors. This book aims to address the main characteristics and driving forces of this model of production and consumption and the paths of entrepreneurship and technological change in attaining a better productive environment and a sustainable development. Particularly, this book explores the meaning and impact of the CE on

L. C. Carvalho
Instituto Politécnico de Setúbal, Setúbal, Portugal
e-mail: luisa.c.carvalho@esce.ips.pt

A. B. Noronha
School of Economics, Business and Accounting, University of São Paulo, São Paulo, Brazil
e-mail: backx@usp.br

entrepreneurship and business models, innovation and value creation, business dynamic capabilities, green and digital economy, lean thinking management and sustainable development goals. It also embraces the CE paradigm focusing on agriculture and agribusiness entrepreneurial ecosystems, sustainability challenges, and structural change for a society and a better productive environment.

This comprehensive and timely publication, which complements other previous publications of the editors (Carvalho et al., 2018, 2019; Sánchez-Hernández et al., 2021a, 2021b), aims to be an essential reference source, building on the available literature in the field of CE in the academic literature devoted to the entrepreneurship, technological change, and business dynamics. This book addresses a range of approaches including conceptual, theoretical, and practical case studies and projects related to these developments. The book contains 14 chapters divided into two parts. The first part regards the advanced debate on the meaning and impact of the CE on enterprises, entrepreneurship, and business models and it includes five chapters. The second part offers insights, barriers, challenges, and new trends on the agriculture and agribusiness sector and includes eight chapters. The next paragraphs detail the book structure, meaning its sections and respective chapters.

Part I Enterprises and Entrepreneurship

The relationship between entrepreneurship, technological change, and the CE is mutually reinforcing, and this first part of the book shows how they interact to drive the green transition. Entrepreneurs are instrumental in promoting circular business models. They identify ways to incorporate principles of the CE into their products and services. This may involve adopting strategies like product-as-a-service, sharing platforms, or designing products with recyclability and reusability in mind. Technological advancements further enable these models by making processes more efficient and enhancing product life cycles. In addition, entrepreneurs and technological advancements can influence policymakers to create a conducive environment for CE practices. Through advocacy and collaboration with governments and industry stakeholders, entrepreneurs can shape policies that support sustainable initiatives and foster a green transition.

The second chapter, "Entrepreneurship, Circular Economy and New Business Models", written by Mansidão and Alves, aims to contribute to advancing the debate on the meaning and impact of the CE on entrepreneurship and business models, presenting the results of a bibliographical research and a theoretical and conceptual framework for this area of investigation.

The third chapter, "Toward the Green Transition of Business Through the Circular Economy: Understanding the Role of Dynamic Capabilities", developed by Baena-Luna, García-Río, and Felipe-Llanos, also conducts a bibliometric analysis of the scientific literature that connects the relationship and impact of the dynamic capabilities on adopting CE business models to foster green transition in companies. The chapter identifies the principal research authors and topics in this area, analyses main conceptual, intellectual, and social structures, and detects future research opportunities.

The fourth chapter, "The Path to a Circular Economy Through a Green and Digital Transition: The Paradigm Shift in Mexican Companies in the State of Hidalgo", made by Gordillo-Benavente and Vega-Hernández, addresses the green and digital CE, analysing how Mexican organizations in the State of Hidalgo perform it considering sustainability features that they must comply. The chapter shows also how imperative is for companies to have a transition to a green and digital CE and the importance of this paradigm shift in organizations.

The fifth chapter, "How Does the Circular Economy Address the Sustainable Development Goals? Evidence from the Brazilian Legal Amazon", proposed by Rosa, Cohen, Campos, Rodrigues, and Avila, verifies whether the circular business model helps in the implementation of SDGs in the Amazon and which objectives foster the CE. The chapter recommends further quantitative research for the formulation of more detailed conceptual models in the future.

The sixth chapter, "Lean and Green: 3Ms Meets 5Rs Creating Environmental value for the Planet", composed by Torres Junior and Nagai, proposes operationalizing the CE in companies through the lean thinking management model. A theoretical discussion is exemplified through a clothing company that applies the concept of lean thinking in its product development and operations with promising results for a green transition.

Part II Agriculture and Agribusiness

The second part of this book is devoted to the specific sector of agriculture and agribusiness showing how *innovation and green technologies* are also transforming the area. Agribusiness must identify environmental challenges and market opportunities, leading to the creation of sustainable solutions. Technological change, driven by agribusiness ventures, are introducing cleaner production processes, energy-efficient technologies, and renewable energy sources, which are crucial components of a CE. In fact, the CE relies on resource efficiency and waste reduction. Agribusiness are playing a vital role in optimizing resource use and finding innovative ways to minimize environmental impacts. Technological change provides the tools and methods to achieve these goals effectively. In addition, entrepreneurship and technological change can drive market transformation by creating demand for green products and services. As consumers become more environmentally conscious, agribusiness must respond with sustainable solutions that align with circular principles. This leads to a shift in consumer behaviour and encourages more agribusinesses to adopt sustainable practices, further promoting the green transition.

The seventh chapter, "An Overview of Circular Business Models in Agribusiness", proposed by Lamolinara, Teixeira, Marreiros, Ferreira, and Pérez-Martínez, conducts a systematic literature review which provide a valuable contribution to the understanding of circular business models in agribusinesses and highlights important aspects useful for entrepreneurs and policymakers.

The eighth chapter, "The Circular Economy as a Creator of Value in Agribusiness 4.0: Proposals for a Life", written by Pimenta-Alonso and González-Rodrigo, deals with four specific aspects of the agricultural industry sector—use of biomass residues and their contribution to the food sector, the generation of biofuel and electricity from biogas for agricultural use, the reuse of treated waste water for irrigation, and the use of regenerated cotton by the textile industry—to improve lives and the sustainability of the planet.

The ninth chapter, "The Social Dimension of Agribusiness and How It Has Been Measured", suggested by Massuça, Marta-Costa, and Lucas, aims to classify agricultural social indicators according to their thematic scope, grouping them in distinct clusters, and subsequently sorted the most frequently referenced indicators in each of the defined clusters. The study is based on a previously extensive literature review and on a comprehensive bibliometric study about agricultural social indicators, between 1999 and 2022.

The tenth chapter, "Urban Green Infrastructure: Circular Socio-economy with Nature-based Solutions", developed by Briz, De Felipe, and Briz, complemented a national online research on social perception of green urban infrastructure with case studies on energy, urban agriculture in Spain, and some projects in Madrid, particularly the project gas recycling harm waste into urban benefits. The chapter highlights the role of CE throughout gas recycling as a useful instrument in activities for social welfare.

The eleventh chapter, "Alentejo Wines and Olive Oil Sustainability Programs: Background of Absorptive Capacity in the Development of Circular Economy Business Models", designed by Prazeres, Silva, Henriques, Ramos, Rego, and Lucas, presented the two cases of wines and olive oil from Alentejo region in Portugal that implemented a sustainability program to investigate the antecedents of absorptive capacity in CE business.

The twelfth chapter, "Circular Economy Business Model—Let's Wine About It: An Exploratory Research Applied to Portuguese Wineries", proposed by Moreira, Carvalho and Travassos, discusses the CE practices adoption by wine sector companies, the managers engaged in a CE strategy, and the companies' main difficulties when switching CE business models. The chapter also reveals that higher education institutions can make an essential contribution to the transition from a linear to a CE.

The thirteenth chapter, "Towards Circular Economy: Barriers, Drivers, Challenges and Potentialities in Angola's Agribusiness", written by Kamutali, Henriques, Lucas, and Marta-Costa analyses the barriers, potentialities, and challenges to agribusiness development in the provinces of Cuanza Sul, Huambo, Bié and Luanda, in Angola, from a perspective of CE and sustainability.

The fourteenth chapter, "Reputational Leadership, Family Business and Sustainability in the Creation of Economic and Social Value: The Case of the Company Agroparque Esperanza", created by Tato-Jiménez, Cabrera-Gala and Carrillo-Durán proposes a research study based on a case methodology examining the Agroparque Esperanza company, an example of entrepreneurship in the CE, within the Mexican scope. The chapter argues that this Agropark is an innovative case of value creation and a clear example of sustainable entrepreneurship, both because it is a family business with a particular leadership style, and because it is an agricultural company that focuses its activity on producing and processing apples under a sustainable context.

This book allows an enlarged perspective, theoretical and empirical, on an issue that is still quite undeveloped and unknown, providing a set of perspectives related to conceptualization, framework, adoption, operationalization, and new trends on the CE.

References

Carvalho, L., Rego, C., Lucas, M. R., Sánchez-Hernández, M. I., & Backx Noronha, A. (Eds.). (2018). Entrepreneurship and structural change in dynamic territories: Contributions from developed and developing countries. In *Structural change and industrial dynamics*. Springer.

Carvalho, L., Rego, C., Lucas M. R., Sánchez-Hernández, M. I., & Backx Noronha, A. (Eds.). (2019). New paths of entrepreneurship development. The role of education, smart cities, and social factors. In *Structural change and industrial dynamics*. Springer.

Sánchez-Hernández, M. I., Carvalho, L., Rego, C., Lucas, M. R., & Backx Noronha, A. (Eds.). (2021a). Entrepreneurship in the fourth sector: Entrepreneurial ecosystems and sustainable business models. In *Studies on entrepreneurship. Structural change and industrial dynamics*. Springer.

Sánchez-Hernández, M. I., Carvalho, L., Rego, C., Lucas M. R., & Backx Noronha, A. (Eds.). (2021b). Social innovation and entrepreneurship in the fourth sector. Sustainable best-practices from across the world. In *Studies on entrepreneurship. Structural change and industrial dynamics*. Springer.

Part I
Enterprises and Entrepreneurship

Entrepreneurship, Circular Economy and New Business Models

Rui Manuel Mansidão and André Bernardo Alves

Abstract In the new paths of sustainability and green transition, the circular economy is a central theme on the agendas of academics, policy-makers, and business leaders. Above all, a way to overcome the negative impacts of a linear economy on the environment and society, in general, is being sought. In this movement of initiation and technological acceleration, due to its ability to add value through the vision, creativity and innovation of entrepreneurs, entrepreneurship stands out as a key concept for changes in society and for sustainable economic growth. The literature emphasizes that circular entrepreneurs can be agents of change in organizational structures and drivers of new business models and processes.

Based on a structured review of the literature on databases of reference, this work aims to contribute to advancing the debate on the meaning and impact of the circular economy on entrepreneurship and business models, presenting the results of the bibliographical research and a theoretical, conceptual framework for this area of investigation.

Keywords Entrepreneurship · Circular economy · Business models · Literature review · Circularity · Green transition

1 Introduction

Since the Industrial Revolution, the economy has been structured in a business model anchored on a linear flow of extraction, processing and transformation of raw materials into goods, consumption and disposal. At the apex of economic development at the end of the last century, this growth model, assuming that resources are

R. M. Mansidão (✉) · A. B. Alves
Campus IPS, Setúbal, Portugal
e-mail: rui.mansidao@estsetubal.ips.pt; andre.lemos.alves@estsetubal.ips.pt

C. Rego et al. (eds.), *Entrepreneurship, Technological Change and Circular Economy for a Green Transition*, Studies on Entrepreneurship, Structural Change and Industrial Dynamics, https://doi.org/10.1007/978-3-031-48079-9_2

abundant, available and without concern for recovering the waste generated or the components of technologically obsolete products, has created an unprecedented depletion of the planet's resources and a problem regarding the environment's capacity to absorb pollution (Bianchini et al., 2019; Geissdoerfer et al., 2018; Veleva & Bodkin, 2018; Ellen MacArthur Foundation, 2019).

In view of this finding, alternatives have emerged for the introduction of socio-environmental criteria in organizations' business models. The concept of the circular economy emerged in opposition to the current linear economic system, based on mostly non-renewable resources, and provides the guidelines for implementing production based on the regeneration and restoration of materials (Sariatli, 2017), and on industrial symbiosis, where a residue from one operation can provide raw material for the manufacturing process (Geissdoerfer et al., 2018).

Naturally, as an emerging concept, the circular economy has captured the interest of researchers in the most diverse areas of knowledge.

The regenerative and restorative model aims at the systemic knowledge of flows, reducing the amount of resources used and limiting the environmental impacts (Centobelli et al., 2020). Thus, it represents several challenges but simultaneously provides an opportunity for companies to become competitive and for regional development (Ormazabal et al., 2018).

The circular economy emphasizes dematerialization and the length of the product's life-cycle, transcends the silo mode of governance (Hofstetter et al., 2021) and encourages the development of cooperative dynamics (Ellen MacArthur Foundation, 2019). This new economic model promotes integration and cooperation between the actors in the value chain process, using the principles of economic entrepreneurship, based on knowledge capital and knowledge assets instead of mostly non-renewable resources developed in a convergence towards technological obsolescence and unsustainable environmental trajectories.

In the current economic model, the role of the consumer and society's decisions are key factors in maintaining the dynamism of the economy and the prevailing market. Business models and aggressive marketing policies carried out by companies have been decisive in influencing customers in the market and leading to unsustainable consumption rhythms for the planet's natural balance.

However, at the same time, the environmental issue's prominence on the discussion agenda of relevant worldwide institutions, such as the United Nations and the European Union, has aroused interest, concern and reaction among consumers who value a more restrained lifestyle and consumption in harmony with the environment (Hofstetter et al., 2021; Sariatli, 2017).

The circular economic model represents a proposal to reverse the current situation, connecting the tips of the linear system. That is, closing the circle, involving manufacturers and consumers, to make the most of resources and the products made from them, minimizing the waste generated to ideally non-existent levels (Lüdeke-Freund et al., 2019).

Companies are a key element in the transition to the circular economy through new strategies and business models to make the process operationally viable. Since the circular model is based on retaining value from resources and the products derived from them, it is necessary to define business models strategically focused on the most varied forms of retaining that value.

In work published in scientific journals, researchers pay special attention to the circular economy in companies having an impact on business models, since the circular economy entails profound changes involving new value propositions and the development of competitive advantages (Centobelli et al., 2020; Urbinati et al., 2017).

As documented in the literature, there are various definitions of the concept and what a business model represents. In general, a business model can be defined as a conceptual tool, which an organization has at its disposal to achieve its objectives and add value.

This value consists of three elements, the value proposition representing the idea of the products and services to be offered, the remuneration of value or economic model and value sharing, which identifies the conventions and convictions of a given sector, i.e., the stakeholders and how value is distributed among them (Geissdoerfer et al., 2018).

The new business models, which emerge in the search for solutions to current problems, are based on the principles of economic circularization (Lahti et al., 2018), i.e., all the dimensions of business activity and supply chain management. This new sustainability paradigm encourages new management practices and reveals new business opportunities, adding value to the organization and customers, in harmony with the environment (Centobelli et al., 2020), in addition to the competitive advantages in the context of a highly dynamic global market.

The circular economy requires the alignment of business activity in accordance with the principles of Reduction, Reuse and Recycling. This must guide its production model, where the preference lies in reuse and reproduction due to aspects of an economic nature, given that a substantial part of the added value in the production of a product remains in its components.

Authors such as Linder and Williander (2017) consider a business model, within the scope of the circular economy, as a representation of the organizational strategic decision of the form of business, centred on efficiency and circularity. Here, the concept of value aggregation is based on the economic utility added to products after applying materials in their production.

According to these authors, the strategic focus of the business model changes significantly, moving from the gains obtained from the commercial transaction to the return from the movement of resources and efficiency of the industrial assets throughout their lifecycle, i.e., reuse, recycling, sharing and repair (Ellen MacArthur Foundation, 2019).

One of the basic principles of the circular economy involves maintaining and preserving resources, value and time while minimizing impacts. The business model must first focus on the product and process design, considering the best form of maintenance, quality, durability and modularization to allow for repairs or technology upgrades.

The expected model is one in which the good can be used for a longer period, be reused, and in the end, be processed again. It is recommended that resources be kept in their cycle, the so-called "closed loop" or "closed loop system" (Dragomir & Duţescu, 2022).

Corroborating this view, Van Renswoude et al. (2015) point out that the strategic method of supporting business in creating value should facilitate activities to maintain products and processes, increasing their life cycle, absorbing derived by-products, reusing resources and reintegrating components, and in a final phase, offer the customer or user the customization and marketing of product functions.

The circular economy, and consequently, the business model is centred on using the good for as long as possible, due to its greater installed capacity. When this does not happen, there is inefficiency (Hofstetter et al., 2021; Sariatli, 2017). Thus, circular economy business models must relate the closed loop economy with re-design thinking, including low energy consumption, low pollutant emissions and high efficiency, to restore the flow of materials in the industrial processes.

The innovation of products and services, productive systems and business models, based on a new paradigm of sustainability to generate economic growth in balance with the environment, is one of the challenges of today's society.

This new Era of Sustainability stimulates, at the organizational level, new management practices and reveals new opportunities that generate value for organizations and business projects. However, the role of entrepreneurs and new ventures is still little explored in the literature (Heshmati, 2017), where countless studies are centred on business organizations which have the technological, financial and human resources to adapt to the circular model.

Entrepreneurs' projects are of great importance. They solve complex problems involving uncertainty and risk because they are able to "generate value from uncertainty" (York & Venkataraman, 2010). The move to a circular economy will require massive innovation and mindset changes that cannot be anticipated, thus presenting a huge opportunity for new, innovative ventures, both large and small, that will affect regional development (York & Venkataraman, 2010; Murray et al., 2017; Veleva & Bodkin, 2018). These factors suggest the need to research situations, dynamics and processes of new innovative ventures and business models on the path to the circular economy. This is the main motivation behind this work.

Based on a structured review of the literature, using databases of reference, and following the suggestion by authors such as Seuring and Müller (2008), Massaro et al. (2015) and Centobelli et al. (2017), the work intends to contribute with theory to advancing knowledge in this field of investigation and to stimulate the debate about the impact of the circular economy on business models and new innovative ventures, presenting the results of bibliographical research and a conceptual framework for this research framework.

This chapter is structured in six parts. After this introduction, the second part presents the arguments in the literature concerning research on entrepreneurship, sustainable innovation and business models in the context of changing to the circular economy. The research design and the method used are explained in the third part. The fourth part presents the results obtained from the literature review. The fifth part contains a critical analysis of the data referring to the search words. Finally, in the sixth part, the conclusions are presented.

2 Theorical Framework

2.1 Entrepreneurship, Innovation and Circular Economy

Entrepreneurship is a polysemic concept with no established definition, which generates many doubts due to the lack of consensus. Recurrently, the term is associated with the creation of a company, but the development and implementation of an entrepreneurial business project is not a determining condition for considering someone an entrepreneur.

In general, and in accordance with the literature, entrepreneurship is understood to be the process, in the broadest sense, by which people develop their idea to become a concrete project. This can have a business objective, be intra-company or result in a social benefit generating innovation and employment.

Historically, the word appeared in the lexicon in France, associated with mercantilism with Richard Cantillon, passing through Jean-Baptiste Say, who added the notion of value generation, until more recently, with Schumpeter and Drucker, the term is associated with the innovation component and the emergence of a business (Drucker, 2014).

In economic theory, entrepreneurship is widely referred to as a production factor, the fourth, preceded by capital, labor and land (Elias et al., 2015). Since its origin as "entreprendre" indicates doing something new, creating and innovating, without being sure of the result, entrepreneurship is seen more as an individual phenomenon, linked to the creation of companies through taking advantage of an opportunity, than a social phenomenon, which can lead an individual or a community to develop resilience, overcoming barriers and seeking to build the future.

The European Commission (2003), in a more comprehensive definition, states that "entrepreneurship is above all a mental and behavioral attitude towards the process of creating and developing economic activities, combining risk and creativity and innovation with rigorous management within the scope of a new or existing organism".

Thus, entrepreneurship comes to represent a dynamic procedural concept that is inherent in the vision of opportunity, the design and realization of a business project, which by combining human resources with process development, transforms ideas into business opportunities. (Kuratko et al., 2021). In this way, entrepreneurship presupposes the existence of two determining elements, Creativity, as a connection and organization of knowledge in order to generate innovative ideas, and Innovation, which is the practical implementation of a creative idea (Ahlin et al., 2014).

Currently, government officials, business decision-makers and academics refer to the concept of entrepreneurship as an element that adds value and is decisive for economic evolution, for the progress of products and services, for improved product quality and changes in society (Ács et al., 2018; Carree & Thurik, 2010; Manullang, 2018; Ratten & Dana, 2019; Schaltegger et al., 2016).

2.2 *Circular Entrepreneurship and Circular Economy*

In the age of digitization and sustainability, the theme of circular entrepreneurship has gained ground in both academic research and public policy. This corresponds to recognition of the economic relevance of entrepreneurship.

In a period marked by economic instability and significant concerns about losses and waste resulting from a linear economy, naturally the creation of new companies and "circular" projects has become more prominent on the agenda of policy-makers, particularly in more developed countries. New companies and their entrepreneurs are recognized for their ability to promote and develop, with the introduction of new products into the economy (Schaltegger et al., 2016).

It starts from the contextual principle that entrepreneurship implies an attitude of taking risks in the face of uncertainties, but at the same time the perspective of obtaining a certain return. The concerns surrounding environmental issues represent opportunities for the emergence of disruptive innovations and the creation of new companies by entrepreneurs. These activities can eliminate the gaps in the economic transition, by identifying and subsequently exploiting the opportunities arising, such as more efficient use of natural resources and promoting the path towards a sustainable economy (Dean & McMullen, 2007).

In addition, sustainable entrepreneurs can be agents of change in organizational structures, by allowing the exploration of sustainable opportunities, increasing the competitiveness of environmentally sustainable behaviors and leveraging new entrepreneurial attitudes (Zucchella & Urban, 2019). This enables connectivity between corporate businesses, creative processes and well-being at a socio-economic and environmental level (Crecente et al., 2021).

Faced with a model based on the premises of regeneration and restoration, the need for materials, energy, underutilization and the emission of pollutants are limited by the slowdown, closing and narrowing material cycles and energy efficiency. Based on the design of sustainable products and processes, the rehabilitation and maintenance of materials and reuse and reintegration, as well as recycling (Geissdoerfer et al., 2017), circular entrepreneurship represents a radical paradigm shift in the governance of resources, in this new economic, social and environmental dimension (Zucchella & Urban, 2019).

In the literature, Circular Entrepreneurship is conceptualized as the process of developing and taking advantage of opportunities in the economic sphere, to address sustainability, with the aim of closing, slowing down and reducing the cycle of material exploitation by regenerating environmental resources (Zucchella & Urban, 2019).

When we consider the circular economy as a strategy for sustainability (Geissdoerfer et al., 2017; Velenturf & Purnell, 2021), circular entrepreneurship represents a concept framed in the identification of business opportunities and creation of value in terms of sustainability. It can be designated as sustainable entrepreneurship and is a fundamental and decisive element in the process of innovation and transformation towards the circular economy.

Still regarding circular entrepreneurship, Zucchella and Urban (2019) say this can be represented by new companies, which are "companies born circular", created to add value in the context of the circular economy. New business projects with a circular scope contribute to solving market needs related to the environment. This is possible through innovation in environmentally sustainable products and services.

The aforementioned authors also refer to another organizational typology, i.e., established organizations that seek to adjust or transform their business in the context of the circular economy, known as "circular growing" companies.

Circular entrepreneurship is necessary and decisive in transformation to the circular economy, where new ventures play a fundamental role in the search for more ambitious efficiency strategies, providing the disruptive innovations necessary for evolution and paradigm shift (Si et al., 2020).

Although entrepreneurship focused on sustainable development has received considerable attention in recent years, authors such as Filser et al. (2019) suggest the need for further studies in this field. Along the same lines, researchers such as Heshmati (2017) state that study of the interface between entrepreneurship in the transition process and the circular economy is still at an incipient stage.

2.3 Business Models and Circular Economy

A business model is defined by Osterwalder and Pigneur (2010) as an economic model that defines an organization's strategic lines of action in the business context, namely how it creates value from the resources used. This means that a business model identifies how organizations develop their economic activity, how the "value" of their product or service portfolio is generated, distributed and remunerated.

Having initially emerged as a communication tool to attract interested parties to investment projects (Zott et al., 2011), the concept quickly attracted the attention of business decision-makers and academics as a relevant conceptual instrument to develop an organization's strategic options to generate, capture and share value over time (Foss & Saebi, 2017; Fraccascia et al., 2019a, 2019b).

The circular economy, representing an initiating and accelerating vector of economic transition, forces business structures to demarcate themselves from the classic value generation model based on ownership and obsolescence of products, and to transform productive and economic systems, with the need to redefine the company's role (Centobelli et al., 2020).

This transformation is particularly relevant in the design and modeling of businesses, because as Zucchella and Urban (2019) mention, the circular economy also leads to new relationships between companies, territories and the surrounding environment.

The development of information and communication technologies makes new policies for the location of production centers possible. The Internet of Things is propelling industry into the new digital era, the "4.0", where physical flows of goods are interconnected and allow efficient supply chain management.

In the light of this new technological revolution (4.0) and the sustainability paradigm, all stages of the value chain, including design, production, logistics, the use of products and recovery of materials and waste, are positioned in the research, development and innovation of new business models.

In this context, the concept of circular business models has emerged as a series of definitions proposed by different researchers. For example, Bocken et al. (2016) state that these business models represent strategic guidelines for the economic transition based on the principles of the circular economy. Linder and Williander (2017), on the other hand, define it as a model based on a conceptual framework for adding value after the end of the product cycle, incorporating its components in the production of other products.

According to other researchers, such as Geissdoerfer et al. (2020), in a circular business, it is assumed that a company adds value in the circular circuit of materials. Zucchella and Previtali (2019) focus on the balance between sustainability and economic business performance, which guarantees survival and prosperity.

They state that the fundamental role of a circular business model is to incorporate the principles of the circular economy in a business activity project, with a balanced structure of costs and revenues in terms of sustainability and profitability.

The contextual principle of circular economy development is based on economic agents' ability to identify new opportunities, to develop and implement their businesses with valuable propositions that provide competitive advantages, in a profitable manner and in line with the principles of circularity (Linder & Williander, 2017; Urbinati et al., 2017; Zucchella & Previtali, 2019).

This puts pressure on companies and entrepreneurs to develop innovations in terms of products, processes, markets and even new methods of organizing economic activities. These radical innovations are necessary to pursue the objectives of the circular economy, and companies and entrepreneurs must reconcile them with a perspective of return on their investment (Zucchella & Urban, 2019).

However, being a relatively recent topic, existing studies are found to be rather dispersed and little systematized. The literature does not explain how business models are planned and designed according to circular economy principles. There is a framework to deepen research into the intersection between the circular economy and the business model to be developed (Fraccascia et al., 2019b; Lüdeke-Freund et al., 2019; Perey et al., 2018; Centobelli et al., 2020).

Zott et al. (2011) point out that innovation of the business model is an element that enhances innovation, which complements the traditional issues of process, product and innovation in organizations, forming a determining and crucial element of circular entrepreneurship.

3 Methodology

The method adopted here is a literature review, which was carried out in order to understand research trends in the framework associated with the theme of entrepreneurship in the circular economy and business models associated with new industries.

In scientific research, a literature review is a methodology to specify what has already been done and found out, and thus identify the limits of knowledge in relation to a given field of research (Maiberger & Sunmola, 2022).

A literature review can be developed in different ways according to the method and the objective to be achieved. However, its main purpose is to collect information that can delineate the research area, planning of the work to be carried out and the collection of summaries of published research (Paul & Criado, 2020).

Being a widely used research method due to the growth in the number and complexity of scientific publications, there are several ways to carry out literature reviews. Despite the proliferation of concepts in the literature review, fundamental characteristics such as research and collection, evaluation and presentation of evidence are common elements shared by all approaches. (Arksey & O'Malley, 2005).

In the present work, a literature review is developed using the structured method. This methodology of synthesis is considered, by some authors, as an adequate approach, because it can identify, evaluate, and synthesize published studies, in the literary reference bases by researchers, academics and professionals.

In particular, some authors point out that due to the wide availability of scientific work, the structured review goes beyond simply summarizing and systematizing previous research and can contribute to identifying research trends and potential future lines of study (Petticrew & Roberts, 2008; Massaro et al., 2016).

At the conceptual level, it is possible to find different steps to develop a literature review (Christoffersen, 2013; Thorpe et al., 2005). The methodology followed here is based on previous studies (Dumay & Cai, 2014; Guthrie et al., 2012) and the approach of Massaro et al. (2015), which comprises the following steps: Definition of the problem and questions to be investigated, a search of the literature with the definition of keywords, databases and criteria for selection of studies, research evaluation with the selection of articles, and analysis and interpretation of the items and discussion of the results as outlined in Fig. 1.

At the conceptual level, as mentioned by some authors, highlighting Thorpe et al. (2005) and Christoffersen (2013), it is possible to find different stages to carry out a literature review.

The methodology followed here is based on previous studies (Dumay & Cai, 2014; Guthrie et al., 2012) and the one defined by Massaro et al. (2015), which comprises five steps: definition of the problem and questions to be investigated, a search of the literature with the definition of keywords, databases and criteria for selection of studies, research evaluation with the selection of articles, and analysis and interpretation of the items and discussion of the results as outlined in the previous figure.

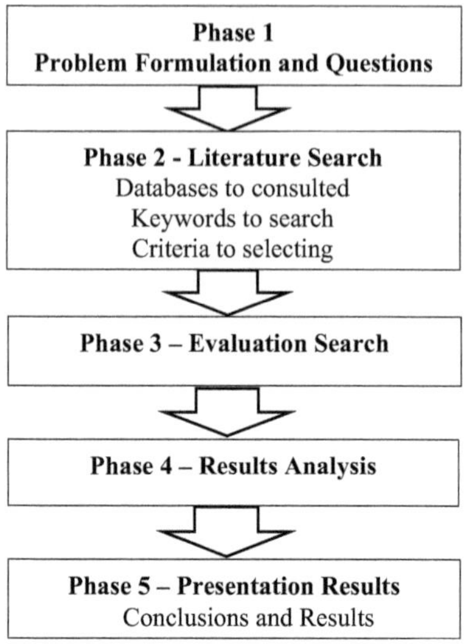

Fig. 1 Literature review approach. (Source: Own elaboration)

3.1 Research Questions

In the introduction and in Chapter. "Toward the Green Transition of Business through the Circular Economy: Understanding the Role of Dynamic Capabilities", a theoretical framework was presented regarding entrepreneurship, the circular economy and new business models. Considering the points developed in this field, three research axes were defined to guide this work and these relate to the following questions:

- RQ1: How is the literature on the circular economy, entrepreneurship and new business models developing?
- RQ2: • What are the trends in the field present in the literature?
- RQ3: • What is the impact of published studies in this framework?

3.2 Search of the Literature

In the second stage, a protocol was defined to determine the methods, information resources and tools to use to explore and summarize the items, as explained by Chapman (2021).

For a search of the literature, various sources of information available on the internet are recommended by some researchers, such as Tranfield et al. (2003).

However, in this research, only publications published in databases of reference were considered, in order to improve the quality of the sample. For the collection of articles, the Web of Science and Scopus databases were used to ensure significant coverage of published work (Paul & Criado, 2020).

The strings that make up the search flow were chosen to ensure the location of articles related to entrepreneurship and business models in the circular economy. For this purpose, in a first phase, the strings of "entrepreneurship", "circular economy" and "business models" were used, but the results obtained were not significant.

In a next step, different conjugations of keywords were defined, and for a more robust sample of papers for analysis, we used the keywords: "entrepreneurship", "circular economy" and "business models". In order to obtain a broader sample of articles for analysis, the keywords of "technology change", "new ventures" and "closed-loop" were used.

The search strategy considered work published up to 2022 and was carried out in the "topic" field, in the case of "Web of Science", and by article title, abstracts and keywords in the case of Scopus.

3.3 *Data Evaluation*

The result of the search in the databases used was: 19 articles found in Web of Science, and 39 articles in Scopus. In the next step, the results found in the databases were compared and duplicated items were extracted, resulting in a total of 47 documents.

Given the polysemic nature of the concepts of entrepreneurship and the circular economy, the title, abstracts, and subsequently, the content of the articles in the sample obtained were submitted to a review process.

To ensure the articles found were pertinent for the study, the following criteria were used: only papers written in English, excluding all papers with a different format from the article type and excluding articles that were not profiled in the line of investigation. From this process, we arrived at a final sample of 36 papers.

The next step, to obtain the results and classification of the articles, was based on the theoretical framework. The papers were fully analyzed to obtain relevant information to characterize the theme under analysis and to define analysis groups.

The categories defined for analysis of the sample, according to codification of the bibliometric information, were the following:

- Publication period: i.e., distribution of the number of articles published over time;
- Analysis of citations, through the number of citations, citations per year;
- Research methods used by the authors;
- Investigation/Research frameworks.

For this study, and considering the emerging nature of some of the concepts analyzed, it was deemed appropriate to develop a descriptive analysis, supported by data such as the distribution of articles among countries and the tendency in the number of publications, aiming to underline how the literature portrays the role of entrepreneurial activity in a context of a new economic, social and environmental dimension.

For keyword analysis, VOSviewer software was used (Eck & Waltman, 2017).

4 Results

This section, which corresponds to the fourth stage of the research methodology, presents the results obtained. The central objective here is to analyse the sample content and provide results/answers/inputs to the research questions.

4.1 Descriptive Analysis

The publications forming the sample cover the period between 2017 and 2022. The annual productivity chart (Fig. 2) shows the interest in the subject analyzed, through the growth in the number of publications per year.

In the final sample, consisting of 36 papers, most of the documents analyzed correspond to articles published in international journals, representing about 86% of the total number of documents in the sample, with the remaining 14% being articles from international conferences.

As shown in Fig. 2, the themes of entrepreneurship, circular economy and business models began to be part of researchers' agendas from 2017 onwards, registering an increase in the following years. To date, the year with the highest number of articles was 2021, with a total of 14 articles published. Despite some oscillation in 2020 and 2022, compared to previous years, overall there is a growing trend in relation to the emergence of these topics on the scientific agenda.

This trend can also be observed through the number of citations reported in the Web of Science database, as illustrated in Fig. 3. This indicates the annual number of citations of the papers, where the sum of citations, in the period under analysis, follows the same tendency as scientific production.

4.2 Bibliometric Analysis

For this study, a bibliometric analysis was performed on the selected articles. This research technique allows researchers to deepen knowledge in a given field of research, through a more meticulous approach, given that bibliometrics uses statistical techniques in the scientific investigation of a particular field of study (Pizzi et al., 2020).

This analysis technique is based on two development axes, represented by the graphical representation of scientific concepts and by performance analysis. The scientific representation or mapping is based on first and second-generation

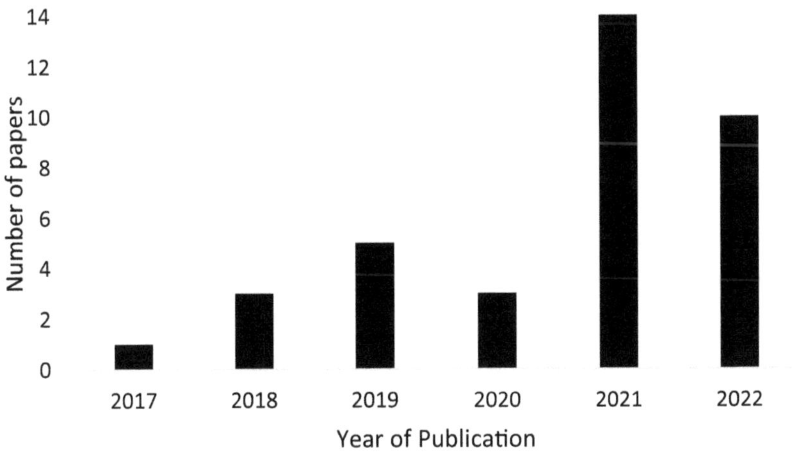

Fig. 2 Number of documents per year

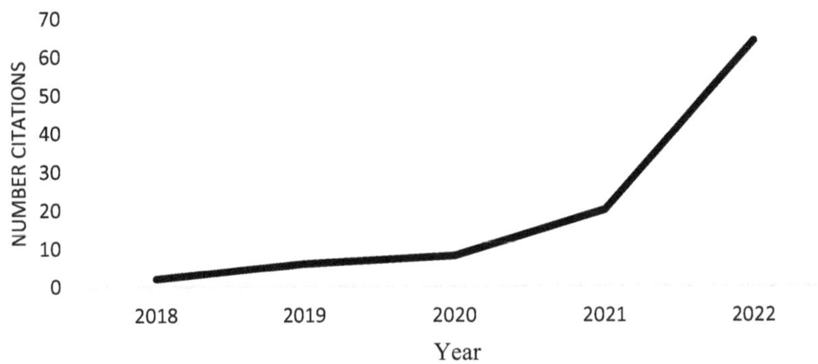

Fig. 3 Number of citations per year

relational metrics, which provide a spatial representation of the relationship between the various scientific elements (Pizzi et al., 2020).

Authors such as Rosato et al. (2021) state that the central objective of this axis is to highlight "the dynamic and structural organization of knowledge" in this area of research.

According to the cited authors, performance analysis is based on metrics of information on scientific production and the impact of publications. For this purpose a set of techniques was used, such as the analysis of citations or the frequency of the words in the publication (Fig. 4).

We chose to use a metric that allows us to determine the main topics and the research trends, through the co-occurrence of the keywords. By taking the authors' keywords as input, this indicator allows us to obtain information regarding the conceptual framework to be investigated.

According to the literature, the analysis of keyword co-occurrence assumes that a given framework can be traced by the similarities identified between these keywords.

As this indicator does not need an intrinsic bias for studies that cover a longer period of time, the relationship between the similarity of keywords has been a widely used method in scientific research (Pizzi et al., 2020).

Considering our objectives, VOSviewer software was used to determine this metric. This software allows the researcher to obtain network maps between similarities and connections with defined words, citation network, density diagrams and sources, among others (Eck & Waltman, 2017).

As mentioned in the literature, VOSviewer has become popular in the scientific community for bibliometric analysis and other areas of research where scientific mapping is used, due to the ease and variety of resources for the user, unlike other software used for this function.

Given its graphical representation, as a bibliometric tool for literature analysis, it has some advantages, such as a global view of the literature and a series of tools to obtain reliable data from the units under analysis and a set of results obtained through a rigorous process (Oladinrin et al., 2022).

Thus, VOSviewer bibliometric mapping software was applied to create a visual network of the most used terms in the connection between entrepreneurship, circular economy and new business models.

4.2.1 Analysis of Keywords in the Sample

Analysis of the research's keywords was carried out to assess the specificities of the discussion on the concepts of entrepreneurship and business models, studied in a context of transition to the circular economy.

For this purpose, the Keywords Plus function was used to determine the keywords referenced by the studies in our sample.

This function revealed that 1176 keywords were used in the titles, abstracts and keyword field of the sample. In this set, five keywords with the highest number of

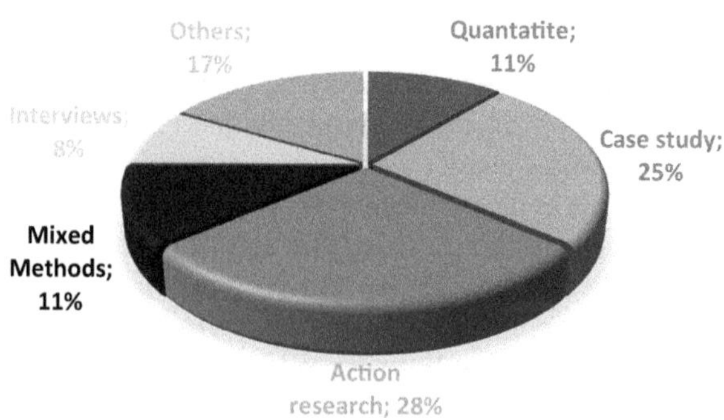

Fig. 4 Articles per research method

occurrences stand out: “circular economy”, “sustainability”, “entrepreneurship”, “business model” and “environmental economics”.

The keyword co-occurrence network map generated by VOSviewer is shown in Fig. 5. Here, the clusters aggregate the nodes and the size of the node illustrates the frequency of the term. “Entrepreneurship”, “circular economy” and “business models” are central concepts in the graphic representation of the bibliometric mapping.

The density of the analysis is shown in Fig. 6. As can be observed in this graphic representation, a high number of articles deal with research into the circular economy with entrepreneurship.

In this regard, this analysis confirms the relevance of the concept and entrepreneurship’s potential to initiate and accelerate the process of transitioning to a more environmentally sustainable economy. Researchers are also concerned about associating the emergence of new circular business models with this new economic dimension.

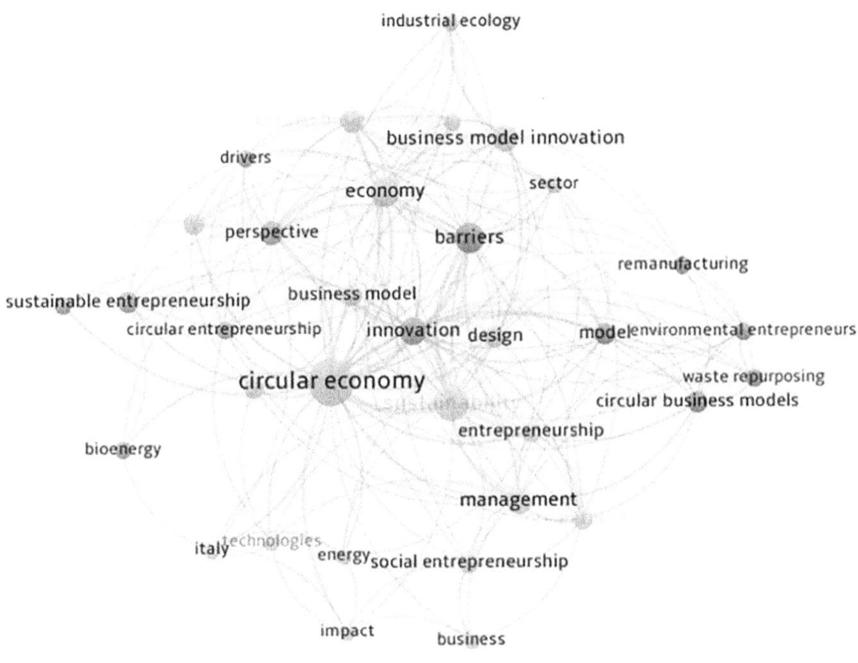

Fig. 5 Diagram of the co-occurrence of keywords

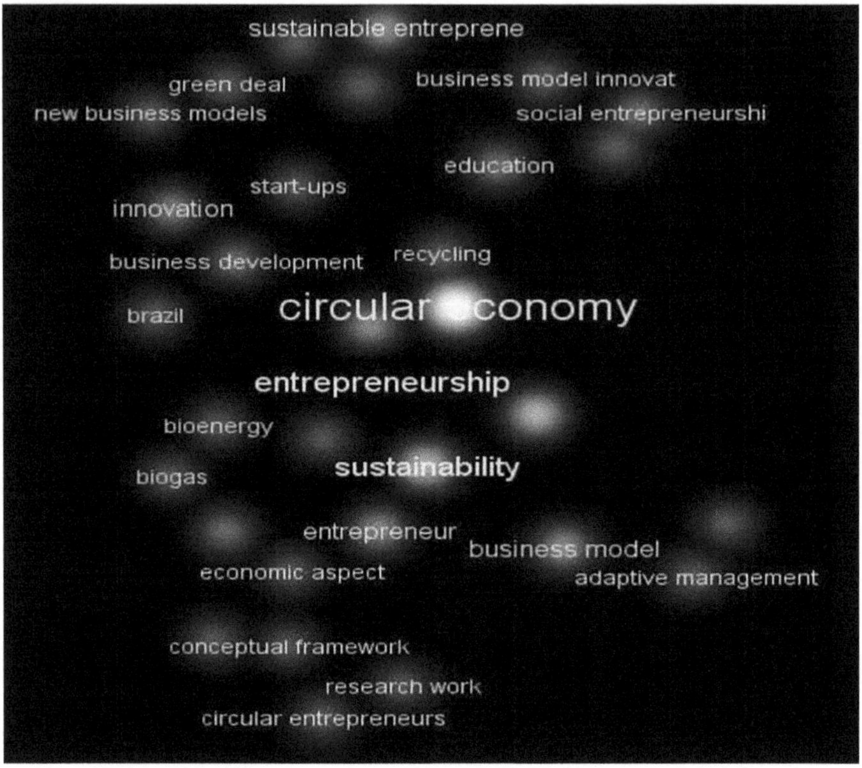

Fig. 6 Density diagram of the co-ocurrence of keywords

4.3 *Content Analysis*

At this point, the main concepts determined through content analysis will be presented, resulting from the simultaneity of the keywords present in the sample under study. The intention, therefore, is to understand the most relevant points in researchers' agendas on entrepreneurship, the circular economy, and new business models.

Based on the output provided by the computer tool used and according to the analysis carried out, the keywords were grouped into four main thematic areas, which were classified in the following clusters: Economy and Sustainability, Circular Entrepreneurship, Business Activity, and Georeferencing.

The first cluster represents the set of publications studying how the circular economy concept is investigated and encouraged in a context of sustainable development.

Some studies pay particular attention to the question of how the circular economy aims at resource efficiency as a determining condition for eliminating waste (Buch et al., 2021; Cullen & De Angelis, 2021; Despeisse et al., 2017), to minimize the use of production materials that do not contribute to environmental sustainability (Elf et al., 2022), and the reduction of waste with an impact on the economy and organizational subsystems (Kanda et al., 2021).

Some authors highlight the interest in investigating the benefits of sustainability as an important and timely topic (Dentchev et al., 2018) and the need and evidence of transition to a more sustainable economy (Lauten-Weiss & Ramesohl, 2021), the holistic view of the relationship between the concepts (Buch et al., 2021) and the importance of new technologies as initiating elements in the change to the circular economy and sustainability (Saura et al., 2022).

Regarding the second cluster, the concept of circular entrepreneurship is conceptualized as a process of exploring and taking advantage of opportunities in the domain of the circular economy (Cullen & De Angelis, 2021). The role of entrepreneurship in economic transformation is mentioned by Saura et al. (2022).

Regarding forms of entrepreneurship, some authors have empirically studied the motivations and identity of entrepreneurs (Henry et al., 2023), and the motives of micro entrepreneurship in the circular economy (Hu et al., 2019).

In a circular economy context, how innovation and sustainable entrepreneurship influence social progress (Manea et al., 2021). Upcycling entrepreneurship (Donner et al., 2020). Staicu and Pop (2018) addressed the role of social entrepreneurship.

Some authors, such as Sehnem et al. (2022), explore the relationship between start-ups and the circular economy, illustrating the interest of entrepreneurial ecosystems. Kanda et al. (2021) developed an ecosystem-based approach to leverage entrepreneurship and innovation in the context of the circular economy.

In the documents present in the sample, industry-level studies paid special attention to waste management (Despeisse et al., 2017), waste management in the textile industry (Staicu & Pop, 2018), food industry (Cullen & De Angelis, 2021), furniture industry (Igelbrink et al., 2019) and waste management cooperatives (Buch et al., 2021).

Concerning other economic activities, studies were carried out in the agricultural sector (Donner et al., 2020; Zhu et al., 2019), the hotel industry (Saura et al., 2022), foodtech industry (Sehnem et al., 2022) and supply chain management (Le et al., 2022).

At a geographical level, the articles in the sample are diverse and provide evidence from certain regions, such as the article by Brendzel-Skowera (2019) on the difficulties in the transition to the circular economy for small and medium-sized companies in the Częstochowa region (Poland). Also in this country, Rok and Kulik (2021) studied innovation in start-up business models.

In Romania, Staicu and Pop (2018) addressed the role of entrepreneurship in developing clothing reuse practices.

Europe is the geographic context of analysis for Krajnc et al. (2022), addressing the concept of the Green Deal and young people's attitudes in five European countries.

Some studies in the sample focus on the characteristics of entrepreneurs from different geographical areas and economies, such as the work of Henry et al. (2023), who analyzed the profile of founders of start-ups in Europe and Australia, having identified distinct characteristics in the profile of entrepreneurs.

In Brazil, Sehnem et al. (2022) studied the relationship between start-ups and business ecosystems in increasing entrepreneurial activity.

Zhu et al. (2019) studied the economic, ecological and social benefits of adapting agriculture to transition to a circular economy in China.

5 Analysis and Discussion

This chapter presents an investigation in the form of a review of the literature on the relationship between entrepreneurship and concerns for a circular economy and business models. Given the scarcity of published literature, this topic lacks depth and breadth in investigations. Thus, there is a gap in research that was explored with this work, presenting a convincing argument and highlighting the need to address this framework of investigation.

Due to the key role entrepreneurship plays in today's society, some theoretical connections were established between this concept, the circular economy and business models, placing a strong emphasis on entrepreneurship as a determining element for the initiation and transition to a new economic dimension.

For the above reasons, it was decided to set three research questions to guide the research, which consisted of investigating, in the first part, how the literature on the relationship between the circular economy and entrepreneurship and new business models is evolving. In a second point, the question of establishing the focus of the published work, and in a third point, to determine the implications for the future.

In response to the first question, we found that despite being a recent topic on research agendas, the number of papers has increased over the past few years. The bibliographical research identified 36 studies relevant to the scope of the work, of which around 67% were published in the last 2 years, showing researchers' growing interest in the subject.

The number of citations, according to the bibliography (Massaro et al., 2015), can also be an indicator of analysis of scientific performance. According to the Fig. 2, the number of citations has been increasing annually since 2017.

This growth in citation has followed the trend of publications, with a considerable growth in the period from 2020 to 2021.

In the bibliographical analysis, the topic identified in the literature as an aggregating element was the circular economy. In this centrality, a set of themes related to the circular economy were identified, i.e., sustainability, technological innovation, value creation, circular business models and circular entrepreneurship.

Waste management in industry was a popular topic among researchers. In published articles, research on the process of reusing materials, low use of natural resources and energy efficiency in industrial processes stands out.

The framework of industries in the new economic era is also studied across economic sectors, namely in agriculture, the furniture industry, fashion industry, start-ups and smart solution companies. In these studies, the approach is oriented towards business and solutions and sustainability issues, with regard to productive activities, which promotes the topic of entrepreneurship to a prominent position in this study.

According to researchers, in initiating and accelerating the economic transition, entrepreneurship, due to its ability to seize an opportunity and add value, stands out as a key concept for sustainable economic growth, for the development of innovative products and services, and for changes in society. Corroborating this view, the term appears associated with concepts such as social entrepreneurship, technological innovation, value creation and new ventures.

Studies highlight the relationship between the circular economy and entrepreneurship. The sample documents confirm that the circular economy is seen as an opportunity to develop new forms of entrepreneurial projects, that is, circular entrepreneurship is defined as the process of exploring opportunities in the field of the circular economy.

Circular entrepreneurship can introduce new forms of consumption, reducing waste, energy consumption and promoting more flexible products, through digitalization and monitoring of production processes. This allows companies to be more disruptive and innovative in technological change processes.

In general, where entrepreneurship meets the circular economy, circular entrepreneurship is based on the contextual principle that entrepreneurial projects tend to incorporate the concepts of the circular economy, originating sustainable business models based on the circularity of resources.

New businesses derived from the green transition aim to innovate, develop and add value at an environmental level. The circular economy is a way to achieve sustainability and entrepreneurial processes, a key and decisive element in this cycle change.

6 Conclusions

In the new paths of green transition and technological change, the circular economy is a central theme on the agendas of policy-makers, academics, and business leaders. Above all, the aim is to overcome the negative impacts of the overuse of natural resources and the depletion of ecosystems that have been inseparable from the linear economic model established for decades.

In this growing interest in a new economic, social, and environmental paradigm, its impact on entrepreneurship and business models is still an open and relevant question for research.

Circular principles contribute to a sustainable and innovative future for humans and nature. The circular economy introduces changes and new relationships between economic agents, in order to develop value-adding products for new needs (Dean & McMullen, 2007; Crecente et al., 2021).

Entrepreneurs are transforming the global economy, and as Zucchella and Urban (2019) mention, they are simultaneously a representative of humanity, pursuing personal goals, and a functional element responsible for an organization's performance.

However, an organization's goals are currently only one facet of the decision-making process. Global, holistic, and environmental considerations appear as key factors in business modeling processes in a sustainability context.

This chapter, which was based on a structured literature review, presents the state of the art on entrepreneurship, the circular economy, and business models. The objective was to contribute with a useful synthesis in this area of investigation. The bibliographic research was carried out in the Scopus and WOS databases, which resulted in a total sample of 36 articles.

In general terms, one of the first conclusions of this study is that entrepreneurship, by integrating the principles of circularity, the so-called circular entrepreneurship, plays a preponderant role in a transversal way, in different technological areas, business sectors and markets, in changing the economic, social and environmental paradigm.

In more specific terms, firstly, the results obtained from the bibliographical research showed that the theme is beginning to arouse interest in the scientific community, as observed through the growth in the number of articles published in the period under analysis.

These studies address topics such as industrial processes, strategies to be developed within the circular scope, the analysis of technologies to support change, and even organizational culture. For these purposes, the researchers resorted to different methodologies such as case studies, action research, quantitative methods, interviews, and mixed methods.

Secondly, it was concluded that the aggregating element of the published studies is the circular economy, and within this, a set of related topics such as sustainability, innovation, business models, and circular entrepreneurship were investigated, emphasizing the role of circular entrepreneurship and business models in the new economy.

In particular, we highlight that for entrepreneurs the circular economy, by implying a systematic change in the way of rethinking production and consumption models, provides opportunities for innovative solutions with a view to disruption and technological change. New companies and start-ups are the main enablers of green and circular business models, necessary for the green transition. Thus, the strategic lines of new business models cannot be limited solely to economic aspects, but they must also equate an ecological dimension, the result of a philosophy of green transformation, based on a new vision and innovative creation and governance methods, which we can call "eco-competitiveness".

In this dawn of the green transition, entrepreneurial capacity is an essential pillar for technological change, as entrepreneurs and their start-ups are catalysts for innovation ecosystems, with responsibility for leveraging the circular economy, contributing to the green transition, the creation of new jobs, the economic development of regions and nations, constituting a fundamental pillar in the creation of a new economic system.

Finally, we find that the relationship between the topics of this research is still in an emerging investigation phase. To expand the sample, it is necessary to expand the keywords and databases used.

From this study, it is possible to conclude on the limited amount of research combining the circular economy with entrepreneurship and circular business models, highlighting the need to explore this area of research further.

References

Ács, Z. J., Szerb, L., Lafuente, E., & Lloyd, A. (2018). The entrepreneurial ecosystem and global prosperity. In *Global entrepreneurship and development index*. Springer.

Ahlin, B., Drnovšek, M., & Hisrich, R. D. (2014). Entrepreneurs' creativity and firm innovation: The moderating role of entrepreneurial self-efficacy. *Small Business Economics, 43*(1), 101–117.

Arksey, H., & O'Malley, L. (2005). Scoping studies: Towards a methodological framework. *International Journal of Social Research Methodology, 8*(1), 19–32.

Bianchini, A., Rossi, J., & Pellegrini, M. (2019). Overcoming the main barriers of circular economy implementation through a new visualization tool for circular business models. *Sustainability, 11*(23), 6614.

Bocken, N. M., de Pauw, I., Bakker, C., & van der Grinten, B. (2016). Product design and business model strategies for a circular economy. *Journal of Industrial and Production Engineering, 33*, 308–320.

Brendzel-Skowera, K. (2019). Circular economy in the development of sustainable entrepreneurship. In *Proceedings of the 33rd international business information management association conference, IBIMA 2019: Education excellence and innovation management through vision* (pp. 8390–8399). International Business Information Management Association.

Buch, R., Marseille, A., Williams, M., Aggarwal, R., & Sharma, A. (2021). From waste pickers to producers: An inclusive circular economy solution through development of cooperatives in waste management. *Sustainability, 13*(16), 8925.

Carree, M. A., & Thurik, A. R. (2010). The impact of entrepreneurship on economic growth. In *Handbook of entrepreneurship research* (pp. 557–594). Springer.

Centobelli, P., Cerchione, R., Chiaroni, D., Del Vecchio, P., & Urbinati, A. (2020). Designing business models in circular economy: A systematic literature review and research agenda. *Business Strategy and the Environment, 29*(4), 1734–1749.

Centobelli, P., Cerchione, R., & Esposito, E. (2017). Environmental sustainability in the service industry of transportation and logistics service providers: Systematic literature review and research directions. *Transportation Research Part D: Transport and Environment, 53*, 454–470.

Chapman, K. (2021). Characteristics of systematic reviews in the social sciences. *The Journal of Academic Librarianship, 47*(5), 102396.

Christoffersen, J. (2013). A review of antecedents of international strategic alliance performance: Synthesized evidence and new directions for core constructs. *International Journal of Management Reviews, 15*(1), 66–85.

Crecente, F., Sarabia, M., & Del Val, M. T. (2021). Sustainable entrepreneurship in the 2030 horizon. *Sustainability, 13*(2), 9.

Cullen, U. A., & De Angelis, R. (2021). Circular entrepreneurship: A business model perspective. *Resources, Conservation and Recycling, 168*, 105300.

Dean, T. J., & McMullen, J. S. (2007). Toward a theory of sustainable entrepreneurship: Reducing environmental degradation through entrepreneurial action. *Journal of Business Venturing, 22*(1), 50–76.

Dentchev, N., Rauter, R., Jóhannsdóttir, L., Snihur, Y., Rosano, M., Baumgartner, R., et al. (2018). Embracing the variety of sustainable business models: A prolific field of research and a future research agenda. *Journal of Cleaner Production, 194*, 695–703.
Despeisse, M., Baumers, M., Brown, P., Charnley, F., Ford, S. J., Garmulewicz, A., et al. (2017). Unlocking value for a circular economy through 3D printing: A research agenda. *Technological Forecasting and Social Change, 115*, 75–84.
Donner, M., Gohier, R., & de Vries, H. (2020). A new circular business model typology for creating value from agro-waste. *Science of the Total Environment, 716*, 137065.
Dragomir, V. D., & Duţescu, A. (2022). New business models in the circular economy. *Proceedings of the International Conference on Business Excellence, 16*(1), 792–804.
Drucker, P. (2014). *Innovation and entrepreneurship*. Routledge.
Dumay, J., & Cai, L. (2014). A review and critique of content analysis as a methodology for inquiring into IC disclosure. *Journal of Intellectual Capital, 15*(2), 264–290.
Eck, N. J., & Waltman, L. (2017). Citation-based clustering of publications using CitNetExplorer and VOSviewer. *Scientometrics, 111*(2), 1053–1070.
Elf, P., Werner, A., & Black, S. (2022). Advancing the circular economy through dynamic capabilities and extended customer engagement: Insights from small sustainable fashion enterprises in the UK. *Business Strategy and the Environment, 31*(6).
Elias, G., Carayannis Elpida, T., & Samara Yannis, L. B. (2015). *Innovation and entrepreneurship: Theory, policy and practice*. Springer.
Ellen MacArthur Foundation. (2019). *Completando a figura: Como a economia circular ajuda a enfrentar as mudanças climáticas*. Ellen MacArthur Foundation. https://www.ellenmacarthurfoundation.org/publications
European Commission. (2003). *Livro Verde*. Espírito Empresarial na Europa.
Filser, M., Kraus, S., Roig-Tierno, N., Kailer, N., & Fischer, U. (2019). Entrepreneurship as catalyst for sustainable development: Opening the black box. *Sustainability, 11*(16).
Foss, N. J., & Saebi, T. (2017). Fifteen years of research on business model innovation: How far have we come, and where should we go? *Journal of Management, 43*, 200–227.
Fraccascia, L., Giannoccaro, I., Agarwal, A., & Hansen, E. (2019a). Business models for the circular economy: Opportunities and challenges. *Business Strategy and the Environment, 28*, 430–432.
Fraccascia, L., Giannoccaro, I., & Albino, V. (2019b). Business models for industrial symbiosis: A taxonomy focused on the form of governance. *Resources, Conservation and Recycling, 146*, 114–126.
Geissdoerfer, M., Morioka, S. N., de Carvalho, M. M., & Evans, S. (2018). Business models and supply chains for the circular economy. *Journal of Cleaner Production, 190*, 712–721.
Geissdoerfer, M., Pieroni, M., Pigosso, D., & Soufani, K. (2020). Circular business models: A review. *Journal of Cleaner Production, 277*, 123741.
Geissdoerfer, M., Savaget, P., Bocken, N. M., & Hultink, E. J. (2017). The circular economy–a new sustainability paradigm. *Journal of Cleaner Production, 143*, 757–768.
Guthrie, J., Ricceri, F., & Dumay, J. (2012). Reflections and projections: A decade of intellectual capital accounting research. *The British Accounting Review, 44*(2), 68–82.
Henry, M., Hoogenstrijd, T., & Kirchherr, J. (2023). Motivations and identities of "grassroots" circular entrepreneurs: An initial exploration. *Business Strategy and the Environment, 32*(3), 1122–1141.
Heshmati, A. (2017). A review of the circular economy and its implementation. *International Journal of Green Economics, 11*, 251–288.
Hofstetter, J. S., De Marchi, V., Sarkis, J., Govindan, K., Klassen, R., Ometto, A. R., et al. (2021). From sustainable global value chains to circular economy—Different silos, different perspectives, but many opportunities to build bridges. *Circular Economy and Sustainability, 1*(1), 21–47.
Hu, S., Henninger, C. E., Boardman, R., & Ryding, D. (2019). Challenging current fashion business models: Entrepreneurship through access-based consumption in the second-hand luxury

garment sector within a circular economy. *Sustainable luxury: Cases on circular economy and entrepreneurship*, 39–54.

Igelbrink, A., Kälin, A., Krajner, M., & Kunič, R. (2019). Cradle to cradle®—Parquet for generations: Respect natural resources and offers preservation for the future. *Sustainable Luxury: Cases on Circular Economy and Entrepreneurship*, 83–106.

Kanda, W., Geissdoerfer, M., & Hjelm, O. (2021). From circular business models to circular business ecosystems. *Business Strategy and the Environment, 30*(6), 2814–2829.

Krajnc, D., Kovačič, D., Žunec, E., Brglez, K., & Kovačič Lukman, R. (2022). Youth awareness and attitudes towards a circular economy to achieve the green Deal goals. *Sustainability, 14*(19), 12050.

Kuratko, D. F., Fisher, G., & Audretsch, D. B. (2021). Unraveling the entrepreneurial mindset. *Small Business Economics, 57*, 1681–1691.

Lahti, T., Wincent, J., & Parida, V. (2018). A definition and theoretical review of the circular economy, value creation, and sustainable business models: where are we now and where should research move in the future?. *Sustainability, 10*(8), 2799.

Lauten-Weiss, J., & Ramesohl, S. (2021). The circular business framework for building, developing and steering businesses in the circular economy. *Sustainability, 13*(2), 963.

Le, T. T., Behl, A., & Pereira, V. (2022). Establishing linkages between circular economy practices and sustainable performance: The moderating role of circular economy entrepreneurship. *Management Decision*, (ahead-of-print).

Linder, M., & Williander, M. (2017). Circular business model innovation: Inherent uncertainties. *Business Strategy and the Environment, 26*(2), 182–196.

Lüdeke-Freund, F., Gold, S., & Bocken, N. M. (2019). A review and typology of circular economy business model patterns. *Journal of Industrial Ecology, 23*, 36–61.

Maiberger, T. W., & Sunmola, F. T. (2022). Effectiveness factors of food safety management systems: A systematic literature review. *British Food Journal*. (ahead-of-print).

Manea, D. I., Istudor, N., Dinu, V., & Paraschiv, D. M. (2021). Circular economy and innovative entrepreneurship, prerequisites for social progress. *Journal of Business Economics and Management, 22*(5), 1342–1359.

Manullang, M. (2018). Determinant the entrepreneurship ULOS Batak weaving shawl sellers revenue in Medan, Indonesia. *Academy of Entrepreneurship Journal, 24*, 1–8.

Massaro, M., Dumay, J., & Garlati, A. (2015). Public sector knowledge management: A structural literature review. *Journal of Knowledge Management, 19*(3), 530–558.

Massaro, M., Handley, K., Bagnoli, C., & Dumay, J. (2016). Knowledge management in small and medium enterprises: A structured literature review. *Journal of Knowledge Management, 20*(2), 258–291.

Murray, A., Skene, K., & Haynes, K. (2017). The circular economy: an interdisciplinary exploration of the concept and application in a global context. *Journal of Business Ethics, 140*, 369–380.

Oladinrin, O. T., Arif, M., Rana, M. Q., & Gyoh, L. (2022). Interrelations between construction ethics and innovation: A bibliometric analysis using VOSviewer. *Construction Innovation, 23*(3), 1471–4175.

Ormazabal, M., Prieto-Sandoval, V., Puga-Leal, R., & Jaca, C. (2018). Circular economy in Spanish SMEs: Challenges and opportunities. *Journal of Cleaner Production, 185*, 157–167.

Osterwalder, A., & Pigneur, Y. (2010). *Business model generation: A handbook for visionaries, game changers, and challengers*. John Wiley & Sons.

Paul, J., & Criado, A. R. (2020). The art of writing literature review: What do we know and what do we need to know? *International Business Review, 29*(4), 101717.

Perey, R., Benn, S., Agarwal, R., & Edwards, M. (2018). The place of waste: Changing business value for the circular economy. *Business Strategy and the Environment, 27*(5), 631–642.

Petticrew, M., & Roberts, H. (2008). *Systematic reviews in the social sciences: A practical guide*. John Wiley & Sons.

Pizzi, S., Caputo, A., Corvino, A., & Venturelli, A. (2020). Management research and the UN sustainable development goals (SDGs): A bibliometric investigation and systematic review. *Journal of Cleaner Production, 276*, 124033.
Ratten, V., & Dana, L. P. (2019). *Societal entrepreneurship and competitiveness*. Emerald Publishing Limited.
Rok, B., & Kulik, M. (2021). Circular start-up development: The case of positive impact entrepreneurship in Poland. *Corporate Governance: The International Journal of Business in Society, 21*(2), 339–358.
Rosato, P. F., Caputo, A., Valente, D., & Pizzi, S. (2021). 2030 agenda and sustainable business models in tourism: A bibliometric analysis. *Ecological Indicators, 121*, 106978.
Sariatli, F. (2017). Linear economy versus circular economy: A comparative and analyzer study for optimization of economy for sustainability. *Visegrad Journal on Bioeconomy and Sustainable Development, 6*(1), 31–34.
Saura, J. R., Ribeiro-Soriano, D., & Palacios-Marqués, D. (2022). Adopting digital reservation systems to enable circular economy in entrepreneurship. *Management Decision*, (ahead-of-print).
Schaltegger, S., Lüdeke-Freund, F., & Hansen, E. G. (2016). Business models for sustainability: A co-evolutionary analysis of sustainable entrepreneurship, innovation, and transformation. *Organization and Environment, 29*, 264–289.
Sehnem, S., Provensi, T., da Silva, T. H. H., & Pereira, S. C. F. (2022). Disruptive innovation and circularity in start-ups: A path to sustainable development. *Business Strategy and the Environment, 31*(4), 1292–1307.
Seuring, S., & Müller, M. (2008). From a literature review to a conceptual framework for sustainable supply chain management. *Journal of Cleaner Production, 16*(15), 1699–1710.
Si, S., Zahra, S. A., Wu, X., & Jeng, D. J. F. (2020). Disruptive innovation and entrepreneurship in emerging economics. *Journal of Engineering and Technology Management, 58*, 101601.
Staicu, D., & Pop, O. (2018). Mapping the interactions between the stakeholders of the circular economy ecosystem applied to the textile and apparel sector in Romania. *Management & Marketing. Challenges for the Knowledge Society, 13*(4), 1190–1209.
Thorpe, R., Holt, R., Macpherson, A., & Pittaway, L. (2005). Using knowledge within small and medium-sized firms: A systematic review of the evidence. *International Journal of Management Reviews, 7*(4), 257–281.
Tranfield, D., Denyer, D., & Smart, P. (2003). Towards a methodology for developing evidence-informed management knowledge by means of systematic review. *British Journal of Management, 14*(3), 207–222.
Urbinati, A., Chiaroni, D., & Chiesa, V. (2017). Towards a new taxonomy of circular economy business models. *Journal of Cleaner Production, 168*, 487–498.
Van Renswoude, K., Wolde, A. T., & Joustra, D. J. (2015). Circular Business Models. Part 1: An Introduction to IMSA's Circular Business Model Scan. https://documents.pub/document/imsa-circular-business-models.html?page=7
Velenturf, A. P., & Purnell, P. (2021). Principles for a sustainable circular economy. *Sustainable Production and Consumption, 27*, 1437–1457.
Veleva, V., & Bodkin, G. (2018). Corporate-entrepreneur collaborations to advance a circular economy. *Journal of Cleaner Production, 188*, 20–37.
York, J., & Venkataraman, S. (2010). The entrepreneur-environment nexus: Uncertainty, innovation and allocation. *Journal of Business Venturing, 25*, 449–463.
Zhu, Q., Jia, R., & Lin, X. (2019). Building sustainable circular agriculture in China: Economic viability and entrepreneurship. *Management Decision, 57*(4), 25–1747.
Zott, C., Amit, R., & Massa, L. (2011). The business model: Recent developments and future research. *Journal of Management, 37*, 1019–1042.
Zucchella, A., & Previtali, P. (2019). Circular business models for sustainable development: A "waste is food" restorative ecosystem. *Business Strategy and the Environment, 28*, 274–285.
Zucchella, A., & Urban, S. (2019). *Circular entrepreneurship*. Springer International Publishing.

Toward the Green Transition of Business Through the Circular Economy: Understanding the Role of Dynamic Capabilities

Pedro Baena-Luna, Esther García-Río, and Carmen M. Felipe-Llanos

Abstract The green transition (GT) is becoming imperative for all economic actors in the business field. In this scenario, the circular economy (CE) is emerging as a new paradigm to address these challenges. However, approaching these novel business models requires that companies reconfigure their resources and organizational capabilities to enable the transition from a traditional linear model to a circular one. The dynamic capabilities (DCs) view provides an exceptional theoretical framework for examining this reality, offering mechanisms that facilitate companies to face these complex processes from the perspective of the research field of business organizations. This chapter aims to conduct a bibliometric analysis of the scientific literature that connects the relationship and impact of DCs on adopting CE models to foster GT in companies. The scientific databases consulted were Web of Science and Scopus, from an initial number of 103 and 60 papers. Once the results were unified, 108 were finally selected for analysis. Bibliometrix tool was used for processing the information, providing results on annual production, and analysis of journals, authors, documents, keywords, etc. These findings will allow us to identify the principal research authors and topics in this area, analyze main conceptual, intellectual, and social structures, and detect future research opportunities.

Keywords Green transition · Circular economy · Dynamic capabilities · Dynamic capabilities view

P. Baena-Luna (✉) · E. García-Río · C. M. Felipe-Llanos
Faculty of Economics and Business Sciences, University of Seville, Seville, Spain
e-mail: pbaenaluna@us.es; egrio@us.es; cfelipe@us.es

C. Rego et al. (eds.), *Entrepreneurship, Technological Change and Circular Economy for a Green Transition*, Studies on Entrepreneurship, Structural Change and Industrial Dynamics, https://doi.org/10.1007/978-3-031-48079-9_3

1 Introduction

Currently, sustainability concerns are at the forefront of the agenda of major political and social institutions. The ever-increasing effects of climate change, with severe impacts, such as extreme weather events, food crises, and biodiversity loss, have led to the awareness of the imperative need to address a global green transition (GT). The Sustainable Development Goals (SDG) of the United Nations are shaping this process, providing a framework for reconciling social, economic, and environmental sustainability, involving contributions from all social agents (Mio et al., 2020).

In such a scenario, the private sector is urged to participate in sustainable development actively (Williams & Blasberg, 2022). On the one hand, companies are crucial players in economic growth, employment generation, and community prosperity. On the other hand, the effects of their operations have contributed significantly, and so it remains, to environmental degradation, worsening the impacts of climate change.

Thus, new paradigms have been emphasized, such as the circular economy (CE), which proposes innovative models to address the issues above and transform them into opportunities (Bocken et al., 2016). CE advocates the design and optimization of products and the valuable recovery of waste based on activities such as reuse, remanufacturing, and recycling (Singh & Ordoñez, 2016). With a linear production model showing significant adverse impacts, CE is presented as a framework of solutions committed to extending the life cycle of goods.

As an example of this importance, the European Commission adopted in March 2020 the "Circular Economy Action Plan" (European Commission, 2020), whose main objectives are developing more sustainable products, reducing waste, and empowering citizens in the GT. The Commission sets a target date of 2050 to achieve full CE in Europe.

Nevertheless, the path from the traditional linear production model to a CE is complex. Implementing circular business practices requires agility in operations, permanent networking, and integration with supply chain partners (Chowdhury et al., 2022). Therefore, companies must leverage those mechanisms that enable the flexibilization, reorganization, and even restructuration of their resources and capabilities to respond to these defiant challenges.

The dynamic capabilities (DCs) view (Teece et al., 1997) provides an excellent lens to observe these emerging issues. This theoretical perspective is a perfect approach to understanding the basis and the functioning of markets and competition in the twenty-first century (Teece, 2020). Teece et al. (1997) define DCs as those capabilities that allow firms' integration, building, and reconfiguration of internal and external competencies to enable rapid responses to change environments.

Given the above, further exploration of the research work in these fields is required to unveil the capabilities-building mechanisms the DCs view provides companies to embrace the GT through CE practices. A bibliometric analysis of the previous literature can illuminate existing and good relationships between DCs and CE research fields. Consequently, the following research questions are formulated:

RQ1. Which authors are at the forefront of the scientific production from the related scientific literature?
RQ2. Which have been the papers with the highest impact in subsequent related works?
RQ3. What are the main topics investigated, which countries lead this scientific production, and which keywords are the most used in the scientific literature analyzed?
RQ4. What conceptual, intellectual, and social structures were generated from the scientific literature analyzed?

This chapter will attempt to provide answers to these questions. As its main contributions and insofar as the authors' knowledge is concerned, this is the first bibliometric analysis linking the DCs view and the CE, as well as one of the scarcest combining two databases (Scopus and Web of Science) to perform a single integrated analysis.

2 Theoretical Framework

In recent years, interest in the GT phenomenon has significantly risen as a critical factor in business management processes (Kannan et al., 2022; Rojšek, 2001). The development of various initiatives, actions, and policies aimed at promoting and favoring GT from public actors is becoming increasingly common. (Wang et al., 2021). As key moments of these commitments, the adoption of the United Nations Sustainable Development Goals for 2030, which the European Union has assumed, and the Paris Agreement on climate change, legally mandatory for 196 parties, can be highlighted.

In the economic and business sphere, innovative activity is generated to favor an adequate context for companies' GT (Marra et al., 2017). Business development and economic growth, essential for a company, should not endanger society's potential long-term prosperity as long as natural capital conservation is maintained (Kemp-Benedict, 2018). As a consequence of the relevance and extent of the GT topic, its study is being addressed in different fields and approaches that explore metrics, drivers and barriers, potential success and failure factors, as well as the analysis of good practices (Kannan et al., 2022) in different organizational contexts.

CE emerges as one of the most outstanding paradigms among the various schemes targeted to companies and organizations to face GT. Although CE had been previously defined by Pearce and Turner (1989), it is only recently that the concept has acquired greater relevance at the business level and for society in general (Herrero-Luna et al., 2022). According to Korhonen et al. (2018), CE is an approach for companies and organizations to tackle current societal and environmental challenges while promoting and fostering sustainable development by exploiting new business opportunities (Holzer et al., 2021).

CE is a feasible option for the current economic and business models managing different resources, which impact performance in its economic, social, and environmental dimensions (Herrero-Luna et al., 2022), the so-called triple bottom line. A worthy definition is offered by Homrich et al. (2018): *"CE is a strategy that emerges to oppose the traditional open-ended system, aiming to face the challenge of resource scarcity and waste disposal in a win-win approach with economic and value perspective"* (p. 534). The work of the Ellen MacArthur Foundation is of particular relevance. This institution is becoming the backbone of a collaborative network of public and private actors for the practical development and enhancement of the CE concept (Geissdoerfer et al., 2017).

The CE's main foundation is that waste generated by production processes in companies should not be discarded, fundamentally promoting circular activities such as reducing, reusing, and recycling (Velenturf & Purnell, 2021). This so-called 3Rs approach has been progressively evolving with subsequent works in the literature until reaching the current 9Rs of CE practices: refusing (preventing the use of new materials), reduction, reusing, repairing, refurbishing, remanufacturing, repurposing, recycling, and energy recovering (van Buren et al., 2016). CE addresses in its actions the different microprocessors that make up a company with the significant aspects of the economy in its macro concept (Geisendorf & Pietrulla, 2018).

The CE's main objective is to maintain the value of different natural, human, and cultural assets, together with compatible value creation, by transforming the resources above. The emerging interest of companies in CE has motivated a growing integration into their sustainability strategy (Walker et al., 2022). However, adopting these new models is not an effortless path for companies. CE is concerned with environmental issues or the production and reuse of materials and encompasses other business aspects, such as innovation, consumption patterns, and customer behavior (Gonçalves Castro et al., 2022). Addressing this shift requires flexible business processes in which circular practices can be integrated and favor constructing dynamic relationships with the environment and the companies' stakeholders (Aranda-Usón et al., 2020).

Connected to environmental conditions and the identification of the company's potential to address the perception of external opportunities and threats systematically, we find the dynamic capabilities (DCs) view (Oliva et al., 2019). This approach, developed by Teece et al. (1997), arose in response to the need of the resources and capabilities-based view (RCBV) to answer how firms can retain competitive advantages in rapidly evolving environments. DCs enable firms to make business-oriented decisions agilely to promote the maintenance of existing competitive advantages while obtaining new ones (Lee et al., 2021). The DCs' outlook adopts a future-oriented view and is responsible for readjusting and transforming firms' existing resources and capabilities into innovative processes and solutions to respond to changes in uncertain and turbulent environments.

Teece et al. (2016) categorize DCs into three groups: (1) sensing capabilities, those that enable the organization to anticipate the unknown future by matching technological opportunities and threats with customer needs; (2) seizing capabilities, which allow exploiting business opportunities by leveraging organizational resources and capabilities; and (3) transforming capabilities, those that drive the organizational transformation through continuous change and renewal.

The integration of actions aimed at GT in companies, more specifically of CE models, requires the implementation of significantly complex changes of a dynamic nature, both at a technical level and, no less significantly, at an organizational level (Bertassini et al., 2021) According to Teece (2020), DCs contribute to the organization of a company's resources, through the changing demands of customers, the evolutionary possibilities offered by new technologies, and the need to respond to emerging changes in the environment. Additionally, the successful adoption of CE practices requires all supply chain partners to be committed to the model and seamlessly aligned in their activities (Riggs et al., 2023). Under such circumstances, DCs enable organizations to coordinate agilely among partners.

At this point, DCs perform a vital role in adopting and deploying CE practices and are the fundamental reason for our research work. DCs may become crucial for transitioning from linear business models to CE ones. At the same time, their theoretical framework arises as an optimum view to explore and understand these complex organizational change processes.

3 Methodology

Bibliometrics is a widespread discipline of scientific literature analysis, which applies mathematics and statistics to evaluate scientific research objectively, offering advantages in quantitative and modeled studies (Xie et al., 2020). For this reason, its use is spreading to many disciplines, particularly suitable for investigating large-volume, fragmented, and potentially controversial research streams (Aria & Cuccurullo, 2017).

In the first phase of the study, scientific databases were selected for extracting information, specifically Web of Science (WoS) and Scopus, which provide secure, reliable, and trustworthy information concerning scientific output and are the most frequently chosen for bibliometric analysis (Janik et al., 2020). WoS is a website-accessible database that provides access to various databases and citation data from a wide variety of disciplines with a number of records in excess of 90 million. WoS is the largest and most comprehensive database of diverse information resources worldwide (Xie et al., 2020). Like WoS, Scopus is a website that facilitates access to databases and citation data in different scientific areas related to the social sciences (Moral-Muñoz et al., 2020).

In the second phase of the research, a list of words related to the research topic and Boolean operators to be used in the search engine of the selected databases was established. The search performed was in WoS, "dynami* capabilit*" or "Resource* view*" (topic) and "circular* econom*" or "green practic*" or "green transiti*"(topic) and in Scopus; (TITLE-ABS-KEY ("dynami* capabilit*" OR

"Resource* view*") AND TITLE-ABS-KEY ("circular* econom*" OR "green practic*" OR "green transiti*")). The results were filtered by document type, "article," of those collected in both databases, obtaining 93 documents in WoS and 52 in Scopus. The data is extracted in txt and BibteX format and then unified in RSutudio to obtain a joint file in excel format in which the duplicates have been eliminated.

In the next step, once the information has been collected from the WoS and Scopus databases, we analyze the type of computer tools that will allow us to perform the bibliometric analysis. This research provides added value by combining the information collected in the WoS and Scopus databases. It was decided to use R Studio to unify the results and the Bibliometrix package with its Biblioshiny interface as software tools for bibliometric analysis. The following steps were followed to use the bibliometric software: first, RStudio was accessed in its web version, and the following command was typed in the console window: install. Packages ("bibliometrix") and library (bibliometrix). To access the software interface, type the command to run the Biblioshiny program: biblioshiny() (Ab Rashid & Aziz, 2022). Once inside, the unified file is loaded with the information extracted from the two databases in excel format to start the analysis and obtain the graphs. This process is summarized in Fig. 1.

Bibliometrix is a package in R language. This tool was developed by M. Aria and C. Cuccurullo of the University of Naples and L. Vanvitelli of the University of Campania (Italy) (Moral-Muñoz et al., 2020). Its use is recommended for bibliometric analysis and visual presentation of the whole process, such as for statistical analysis, data preprocessing, construction of co-occurrence matrices, co-citation analysis, co-word analysis, and cluster analysis. Biblioshiny allows analysis through graphs in which source, author, and document can be combined by means of three knowledge structures (conceptual, intellectual, and social). (Moral-Muñoz et al., 2020). In the present study, Bibliometrix has enabled the performance of a series of specific analyses, which contribute to answering the research questions defined above, as follows: annual distribution documents, and analysis of citation and analysis of main researcher (response to RQ1); analysis of citations (response to RQ2); analysis of high-frequency keywords (responses to RQ3 and RQ4); cluster analysis

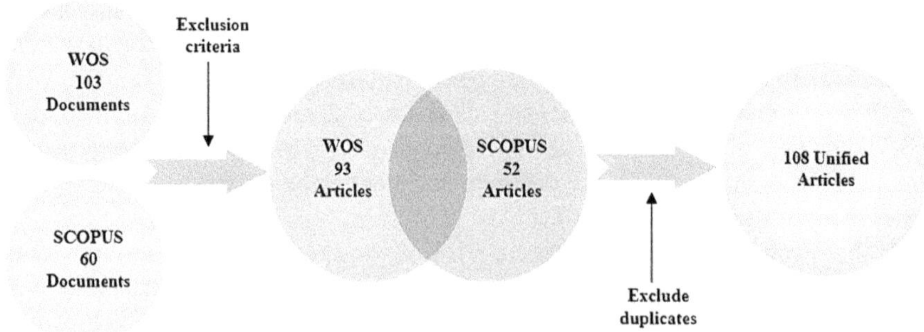

Fig. 1 Flowchart research methodology. *Source*: Author's elaboration

and multiple correspondence analysis of high-frequency keywords (response to RQ4); thematic evolution analysis (response to RQ4); thematic map and co-occurrence network mapping (response to RQ4); and article productivity mapping per country and across countries (response to RQ3).

Only a few bibliometric studies merge the two databases to perform a single integrated analysis (Caputo & Kargina, 2022). The most relevant value provided by the methodology followed in this bibliometric analysis is that it jointly analyzes the documents obtained in the WoS and Scopus databases.

The data collected from the 108 selected papers were subjected to inclusion and exclusion criteria according to Iden and Eikebrokk (2013). They establish data for processing the following: Journal name and reference, author/s and affiliations, countries of affiliations, keywords, research methods used and classification, supporting theoretical framework for the different theories addressed, research questions posed, and study summary.

In addition to using the Bibliometrix tool, the papers were critically analyzed (Bellucci et al., 2021) to establish the derived thematic clusters. As Paul et al. (2021) pointed out, the in-depth review of a topic favors advancing the related scientific theory.

4 Results

The results are initially presented by providing general information on the total "Dataset" and the annual scientific evolution, continuing with the analysis of citations and authors. Next, the documents are analyzed through high-frequency keywords to address finally a conceptual and social study.

4.1 *Annual Distribution Documents*

The results after data processing show a statistical and factorial analysis. Metrics have been obtained for the performance study; total publications, number of authors (single author and co-authorship), number of active years of publication and productivity per year of publication, metrics related to total and average citations, and the collaboration index, as shown in Table 1.

This technique has been used to determine the trend or trajectory of DCs, CE-related research in the GT framework, and map the intellectual structure. These results provide information on related past, present, and future research. The data used have been collected from the WoS and Scopus databases and analyzed jointly, which allows complete information to be obtained more than in cases diagnosed individually.

Regarding the first related publications, it can be seen that these are from 2014, but it was not until 2020 that significant growth was observed in the study area (see

Table 1 Main information

Description	Results
Timespan	2014:2022
Sources (journals, books, etc.)	49
Documents	108
Annual growth rate %	60.02
Document average age	2.37
Average citations per doc	26.52
References	9223
Keywords plus (ID)	442
Author's keywords (DE)	361
Authors	302
Authors of single-authored docs	6
Single-authored docs	6
Co-authors per doc	3.52
International co-authorships %	24.07
Article	102
Conference paper	4
Review	2

Source: Author's elaboration

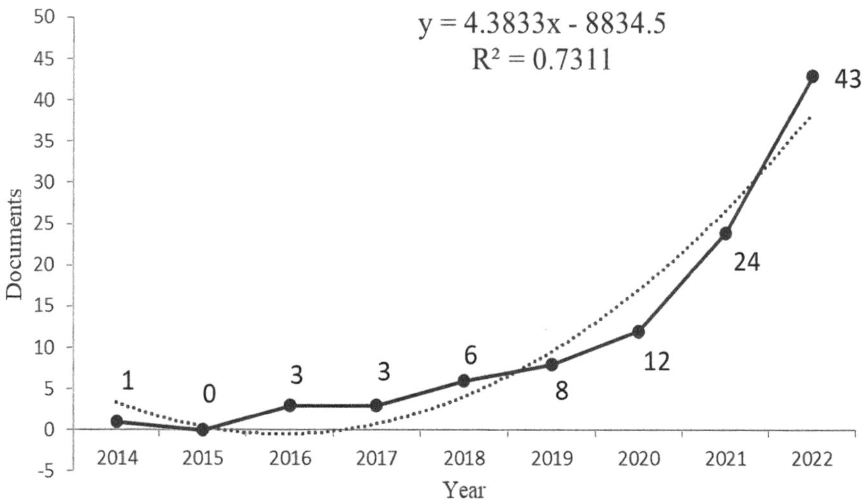

Fig. 2 Growth in scientific production. Note: Dotted line is the polynomial trend line. *Source*: Author's elaboration

Fig. 2). As a result of the low number of published articles, it can be determined that research in this field is still in its initial stage, although it has begun to develop in recent years. This fact may be related to the regulations adopted by the European Commission in the "European Green Pact."

4.2 Analysis of Citation and Analysis of Main Researcher

The average number of citations per year can be seen in Fig. 3. The highest number per year occurred in 2019, with 23 citations, followed by 2020, with 19.17 citations. After this period (2019–2020), in the year 2022, a decrease is shown. However, this is an ordinary and insignificant fact due to being close to the current period.

The authors of most publications and the evolution of their production by year are shown in Fig. 4. This figure shows the authors with the highest number of publications and their development over time, so it is possible to establish periods of higher concentration globally and by authors. S. Scarpellini is in the first position with seven publications. In particular, 2018 and 2020 are the years that produce a higher concentration of scientific works. S. Scarpellini is co-authoring several research pieces from a business approach, which connect CE with DCs. In particular, her work focuses on the transition of companies towards CE models and the critical role of DCs. More specifically, S. Scarpellini highlights how DCs enable the implementation of eco-innovative processes and environmental management in companies that foster the adoption of circular business models. P. Portillo-Tarragona is also a co-author of several related research papers. Most of her research focuses on corporate finance for the CE. She approaches the DCs' view as a framework for studying different environmental capabilities companies leverage to deploy circular practices. However, the first author to publish on the related research area is S.P. Gunasekaran in a work by Ganapathy et al. (2014), although he will not publish again until 2022. The University of Zaragoza (Spain) ranks first in related publications, with Spain having the highest related scientific production.

Also relevant is the analysis of the impact of the works addressing the related topics, measured through the number of citations generated in subsequent works. As shown in Table 2, and because of the novel academic interest in the GT, CE, and DCs concepts and their connecting links, 2019 has the most significant number of

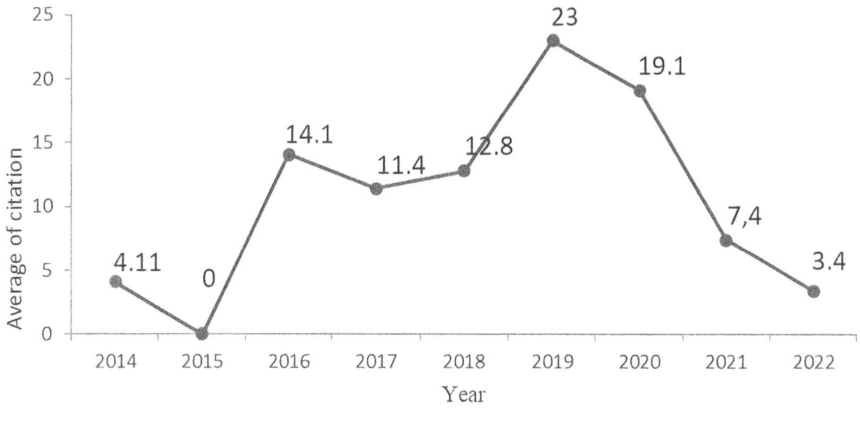

Fig. 3 Average citations per year. *Source*: Author's elaboration

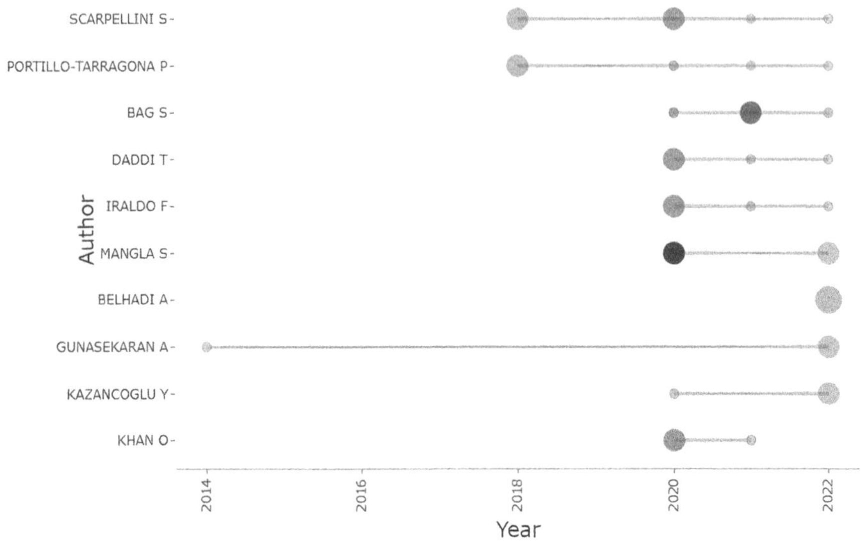

Fig. 4 Authors' Production over Time. *Source*: Bibliometrix Tool

citations derived from a single published paper. This work by Pieroni et al. (2019) currently accumulates 353 citations in subsequent works.

Other relevant information from the data in Table 2 is the scientific journals with the most citations. In particular, Elsevier's "Journal of Cleaner Production" stands out regarding the number of citations derived. This is not a surprising fact, given the scope of the journal. It focuses on sustainable production processes, the environment, sustainable practices in manufacturing business environments, and supply chain management. Thus, seven of the 20 papers with the highest number of citations derived have been published in this journal.

The historiographic analysis shows the chronological map of the most relevant citations or the historical development of the most influential documents in chronological order. Figure 5 shows how two citation networks are constituted from the first publication in 2016 to the last articles published in 2022. The leading network encompasses most of the authors from 2016 to 2022, while the second one presents only two authors: T. Kankonen, cited by A. Arora. A key node is observed in M.P.P. Pieroni, who leads several research branches, from his review of approaches to business model innovation for CE and sustainability, based on a systematic review of the academic literature and practitioner-based methodologies.

Table 2 Most cited papers within scientific production/output

Ranking	Author/s	Year	Title	Source	Total citations
1	Pieroni et al.	2019	Business model innovation for circular economy and sustainability a review of approaches	Journal of cleaner production	353
2	Albort-Morant et al.	2016	The antecedents of green innovation performance. A model of learning and capabilities	Journal of business research	234
3	Hong et al.	2018	Sustainable supply chain management practices, supply chain dynamic capabilities, and enterprise performance	Journal of cleaner production	209
4	Yadav et al.	2020	A framework to overcome sustainable supply chain challenges through solution measures of industry 40 and circular economy: An automotive case	Journal of cleaner production	183
5	Bag et al.	2021	Industry 4.0 adoption and 10R advance manufacturing capabilities for sustainable development	International journal of production economics	148
6	Bag et al.	2020	Procurement 4.0 and its implications on business process performance in a circular economy	Resources, conservation, and recycling	125
7	Kiefer et al.	2019	Drivers and barriers of eco-innovation types for sustainable transitions. A quantitative perspective	Business strategy and the environment	110
8	Lopes et al.	2019	Circular economy business models and operations management	Journal of cleaner production	110
9	Weissbrod et al.	2017	Developing sustainable business experimentation capability, a case study	Journal of cleaner production	99
10	Bai et al.	2017	An implementation path for green information technology systems in the Ghanaian mining industry	Journal of cleaner production	87
11	Khan et al.	2020	Microfoundations of dynamic capabilities: Insights from circular economy business cases	Business strategy and the environment	84

(continued)

Table 2 (continued)

Ranking	Author/s	Year	Title	Source	Total citations
12	Prieto-Sandoval et al.	2019	Key strategies, resources, and capabilities for implementing circular economy in industrial small and medium enterprises	Corporate social responsibility and environmental management	82
13	Katz-Gerro et al.	2019	Mapping circular economy activities in the European Union: Patterns of implementation and their correlates in small and medium-sized enterprises	Business strategy and the environment	69
14	Kumar et al.	2018	Missing link between sustainability collaborative strategy and supply chain performance role of dynamic capability	International journal of production economics	54
15	Scarpellini et al.	2020	Dynamic capabilities and environmental accounting for the circular economy in businesses	Sustainability accounting, management, and policy journal	54
16	Scarpellini, Valero-Gil et al.	2020	Environmental management capabilities for a "circular eco-innovation"	Business strategy and the environment	54
17	Perez-Valls et al.	2016	Green practices and organizational design as sources of strategic flexibility and performance	Business strategy and the environment	44
18	Melkonyan et al.	2020	Sustainability assessment of last-mile logistics and distribution strategies. The case of local food networks	International journal of production economics	43
19	Scarpellini et al.	2018	Defining and measuring different dimensions of financial resources for business eco-innovation and the influence of the firms' capabilities	Journal of cleaner production	40
20	Kahkonen et al.	2018	Sustainable supply management practices: Making a difference in a firm's sustainability performance	Supply chain management-an international journal	37

Source: Author's elaboration

4.3 Analysis of High-Frequency Keywords

Regarding the keywords, it is noteworthy that after the analysis of all the papers from the homogenization of the queries in the WoS and Scopus databases, the terms

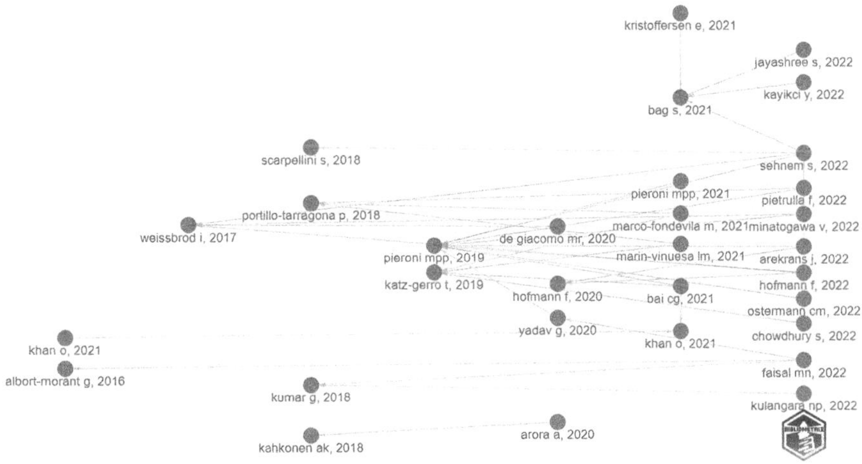

Fig. 5 Historiograph. Historical direct citation network. *Source*: Bibliometrix Tool

Fig. 6 Word Cloud. *Source*: Author's elaboration

with the highest number of repetitions are, as can be seen in Fig. 6: "dynamic capabilities," "circular economy," and "sustainability." These results align with this work's objectives in that they confirm how, within the related literature, these concepts are a recurrent fact and evidence of the relevance of their connection. Another relevant finding is the appearance in several of its formulations of the innovation, one of the dynamic capabilities most supported by previous literature. Additionally, performance is shown in different aspects (financial, company, industrial), a sign of the effects analyzed mainly by the studies connecting the two fields of research we are approaching.

In addition to the relevance of the keywords used in the papers, a result of their bibliometric analysis is the relationship among keywords, authors, and journals in which they were published (see Fig. 7). The value of this analysis lies in the study of the keywords in a connected way in the related works, their authors, and the journals in which these works are published. To this end, a Sankey diagram is performed (Fig. 7).

The Keywords "circular economy" and "dynamic capabilities" occupy the central place. These two Keywords are found in all the works of the most prolific authors in the related topics, mainly deriving from the scientific journals with the highest related production.

The journals with the highest number of papers are "Business strategy and the Environment" and "Journal Cleaner Production," which are very distributed in the authors' case, although S. Scarpellini and P. Portillo-Tarragona stand out in terms of the number of papers.

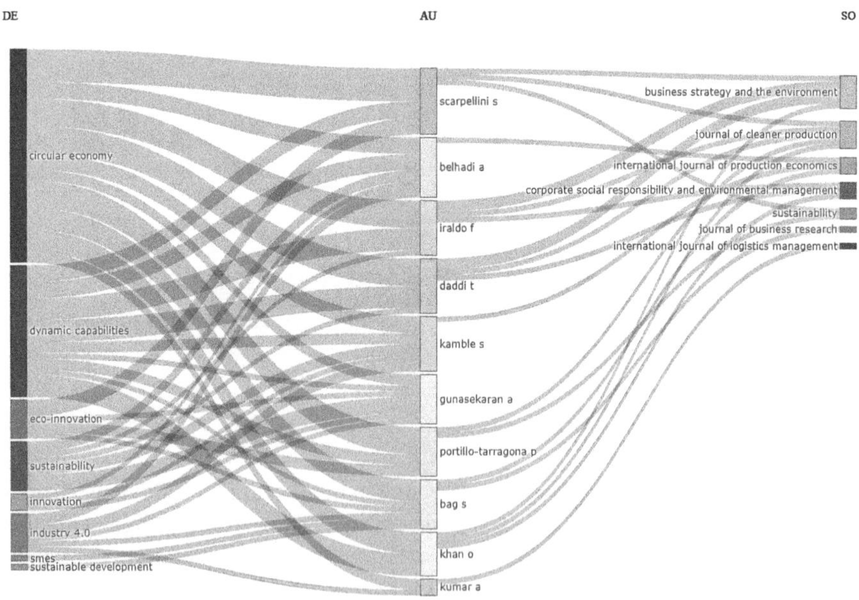

Fig. 7 Three-Field Plot (authors, keywords, and sources). *Source*: Bibliometrix Tool

4.4 Cluster Analysis and Multiple Correspondence Analysis of High-Frequency Keywords

The factor analysis identifies the level of association between the keywords of the articles studied, as shown in Fig. 8. Specific thematic associations, such as innovation, CE, design, model, strategy, and sustainability, are observed in the red region. The blue area shows the critical points of these associations, such as business model, planning and sustainable development, etc.

The Bibliometrix package of R allows the conceptual function Structure to perform a multiple correspondence analysis (MCA). This establishes a conceptual structure of the field and a clustering of K-means that identifies groups of documents that address common concepts (Aria & Cuccurullo, 2017). In MCA analysis, words are represented on a two-dimensional map, and the results are interpreted from the relative positions of the points and their distribution along the dimensions (Cuccurullo et al., 2016). The use of MCA has been extended as an analysis with a sociological approach that compresses large amount of data with multiple variables in a low-dimensional space allowing it to be represented in a two-dimensional graph. This graph uses the distance from the plane to reflect the similarity between keywords approaching the central point according to the attention received in recent years (Xie et al., 2020).

The objective of the MCA technique is to find the latent factors that create a common element according to the data records. In this case, the factor analysis has

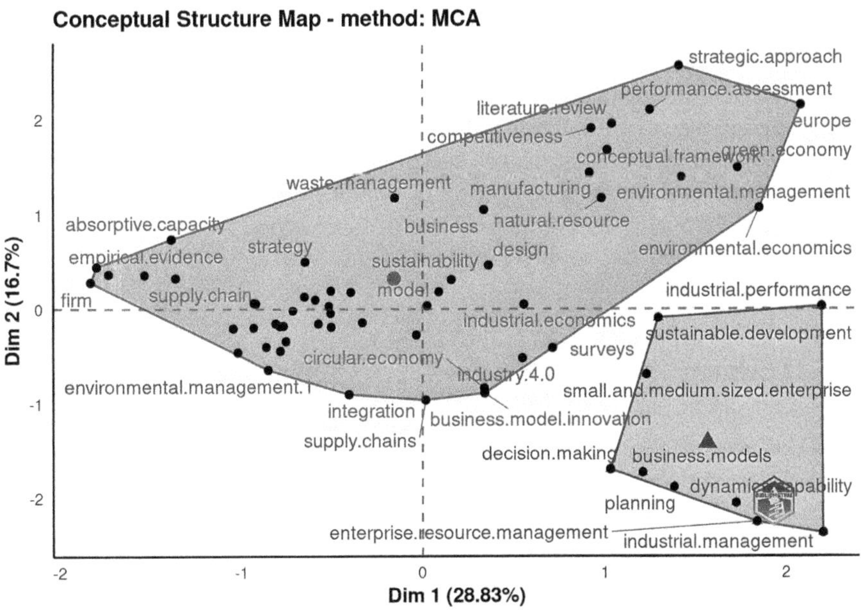

Fig. 8 Conceptual Structure Map (MCA). *Source*: Bibliometrix Tool

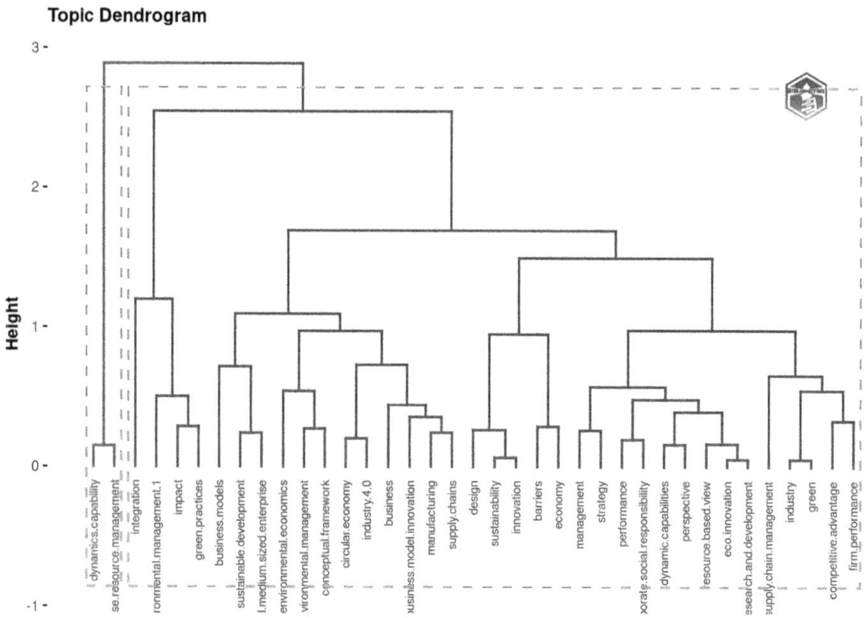

Fig. 9 Dendrogram of hierarchical keyword cluster analysis tree. *Source*: Bibliometrix Tool

generated two keyword classifications. On the one hand, the red classification represents more general keywords, such as "model," "economy," "sustainability," "innovation," "design," etc. On the other hand, the blue sort means more specific keywords such as "business models," "dynamics capability," "planning," "decision making," etc. Those words closest to the edges, such as "industrial management", "enterprise resource management," or "strategic approach," show a transition to other topics in the graph.

Similarly, the dendrogram shows the groups formed by creating clusters of observations at each step and their levels of similarity. The similarity story is measured on the vertical axis (alternatively, the distance level can be shown), and the different observations are specified on the horizontal axis (Fig. 9).

4.5 *Thematic Evolution Analysis*

Evolution analysis focuses on the evolution over time of changing norms, relationships, trajectories, and trends in content, strength, and thematic structures (Xie et al., 2020). The evolution of the terms is represented by a Sankey diagram (see Fig. 10), which shows how the development of the words "dynamic capabilities" began in the early years when related research emerged, evolving in recent years to the concept of "circular economy."

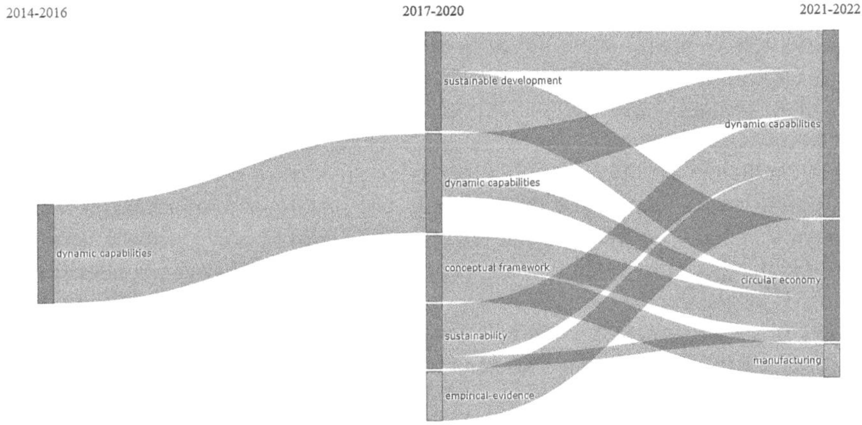

Fig. 10 Thematic Evolution. *Source*: Bibliometrix Tool

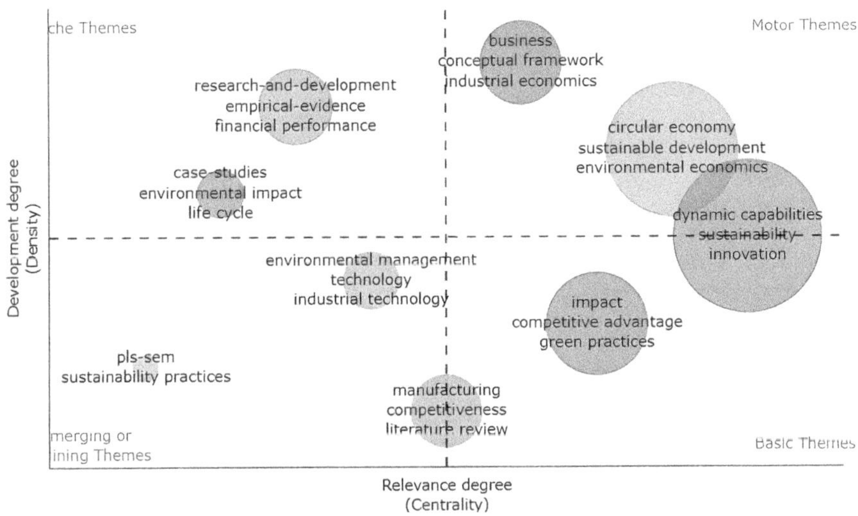

Fig. 11 Thematic Map. *Source*: Bibliometrix Tool

4.6 Thematic Map and Co-occurrence Network Mapping

A two-dimensional conceptual map can be drawn on two axes from the Keywords. On the ordinate axis is the density, which indicates the degree of development of the themes measured by the internal associations between the words or the development of the theme. On the abscissa axis, centrality is represented. This reflects the relevance of the topics, measured by the external associations between the words or the importance of the specific issue within the general scope. As shown in Fig. 11, it is divided into four quadrants: high density and centrality offer the driving themes.

Low density and high centrality represent the essential topics addressed. High density and low centrality show the niche topics. Finally, low density and centrality show the emerging issues. The driving themes, derived from the results, most developed are "business" and "industrial economics," followed by the themes "circular economy" and "sustainable development," which present a higher centrality. The emerging themes are "PLS" and "sustainability practices."

A bibliometric mapping of the keywords was performed to understand the conceptual structure derived from the results. Using a word co-occurrence network, each keyword is represented as a node, and each match of two words is described as a link (Radhakrishnan et al., 2017). The number of times a pair of words coexist in different articles constitutes the link weight connecting the pair. The constructed network, therefore, represents the cumulative knowledge from which information concerning the knowledge patterns and robustness of the links between these keywords can be obtained. This co-occurrence analysis is shown in Fig. 12. The words that stand out are "dynamic capabilities" and "circular economy," presenting larger nodes and, consequently, the higher frequency of keywords. The distance between the concepts shows how much attention has been paid to the effects of these concepts on each other (Sedighi, 2016). The location of this cluster in the central part

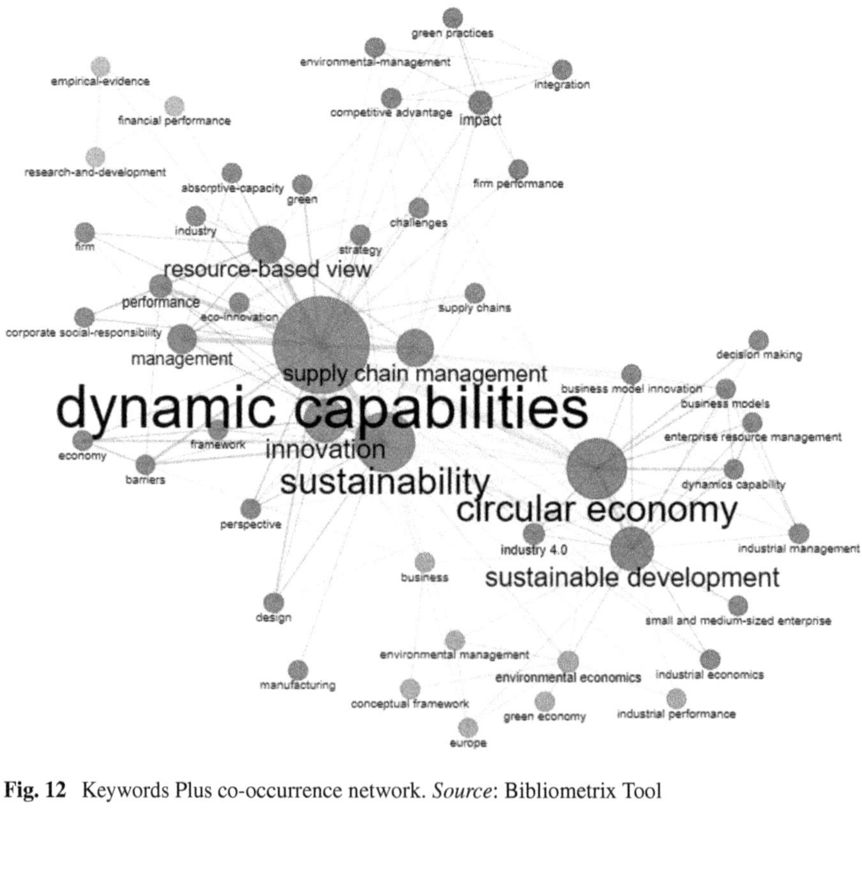

Fig. 12 Keywords Plus co-occurrence network. *Source*: Bibliometrix Tool

of the figure shows the most excellent contact it has with the rest, being subject-related to all of them.

The centrality indicator is essential to network analysis (Sedighi, 2016). The study of intermediation centrality shown by the keywords with the highest degree of intermediation centrality includes "dynamic capabilities," "circular economy," "sustainable development," "sustainability," and "supply chain management." These results show that these concepts have a relevant role in data transmission within the network and are considered subjects with an interdisciplinary approach. Table 3 shows the 20 words with the highest degree of betweenness centrality and closeness centrality.

Betweenness centrality quantifies the number of times a node acts as a bridge along the shortest path between two other nodes. Closeness centrality indicates the distance of a node from all other nodes in the network and is calculated as the average of the shortest path length from the node to all other nodes.

Table 3 Ranking of keywords with the highest degree of intermediation centrality and closeness centrality

Node	Betweenness	Node	Closeness
Dynamic capabilities	567.1949905	Dynamic capabilities	0.016393443
Circular economy	161.6610168	Circular economy	0.013888889
Sustainable development	156.0490758	Sustainability	0.013513514
Sustainability	130.3017306	Sustainable development	0.013157895
Supply chain management	64.44615952	Innovation	0.012500000
Innovation	40.95496995	Supply chain management	0.012048193
Environmental economics	30.41266796	Resource-based view	0.011363636
Business	23.84413373	Management	0.011235955
Impact	22.31408831	Competitive advantage	0.010989011
Resource-based view	17.63996717	Industry 4.0	0.010989011
Competitive advantage	15.62765189	Performance	0.010989011
Management	10.92559738	Business models	0.010869565
Environmental-management	6.898689544	Business	0.010752688
Green practices	5.034007346	Impact	0.010752688
Business models	3.822425015	Business model innovation	0.010638298
Industry 4.0	1.904033584	Perspective	0.010638298
Dynamics capability	1.870542553	Design	0.010416667
Performance	1.559132658	Framework	0.010416667
Barriers	1.111906663	Environmental-management	0.010309278
Firm performance	0.423230516	Barriers	0.010309278

Source: Author's elaboration

4.7 *Article Productivity Mapping Per Country and Across Countries*

The countries with the highest related scientific productivity are (see Fig. 13) Spain, the United Kingdom, and Italy, with fifteen, twelve, and eight papers, respectively. Denmark, India, and Malaysia follow them with six Notably, the United Kingdom has a much higher inter-country collaboration than Spain, although Spain has the highest number of papers.

Figure 14 shows the countries' relationship with the keywords and the affiliation of their authors. The main keywords "circular economy" and "dynamic capabilities" have a more significant relationship with all the countries, although to a greater extent in the case of Spain, India, the United Kingdom, and USA, as well as with almost all the universities. However, the University of Zaragoza stands out above the rest.

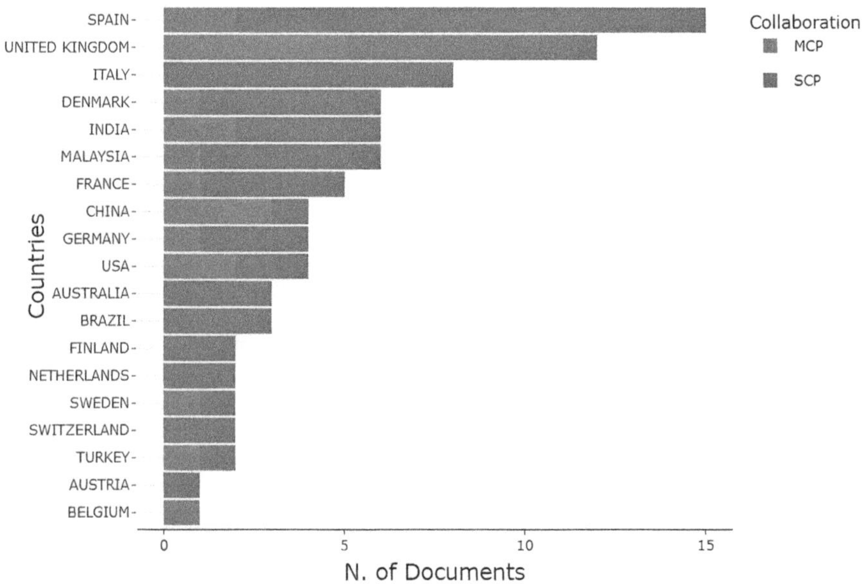

Fig. 13 Corresponding author's country. Intra-country (SCP) and inter-country (MCP) collaboration during 2015–2019. *Source*: Bibliometrix Tool

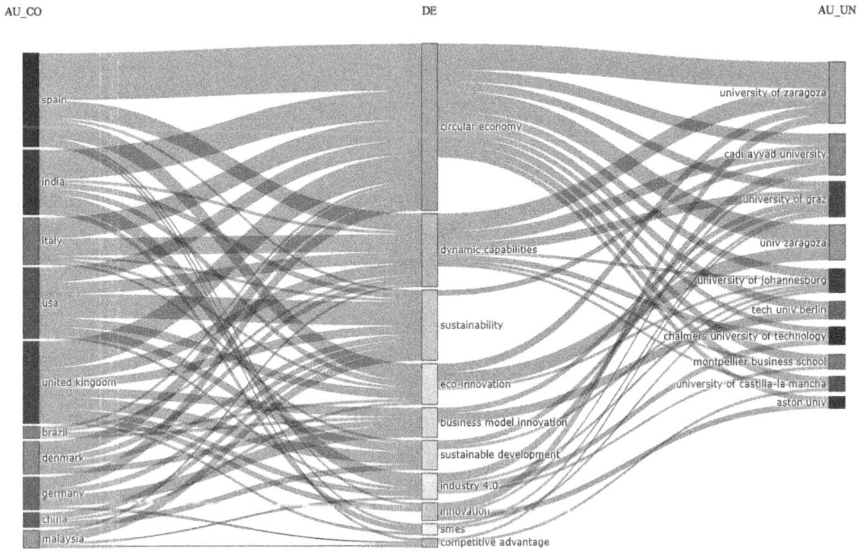

Fig. 14 Relationship between countries, Keywords, and affiliations. *Source*: Bibliometrix Tool

5 Discussion

From analyzing the results after applying the bibliometric techniques, answers can be given to the research questions initially raised in this paper.

RQ1. Which authors are at the forefront of the scientific production from the related scientific literature? The results that provide an answer to this question derive from Sect. 4.1, which describes the annual distribution of the documents, and from Sect. 4.2, in which the analysis of citations and the analysis of main researchers are deployed. Although research in a linked way between the realities of DCs and the GT of companies through CE is a relatively recent field of study (the first works date back to 2014) and therefore emerging, it can be stated that there already exists a sound body of works to identify the most prolific authors in their scientific production addressing these connected realities.

S. Scarpellini, affiliated with the University of Zaragoza (Spain), is the author with the highest number of scientific papers published, followed by P. Portillo-Tarragona, also from the University of Zaragoza (Spain). This shared affiliation undoubtedly shows the preponderant role of Spain in terms of related scientific research and, in particular, Zaragoza and its university. These two authors stand with the highest related production despite this being concentrated over four years, specifically 2018–2022. The author A. Gunasekaran stands out because, although he was the first to address these issues in 2014, his next published related work refers to 2022.

RQ2. Which have been the papers with the highest impact in subsequent related works? The foundation to answer this question is based on the results of the analysis of citations in Sect. 4.1. Despite the emergence of these issues in scientific research, some works with a clear relevance and impact on the following related published works can already be identified. Measuring this potential impact by the number of citations in subsequent works, the result of Pieroni et al. (2019) stands out in the first place. In the case of this work, 353 citations appear in subsequent works. A relevant fact about this work is its relative proximity in time, as it was published in 2019. The second work with the highest number of citations is that of Albort-Morant et al. (2016), which has 234 citations in subsequent related works. This work is one of the first research works that connected these fields, which is essential to highlight its influence on the following works. Next is the work of Hong et al. (2018), which has generated 209 citations in the following works. A relevant finding about these papers is shown when analyzing the ten articles that have generated more citations in subsequent works: six published in the Journal of Cleaner Production. This result confirms this publication's outstanding role in the number of related published papers and its impact on subsequent research.

RQ3. What are the main topics investigated, which countries lead this scientific production, and which keywords are mostly used in the scientific literature analyzed? Results obtained through the analysis of main researchers (Sect. 4.2) clearly show that, after the bibliometric and content analysis of the works under study, it is verifiable that the relationships of DCs and CE to encourage GT in companies are addressed in the scientific literature, mainly under the scope of production and operations areas. This finding perfectly aligns with the requirement of complex cooperation and integration of all the value chain actors for the feasible adoption of CE models. The emerging research sees DCs as an enabler of these mechanisms, allowing the rapid configuration of organizational resources and capabilities, mainly by sensing and seizing environmental threats and opportunities and transforming organizational routines and processes. DCs are facilitating elements, even antecedents, for the CE practices in companies. Another relevant finding regarding the main topics investigated is the connection of DCs and CE with all aspects of the so-called Industry 4.0. This is a logical result, given the increasing (and even dependence) of circular actions in companies with information and communication technologies, which are undoubtedly connected to the DCs view. These two concepts can only be fully comprehended with the foundation of recent IT developments.

Regarding the countries that lead the related scientific production, the article productivity mapping per country and across countries included in Sect. 4.7 shows that Spain undoubtedly occupies the first position, fundamentally with the University of Zaragoza at the head. Two of the most prolific authors of scientific papers addressing both research fields in a related way are affiliated with this university: S. Scarpellini and P. Portillo-Tarragona. Behind Spain, the United Kingdom and Italy have the highest related scientific production. This fact is relevant since most countries identified correspond to those that have experienced an increase in production and where the affiliations of authors with the most significant impact on the number of citations in subsequent related works are found.

Finally, focusing on the keywords with the highest presence in the analyzed works, which are extracted through the analysis of high-frequency keywords (Sect. 4.3), these are undoubtedly directly related to the topics investigated in the associated works. Thus, the obvious Keywords "Circular Economy," "Dynamic Capabilities," and "Sustainability" stand out above the rest. However, Keywords also appear related to production processes and operations, just in line with the previous finding related to the main scope of the papers analyzed. This is the case with the Keywords: "Supply Chain Management" and "Industry 4.0". Another relevant finding is the appearance in several of its formulations of the keyword "innovation", as one of the dynamic capabilities most supported by the previous literature, as a unique link between the DCs view and the CE models. Finally, "performance" is shown under different aspects (financial, company, industrial), reflecting the effects mainly analyzed by the studies connecting the two research fields we are approaching. The relative reduced importance of those keywords related to the strategic field is remarkable, clearly showing future opportunities for upcoming studies.

RQ4. What conceptual, intellectual, and social structures were generated from the scientific literature analyzed? The main results obtained from the thematic map and co-occurrence network mapping in Sect. 4.6, together with the cluster analysis and multiple correspondence analysis of high-frequency keywords performed in Sect. 4.4, and the thematic evolution analysis in Sect. 4.5 conclude on two major conceptual areas related to scientific production. Firstly, and as a significant conceptual area, the relevance of the DCs for the GT of companies through the CE is directly connected in the different works to diverse components of product development and operations processes. This result highlights the need for companies that wish to implement actions favoring the GT, not just as a sole managerial decision but as a participative decision that must be taken, requiring the active implication of various companies' core processes and their related personnel. Secondly, connecting these realities with the concepts and facts of generating new business models based on new business approaches is no less critical. This is undoubtedly not only a decision applicable in the case of large companies with essential production processes, but it is also possible to implement it in different types of companies.

Regarding the theoretical implications of this work, the present analysis offers the first bibliographic approach to the relationship and links between these two concepts -DCs and CE- as well as one of the scarcest bibliometric analyses that merge two different databases. This paper has shown at a glance the current state of the scientific literature and how the companies' GT processes, through the adoption of CE practices and models, are closely linked to the existence and development of a series of DCs, which can even act as enablers or antecedents. This work is a starting point for other researchers who want to address these issues that still need to be well-connected in the scientific literature.

The potential practical implications of the results derived from this work are also important, mainly for the managers of companies that are immersed in or want to plan processes of GT through the CE. Being conscious of the value that the existence and development of diverse DCs in the company will have for successfully deploying these processes is crucial for achieving the proposed objectives.

As with all research work, this has been subject to several limitations. In this case, the most critical hurdle has been determined by the possible dispersion of the information and, on occasions, limited to specific research questions that may have yet to favor a complete overview of the issues relatedly.

6 Conclusions

The most relevant conclusions can be drawn from the analysis and discussion of results obtained in this research work, which have been set out in the section above.

Research about the role that DCs can play to ease the adoption of CE models by companies to tackle the defiant challenges of the GT is relatively novel, as the first paper dates to 2014. Nevertheless, based on the results in scientific output and the impact of the work in subsequent research, it can be stated that this is becoming an exciting emerging field.

Although the scientific works that address these issues are linked to authors with diverse affiliations in a global scope, the prominent role in the related scientific production of European countries is noteworthy. This is the case of Spain, with the University of Zaragoza, and other EU countries. This circumstance is undoubtedly evidence of two complementary trends: (1) the regulatory impulse from the European Union in the last decade; and (2) the growing awareness of companies to develop more resilient and sustainable business models based on innovation and more regenerative production and consumption systems. Therefore, both facts have also gained the scientific community's attention to studying the underlying mechanisms that can play a relevant role in these changes, namely DCs and CE.

The achievement of GT through CE practices is spreading among all management and business activity areas. However, these realities still have a significant weight within business production processes and operations. More specifically, a clear orientation for their integration within the industry 4.0 phenomenon is patent. The development of DCs and contributions to implementing circular practices can only be conceived with the co-existence of advanced information and communication technologies and their related managerial capabilities in firms.

Derived from this last conclusion, the study and analysis of these realities in the context of Industry 4.0 and the fundamental role that technology can play in addressing new related challenges and establishing the necessary measures to achieve the objectives proposed by companies and organizations can be found as future lines of research based on this work. Additionally, the focus on general strategic aspects that complement the current operational majority view could unveil exciting opportunities to understand and exploit the relationships between DCs and CE regarding performance.

Finally, future research possibilities this chapter opens include a review of the literature that could, in greater depth, uncover more specific research trends and gaps for future extensions.

References

Ab Rashid, M. F., & Aziz, M. A. A. (2022). A comprehensive overview of world mapping analysis research trends on impact of artificial intelligence in tourism from 2000 to 2022: A literature review and bibliometric analysis. *ICRRD Quality Index Research Journal, 3*(3), 136–160. https://doi.org/10.53272/icrrd.v3i3.4

Albort-Morant, G., Leal-Millán, A., & Cepeda-Carrión, G. (2016). The antecedents of green innovation performance: A model of learning and capabilities. *Journal of Business Research, 69*(11), 4912–4917. https://doi.org/10.1016/j.jbusres.2016.04.052

Aranda-Usón, A., Portillo-Tarragona, P., Scarpellini, S., & Llena-Macarulla, F. (2020). The progressive adoption of a circular economy by businesses for cleaner production: An approach from a regional study in Spain. *Journal of Cleaner Production, 247*, 119648. https://doi.org/10.1016/j.jclepro.2019.119648

Aria, M., & Cuccurullo, C. (2017). Bibliometrix: An R-tool for comprehensive science mapping analysis. *Journal of Informetrics, 11*(4), 959–975. https://doi.org/10.1016/j.joi.2017.08.007

Bellucci, M., Marzi, G., Orlando, B., & Ciampi, F. (2021). Journal of intellectual capital: A review of emerging themes and future trends. *Journal of Intellectual Capital, 22*(4), 744–767. https://doi.org/10.1108/JIC-10-2019-0239

Bertassini, A. C., Ometto, A. R., Severengiz, S., & Gerolamo, M. C. (2021). Circular economy and sustainability: The role of organizational behavior in the transition journey. *Business Strategy and the Environment, 30*(7), 3160–3193. https://doi.org/10.1002/bse.2796

Bocken, N. M. P., de Pauw, I., Bakker, C., & van der Grinten, B. (2016). Product design and business model strategies for a circular economy. *Journal of Industrial and Production Engineering, 33*(5), 308–320. https://doi.org/10.1080/21681015.2016.1172124

Caputo, A., & Kargina, M. (2022). A user-friendly method to merge Scopus and web of science data during bibliometric analysis. *Journal of Marketing Analytics, 10*(1), 82–88. https://doi.org/10.1057/s41270-021-00142-7

Chowdhury, S., Dey, P. K., Rodríguez-Espíndola, O., Parkes, G., Tuyet, N. T. A., Long, D. D., & Ha, T. P. (2022). Impact of Organisational factors on the circular economy practices and sustainable performance of small and medium-sized Enterprises in Vietnam. *Journal of Business Research, 147*, 362–378. https://doi.org/10.1016/j.jbusres.2022.03.077

Cuccurullo, C., Aria, M., & Sarto, F. (2016). Foundations and trends in performance management. A twenty-five years bibliometric analysis in business and public administration domains. *Scientometrics, 108*(2), 595–611. https://doi.org/10.1007/s11192-016-1948-8

European Commission. (2020). *A new Circular Economy Action Plan. For a cleaner and more competitive Europe.* https://Eur-Lex.Europa.Eu/Legal-Content/EN/TXT/HTML/?Uri=CELEX:52020DC0098&from=EN

Ganapathy, S. P., Natarajan, J., Gunasekaran, A., & Subramanian, N. (2014). Influence of eco-innovation on Indian manufacturing sector sustainable performance. *International Journal of Sustainable Development & World Ecology, 21*(3), 198–209. https://doi.org/10.1080/13504509.2014.907832

Geisendorf, S., & Pietrulla, F. (2018). The circular economy and circular economic concepts—A literature analysis and redefinition. *Thunderbird International Business Review, 60*(5), 771–782. https://doi.org/10.1002/tie.21924

Geissdoerfer, M., Savaget, P., Bocken, N. M. P., & Hultink, E. J. (2017). The Circular economy – A new sustainability paradigm? Journal of Cleaner Production,143, 757–768. https://doi.org/10.1016/j.jclepro.2016.12.048. Elsevier Ltd.

Gonçalves Castro, C., Trevisan, A. H., Pigosso, D. C. A., & Mascarenhas, J. (2022). The rebound effect of circular economy: Definitions, mechanisms and a research agenda. *Journal of Cleaner Production, 345*, 131136. https://doi.org/10.1016/j.jclepro.2022.131136

Herrero-Luna, S., Ferrer-Serrano, M., & Latorre-Martinez, M. P. (2022). Circular economy and innovation: A systematic literature review. *Central European Business Review, 11*(1), 65–84. https://doi.org/10.18267/j.cebr.275

Holzer, D., Rauter, R., Fleiß, E., & Stern, T. (2021). Mind the gap: Towards a systematic circular economy encouragement of small and medium-sized companies. *Journal of Cleaner Production, 298*, 126696. https://doi.org/10.1016/j.jclepro.2021.126696

Homrich, A. S., Galvão, G., Abadia, L. G., & Carvalho, M. M. (2018). The circular economy umbrella: Trends and gaps on integrating pathways. *Journal of Cleaner Production, 175*, 525–543). Elsevier Ltd. https://doi.org/10.1016/j.jclepro.2017.11.064

Hong, J., Zhang, Y., & Ding, M. (2018). Sustainable supply chain management practices, supply chain dynamic capabilities, and enterprise performance. *Journal of Cleaner Production, 172*, 3508–3519. https://doi.org/10.1016/j.jclepro.2017.06.093

Iden, J., & Eikebrokk, T. R. (2013). Implementing IT service management: A systematic literature review. *International Journal of Information Management, 33*(3), 512–523. https://doi.org/10.1016/j.ijinfomgt.2013.01.004

Janik, A., Ryszko, A., & Szafraniec, M. (2020). Scientific landscape of smart and sustainable cities literature: A bibliometric analysis. *Sustainability (Switzerland), 12*(3), 779. https://doi.org/10.3390/su12030779

Kannan, D., Shankar, K. M., & Gholipour, P. (2022). Paving the way for a green transition through mitigation of green manufacturing challenges: A systematic literature review. *Journal of Cleaner Production, 368*, 132578. https://doi.org/10.1016/j.jclepro.2022.132578

Kemp-Benedict, E. (2018). Investing in a green transition. *Ecological Economics, 153*(August), 218–236. https://doi.org/10.1016/j.ecolecon.2018.07.012

Korhonen, J., Nuur, C., Feldmann, A., & Birkie, S. E. (2018). Circular economy as an essentially contested concept. *Journal of Cleaner Production, 175*, 544–552. https://doi.org/10.1016/j.jclepro.2017.12.111

Lee, J. M., Narula, R., & Hillemann, J. (2021). Unraveling asset recombination through the lens of firm-specific advantages: A dynamic capabilities perspective. *Journal of World Business, 56*(2), 101193. https://doi.org/10.1016/j.jwb.2021.101193

Marra, A., Antonelli, P., & Pozzi, C. (2017). Emerging green-tech specializations and clusters – A network analysis on technological innovation at the metropolitan level. *Renewable and Sustainable Energy Reviews, 67*, 1037–1046. https://doi.org/10.1016/j.rser.2016.09.086

Mio, C., Panfilo, S., & Blundo, B. (2020). Sustainable development goals and the strategic role of business: A systematic literature review. *Business Strategy and the Environment, 29*(8), 3220–3245. https://doi.org/10.1002/bse.2568

Moral-Muñoz, J. A., Herrera-Viedma, E., Santisteban-Espejo, A., & Cobo, M. J. (2020). Software tools for conducting bibliometric analysis in science: An up-to-date review. *El Profesional de La Información, 29*, 1–20. https://doi.org/10.3145/epi.2020.ene.03

Oliva, F. L., Gomes Couto, M. H., Fernandes Santos, R., & Bresciani, S. (2019). The integration between knowledge management and dynamic capabilities in agile organizations. *Management Decision, 57*(8), 1960–1979. https://doi.org/10.1108/MD-06-2018-0670

Paul, J., Merchant, A., Dwivedi, Y. K., & Rose, G. (2021). Writing an impactful review article: What do we know and what do we need to know? *Journal of Business Research, 133*, 337–340. https://doi.org/10.1016/j.jbusres.2021.05.005

Pearce, D., & Turner, R. (1989). *Economics of natural resources and the environment*. Johns Hopkins University Press.

Pieroni, M. P. P., McAloone, T. C., & Pigosso, D. C. A. (2019). Business model innovation for circular economy and sustainability: A review of approaches. *Journal of Cleaner Production, 215*, 198–216. https://doi.org/10.1016/j.jclepro.2019.01.036

Radhakrishnan, S., Erbis, S., Isaacs, J. A., & Kamarthi, S. (2017). Novel keyword co-occurrence network-based methods to foster systematic reviews of scientific literature. *PLoS One, 12*(3). https://doi.org/10.1371/journal.pone.0172778. Public Library of Science.

Riggs, R., Roldán, J. L., Real, J. C., & Felipe, C. M. (2023). Opening the black box of big data sustainable value creation: The mediating role of supply chain management capabilities and circular economy practices. *International Journal of Physical Distribution & Logistics Management, 53*(7/8), 762–788. https://doi.org/10.1108/IJPDLM-03-2022-0098

Rojšek, I. (2001). From red to green: Towards the environmental management in the country in transition. *Journal of Business Ethics, 33*(1), 37–50. https://doi.org/10.1023/A:1011977116496

Sedighi, M. (2016). Application of word co-occurrence analysis method in mapping of the scientific fields (case study: The field of Informetrics). *Library Review, 65*(1–2), 52–64. https://doi.org/10.1108/LR-07-2015-0075

Singh, J., & Ordoñez, I. (2016). Resource recovery from post-consumer waste: Important lessons for the upcoming circular economy. *Journal of Cleaner Production, 134*, 342–353. https://doi.org/10.1016/j.jclepro.2015.12.020

Teece, D. J. (2020). Fundamental issues in strategy: Time to reassess? *Strategic Management Review, 1*, 1–45.

Teece, D. J., Peteraf, M., & Leih, S. (2016). Dynamic capabilities and organizational agility: Risk, uncertainty, and strategy in the innovation economy. *California Management Review, 58*(4), 13–35.

Teece, D. J., Pisano, G., & Shuen, A. (1997). Dynamic capabilities and strategic management. *Strategic Management Journal, 18*(7), 509–533.

van Buren, N., Demmers, M., van der Heijden, R., & Witlox, F. (2016). Towards a circular economy: The role of Dutch logistics industries and governments. *Sustainability (Switzerland), 8*(7), 647. https://doi.org/10.3390/su8070647

Velenturf, A. P. M., & Purnell, P. (2021). Principles for a sustainable circular economy. *Sustainable Production and Consumption, 27*(3), 1437–1457. https://doi.org/10.1016/j.spc.2021.02.018

Walker, A. M., Opferkuch, K., Roos Lindgreen, E., Raggi, A., Simboli, A., Vermeulen, W. J. V., Caeiro, S., & Salomone, R. (2022). What is the relation between circular economy and sustainability? Answers from frontrunner companies engaged with circular economy practices. *Circular Economy and Sustainability, 2*(2), 731–758. https://doi.org/10.1007/s43615-021-00064-7

Wang, C., Li, X., Wen, H. X., & Nie, P. Y. (2021). Order financing for promoting green transition. *Journal of Cleaner Production, 283*, 125415. https://doi.org/10.1016/j.jclepro.2020.125415

Williams, A., & Blasberg, L. A. (2022). SDG platforms as strategic innovation through partnerships. *Journal of Business Ethics, 180*, 1041–1057. https://doi.org/10.1007/s10551-022-05194-y

Xie, H., Zhang, Y., Wu, Z., & Lv, T. (2020). A bibliometric analysis on land degradation: Current status, development, and future directions. *Land, 9*(1), 28. https://doi.org/10.3390/LAND9010028

The Path to a Circular Economy Through a Green and Digital Transition. The Paradigm Shift in Mexican Companies in the State of Hidalgo

Liliana de Jesús Gordillo Benavente and Claudia Vega Hernández

Abstract This chapter addresses aspects related to the green and digital circular economy, which are important factors that must be considered nowadays. The objective of this research is to analyze how Mexican organizations in the State of Hidalgo perform the circular economy and digitization, taking into account sustainability features that they must comply with. This is a quantitative study, with a descriptive, correlational and explanatory research design and a systemic approach, the sample size consists of 545 micro and small companies from Hidalgo, the variables analyzed were ISO 26000 principles and features related to environmental management, and digital tools. This research main findings show that companies in the State of Hidalgo conduct activities that contribute to environmental care in a traditional way, such as: responsibility with clients without deceiving information, malpractice avoidance, regulation that promotes ethical behavior, a safe organizational environment, a safety impact of activities, a socially responsible behavior, pollution prevention and control production waste that can positively impact the environment, by recycling and being environmentally friendly. All these variables are carried out in a traditional manner, the results obtained with Likert scale of *agreement*; but the stages of the circular economy were not considered, other findings are that companies do not seek for business digitalization, preferring traditional ways to realize the activities, the items on digital technologies displays that companies don't know about technological alternatives therefore they don't use them, this last statement shows how imperative is for companies to have a transition to a green and digital circular economy and the importance of this paradigm shift in organizations.

Keywords Circular economy · Digitization · Paradigm shift

L. de Jesús Gordillo Benavente (✉) · C. V. Hernández
Universidad Politécnica de Tulancingo, Tulancingo, Hidalgo, Mexico
e-mail: liliana.gordillo@upt.edu.mx; claudia.vega@upt.edu.mx

C. Rego et al. (eds.), *Entrepreneurship, Technological Change and Circular Economy for a Green Transition*, Studies on Entrepreneurship, Structural Change and Industrial Dynamics, https://doi.org/10.1007/978-3-031-48079-9_4

1 Introduction

During the twenty-first century, it's very important that companies can be sustainable through a green and digital circular economy. The world is currently experiencing disasters caused by humans, these disasters can be prevented by improving social and governmental strategies as well as business strategies, and also by avoiding mistreating the environment around us. From this perspective, this chapter analyzes the performance of Mexican companies in the State of Hidalgo and how they face circular economy and digitalization. After this analysis a paradigm change that reinforces sustainability can be proposed.

According to the Organization for Economic Cooperation and Development (OECD), cited in Becker (2018), an economy that emits little carbon, uses resources efficiently and is socially inclusive. He argues that green growth means fostering economic growth and development while ensuring that natural assets continue to provide the resources and environmental services on which our well-being relies. Under this premise, the Development Bank of Latin America (2022) establishes that climate change could significantly worsen economic prospects and exacerbate inequalities in the long term for Latin America and the Caribbean. On the other hand, an effective green transition could potentially bring 10.5% more new jobs by 2030.

The Economic Commission for Latin America and the Caribbean (ECLAC), of the United Nations Organization [UN], (2022), estimates that the world population will go from 7.5 billion people to 10.2 billion in 2060, there will be an increase in the middle class of 1.8 billion individuals from 2010 to 2025, which explains the increase in the demand for goods and services that is being experienced lately, this will keep raising, with its corresponding ecological footprint. The importance of knowing this data, allows us to say that countries need a green transition, if it doesn't occur they will face greater problems. Mexico is a country with an urgent need for a green transition, otherway the country can collapse affecting the economic performance if following the same path.

Under the previous context, Mexican companies should move into a circular economy that through a green and digital transition, since the current linear model of production and consumption has generated environmental difficulties such as pollution, the excessive use of the land and fields, water and air pollutios the loss of biodiversity, amongst other factors that have caused the deterioration of the Earth, Geissdoerfer et al., (2017). Circular Economy generates environmental advantages, social benefits and added value for companies; necessary aspects to guarantee resource sustainability and diversity in a globalized, complex and competitive planetary context (Espalialt, 2017, cited in Falappa et al., 2019). It is necessary to mention that, the circular economy consists of a continuous cycle of positive development that preserves and improves natural capital, optimizes the use of resources and minimizes system risks by managing a finite amount of stocks and renewable flows (Ellen MacArthur Foundation, 2019, as cited in Falappa et al., 2019, p. 9).

Given the relevance of the topic, it is essential to talk about circular economy and the country green and digital transition according to Esposito et al. (2018), circular economy emerges as an alternative to improve the linear form of production, because it promotes maximizing the use of products during their life cycle, from raw materials, supply chain, consumption and after use, converting waste into new raw materials. In this sense, it is intended to implement a cyclical flow from the extraction, transformation, distribution, use and recovery of the materials of products and services available in the market (Pietro-Sandoval et al., 2017), having under this perspective, an alternative to improve the ecosystem in our country. In this context, companies which are the axis of the economy can generate value by improving sustainable development, knowing that circular economy guarantees sustainable development, a process by which environmental resources are preserved, maintained and stored in a natural way, achieving the well-being of current society with the objective of preserving for future generations (Jiménez, 2020), attaining the well-being of a nation.

Under the previous context in Mexico, the General Law of Circular Economy was approved in 2021, the key objectives of this law are to recycle waste and promote reuse and industrial association, transforming the waste generated by a certain industry into the raw materials from another, supporting recovery and recycling regimes (Senate of the Mexican Republic, 2021). It's important to highlight the presentation of Euroclima+ action plan in Mexico, within the framework of the Europe Team for a green transition model in Mexico initiative, where within the framework of this initiative the Delegation of the European Union in Mexico, the Mexican Agency for International Development Cooperation (AMEXCID), the Undersecretariat for Multilateral Affairs and Human Rights of the Ministry of Foreign Affairs and the Ministry of the Environment and Natural Resources (SEMARNAT), participated and reaffirmed the association and solidarity between partners to work in an ecological, fair, sustainable and inclusive manner while responding to human development priorities, (Delegation of the European Union to Mexico, 2022).

According to the National Institute of Statistics and Geography [INEGI] (2020a, 2020b), in Mexico an average of 107,056 tons of garbage were collected on a daily basis during 2018, which is equivalent to 854 g per person, of which 83.93% are collected and only 9.63 are recycled. In Mexico predominates the basic management of urban solid waste, which consists of collecting and disposing waste in sanitary landfills, wasting residues that potentially can be reincorporated into the productive system (Ministry of Environment and Natural Resources, 2017). A research performed by Ceballos et al. (2022), states that according to INEGI (2020a), Hidalgo municipalities produce the largest amount of solid waste, an amount of 1.4 kg per capita are produced on a daily basis, above the national average. According to Saucedo et al. (2019), the generation of Solid Urban Waste is linked to urban and population growth. It is a problem that is still latent in the State, so it is necessary to provide scientific proposals to support companies in this paradigm shift that can help in the transition to a green and digital circular economy.

Following the arguments above, we can say that the circular economy is fundamental, but digital technology is also necessary in a modern and fast changing world. Particularly in Mexico, since the COVID-19 pandemic, it has been brought to the attention that companies need a digital transformation, organizations now are giving the importance to digital technologies in order to be included in the general economy. Under this context, the European Commission (2022) maintains that, to promote the digital transition, additional investments are required. Because an ethical digital transformation has the potential to increase innovation and productivity in the economy and also to offer new opportunities to people and companies. Digital transition contributes to green goals, with synergies in many areas of a smart circular economy.

Having all the information above, it's possible to say that the State of Hidalgo is lagging behind in the circular economy, green and digital transition. There are no studies related to this matter, just a proposal for an ecological transformation, performed by Morales (2016), where he specified that the federal executive, announced the technology participation of a private company with specialized machinery for crushing garbage, melting it and converting it into electricity, but this project was suspended and currently there are no other immediate alternatives. That is why we developed the research presented in this chapter. It enables us to analyze the current state of companies regarding their performance under the circular economy parameters and if there is any digitalization in organizations.

2 Theoretical Framework

According to Espalialt (2017), "the Circular Economy is the antithesis of the linear model. It is, conceptually, a "holistic", "restorative" and "regenerative" model " (p. 8). Products, components and materials maintain their value and usefulness permanently throughout the entire production and use cycle. It generates indisputable environmental advantages, social benefits and added value for companies, necessary features to guarantee the sustainability of resources and ecological diversity in a globalized, complex and often unpredictable planetary context.

Thus, the Circular Economy generates environmental advantages, social benefits and added value for companies; necessary aspects to guarantee the sustainability of resources and diversity in a globalized, complex and competitive planetary context (Espalialt, 2017, cited in Falappa et al., 2019, p. 8). The circular economy consists of a continuous cycle of positive development that preserves and improves natural capital, optimizes the use of resources and minimizes system risks by managing a finite amount of stocks and renewable flows (Ellen MacArthur Foundation, 2019, as cited in Falappa et al., 2019, p. 9).

This is how a circular economy is restorative and regenerative by design, and aims to always keep products, components and materials at their highest usage levels. The concept distinguishes between biological cycles and technical cycles. As envisioned by its creators, a circular economy is a continuous positive development cycle that preserves and increases natural capital, optimizes resource returns, and minimizes system risks, managing finite stocks and renewable flows. In this way it works effectively at any scale (Cerdá & Khalilova, 2016).

For their part, Esposito et al. (2018), mention that the circular economy constitutes an option to improve the linear form of production, since it promotes maximizing the use of products during their life cycle, from raw materials, the supply chain, consumption and after use, converting waste. in new raw materials. In this sense, it is intended to implement a cyclical flow for the extraction, transformation, distribution, use and recovery of the materials of products and services available in the market (Pietro-Sandoval et al., 2017).

Citing Balboa and Domínguez (2014), they mention that the Circular Economy is considered a philosophy of organization systems based on living beings, it seeks the change to a linear economy whose purpose is to produce, use and discard, which becomes increasingly difficult to implement due to the depletion of resources, towards a circular and regenerative model, just as it occurs in nature, also being a great opportunity for business. In addition to the fact that the concept has been known since 1966, its practical applications have been progressively increasing both in economic systems and in industrial processes in recent years. In addition to the fact that the concept has been known since 1966, its practical applications have progressively been increasing in both, economic systems and in industrial processes during the last years.

The circular economy understands a responsible interrelation between economic and ecological cycles, preserving natural capital by improving productive efficiency, also reducing the negative externalities generated by the economic system. This responsible interrelation implies a decrease of the use of materials and resources, the renewability of physical flows and the recovery of waste, turning them into new inputs for the same or other industries on a programmed basis, (Cortés, 2020).

According to Cerdá & Khalilova, (2016), the fundamental principles on which a circular economy is based are: Principle 1, preserve and increase natural capital, by controlling finite stocks and balancing the flows of renewable resources. When resources are needed, the circular economy system selects them wisely and chooses technologies and processes that use renewable or highest-yield resources, whenever possible. A circular economy also increases natural capital by encouraging nutrient flows in the system and creating the conditions for soil regeneration.

As principle number 2, they point out to optimizing the resources performance, circulating products, components and materials at their highest utility level, in the technical and biological cycles. In other words, design to remake, renew and recycle in order to keep materials and components circulating in and contributing to the economy. Finally, principle 3 is to promote the effectiveness of the system, making patents and projecting how to eliminate negative externalities. This includes reducing the damage caused to systems and areas that affect people, such as food, mobility, housing, education, health or entertainment, and managing externalities such as air, water, land and noise pollution, emissions of toxic substances and climate change (Cerdá & Khalilova, 2016).

In this sense, Jaca et al. (2018), mention four principles of the circular economy given its own nature: 1. Garbage is food, 2. Diversity strengthens, 3. Energy from renewable resources and 4. Systemic thinking. In this sense, the use of resources in organizations that can convert sustainable development into the economy is required.

Following the statements, it's important to consider that we are in a stage where the economy must move towards a sustainable change; based on this idea, the objective of the circular economy is to preserve the value of materials and products for as long as possible, seeking to reintegrate them into the production system for reuse and maximum use (Deckymn, 2018; Solórzano, 2018; Ellen MacArthur Foundation, 2019). This is how its life cycle is closed by reducing the generation of waste to a minimum, in such a way that the waste is not seen as waste but as usable resources (Zaman, 2010).

In this context, a new form of economy can be considered, an economy that favors the entire country, allowing sustainability in organizations, based on the fact that all economic entities must produce sustainably, this is how the companies that are the axis of the economy could generate value by improving sustainable development since circular economy guarantees sustainable development, reason why it's considered as the process by which environmental resources are preserved, maintained and stored in a natural way, able to secure the well-being of today's society with the objective of its preservation for the future Jiménez (2020), achieving the entire country well-being. Under this premise, it is possible to talk about ISO 26000 in relation to the environment and the impact of an organization towards sustainable development.

Considering that ISO 26000 standard seeks to contribute to the improvement of the environment, considering as the natural environment in which an organization operates, including air, water, soil, natural resources, flora, fauna, humans, outer space and their interrelationships, including the environment, which extends from within an organization to the global system (ISO 26000, 2010, p. 2). Another very important connotation is that the impact of an organization, which indicates the positive or negative impact that is generated in the society, the economy or the environment, produced as a whole or some sections, as a consequence of past or present decisions and activities taken in the organization.

Citing Ardagoña and Isea (2011), state that in the day to day working activities, it's inevitable that any organization generates impacts on the surrounding environment. The use of resources, the physical location and the production of waste and polluting agents are factors that affect the stability of ecosystems. ISO 26000 and ISO 14000 recommend adopting a holistic approach to address the environmental challenges wich, additionally to recognizing the importance their actions on economic, social, health and environmental factors, also consider the unavoidable interrelation that exists amongst them, this is how ISO 14000 basic objective is to promote the standardization of production methods and providing services that can protect the environment, decreasing harmful effects that organizational activities can cause.

The authors mention that the environmental principles that every organization should take into account are:

- Pollution prevention: The first thing an organization should do is try to prevent as much as possible the generation of polluting agents, such as air emissions, water discharges, waste management and the use of high-risk chemical products, as well as other forms of contamination (radiation, infectious agents, invasive species, noise, odors, etc.).
- Sustainable use of resources: The idea is that the organization contributes to guarantee resources availability in the future, through its responsible use in the present. To do this, ISO 26000 recommends using renewable resources at a lower rate than their renewal rate and, in the case of non-renewable resources, adopting appropriate action to use them less, through technological innovations.
- Climate change adaptation and mitigation: Considering that one of the main causes of climate change are greenhouse gas emissions caused by human activity, ISO 26000 invites organizations to do everything in their power to minimize the emission of this, which requires controlling the use of fossil fuels. It's important to take measures to reduce vulnerability to climate change through, for example: planning the use of natural resources; develop infrastructures and technologies to curb possible effects (floods, droughts, etc.); and contribute to raising awareness of the seriousness of this subject.
- Environment protection, biodiversity and restoration of natural habitats: ISO 26000 highlights the importance of reversing the negative impact of human activity on the stability of ecosystems through a socially responsible use of natural resources. The protection and restoration of biodiversity and ecosystems, the sustainable use of land and natural resources, and the promotion of urban and rural development in harmony with the environment are key factors to take into account (Ardagoña & Isea, 2011, p. 21).

Under the above context, it is important to move towards a green and digital economy, defining green economy as the sustainable development through an appropriate economy, which main objectives are to improve humanity's well-being, social equity, reduce environmental risks and scarcity, reduce carbon emissions, use resources efficiently and be socially inclusive, increase income and creating jobs as a result of public and private investments aimed, promoting energy efficiency as well as the efficient use of resources and avoid the loss of biological diversity and ecosystem services (Herrán, 2012, p. 2).

The green economy is considered as the economy that leads to the improvement of human well-being and social equality. While environmental risks and ecological scarcity are significantly reduced. At the same time, it recognizes that there is an inseparability of the three aspects of sustainability: social, economic and environmental (Estevez, 2016. p. 1).

Considering how essential the circular economy is, becomes necessary to emphasize the importance of digital technology within a fast changing world, especially in Mexico, since the COVID-19 Pandemic, the nation has become aware that companies need a digital transformation, which means that organizations give importance to digital technologies in order to be included in the economy. In this context, the European Commission (2022) maintains that, to promote the digital transition, additional investments are required. This is how a fair digital transformation has the potential to increase the innovation and productivity of the Economy and to offer new opportunities to people and companies. The digital transition also contributes to green goals, with synergies in many areas of a smart circular economy (p. 1).

At societal level, digital disruption generates changes in communication, interaction and consumption models that are reflected in a greater demand for devices, software with more functionalities, cloud computing services and data traffic, as well as basic digital skills for the use of associated technologies (Economic Commission for Latin America and the Caribbean [ECLAC], 2022, p. 11). In turn, the digital economy is becoming more agile and efficient by representing for consumers the possibility of accessing information and knowledge of all kinds in various formats, as well as goods and services, and more agile non-face-to-face forms of consumption. This evolution towards the digitized economy aims to satisfy consumers with intelligent products, with a high degree of customization. All this means an increase in consumer well-being, accompanied by a reconfiguration of the digital skills necessary to carry out more advanced digital consumption and face the new labor demands resulting from the new production models. On the other hand, new forms of consumption are associated with potential benefits derived from reduced materialization and more sustainable environmental decisions, to the extent that they are based on more and better information (for example, that related to the environmental footprint of a product) or reward more environmentally friendly practices (ECLAC, 2022 p.11).

Citing Raya (2020), green technologies -or GreenIT or Green Computing- try to make use of computational resources in an environmentally friendly way in the productive fabric and the economy in general. It also mentions that according to the European digital strategy 2020, there are three fundamental pillars that must be considered in an open, democratic and sustainable digital society, in this sense ICTs can contribute to eradicating climate change, because they have the hability of monitoring consumption, based on meters and smart networks, which is essential for individuals and organizations, by providing information that helps them reduce their carbon footprint (p. 5).

It is important to mention that traditional economies have difficulties adapting to constant changes, which increases their resistance to current paradigms and therefore they assume that traditional production methods are the ones that have worked for them. and they wish to continue that way. Therefore, it is necessary for organizations to transition to a paradigm shift in which production methods are based on a circular economy for sustainability, supported by green technologies. At the moment there is no culture focused on environmental protection which can allow organizations to coexist in a greener development.

According to Pérez (2016), organizations must be conceived based on the basic processes of institutionalization and change. The notion of paradigms is central, since it allows explaining the behavior of any organization. The presence of these paradigms can be appreciated strongly in the formation of the consensus that is established, whether it is around the objectives and goals of the company, as well as in daily aspects of dealing with clients, access to markets or acquisition. of technologies. While collective consensus tends to be the conservative force, change, promoted by forces internal and external to the organization, emerges as the disruption of these consensus. In Kuhn's terms, prevailing beliefs are no longer sufficient to stem the tide of new ideas or practices. It is important to mention the case of behavior change in clients, in market patterns or in technology management. For this reason, the crisis of the organizational paradigm demands a change of knowledge in action, the ways of perceiving and acting on reality must be recreated. The new paradigm brings with it the seeds of a new order, which restores the effectiveness and efficiency in the performance of the organization. In this way, the organizational processes, mediated by the dominant paradigm, are framed in a dynamic cycle of conservation—crisis—change, necessary to increase the ability of any company to survive and grow. Hence the importance that organizations have this trend of the path of a circular economy through a green and digital transition, and that they really make a paradigm shift in Mexican companies in the State of Hidalgo, since in Mexico there is currently no commitment for this paradigm shift.

3 Contextual Framework

3.1 Hidalgo Mexico

According to the National Institute of Statistics and Geography [INEGI] (2020a, 2020b), the state of Hidalgo is made up of 10 regions, among which is the Valley of Tulancingo, formed among other municipalities by Tulancingo de Bravo, Santiago Tulantepec de Lugo Guerrero and Cuautepec de Hinojosa, which is located northeast of this federal entity. The municipality of Tulancingo de Bravo is located between 2200 and 2400 meters above sea level (masl), at the coordinates 20° 04′ 53″ north latitude and 98° 22′ 07″ west longitude. Which has 145 neighborhoods, subdivisions and towns.

The state of Hidalgo has a total population of 3,082,841. Hidalgo ranks 16th nationally for its number of inhabitants. (National Institute of Statistics and Geography [INEGI], 2020a, 2020b). With regard to commerce and services in the State of Hidalgo, it has had a stable behavior in recent years; its contribution to the Gross Domestic Product (GDP) has remained between 11.4% and 13.6%. (Municipal Development Plan, 2016). In 2012 it was the second sector with the highest contribution to the state Gross Domestic Product (GDP); however, in its contribution to the country's commercial Gross Domestic Product (GDP), Hidalgo was ranked 24th (INEGI, 2014). In 2016, Hidalgo's GDP represented 1.6% of the National GDP, occupying the 21st place (Ministry of Economíc, 2016).

According to National Institute of Statistics and Geography [INEGI] (2020a, 2020b) there are 118,821 establishments, which represents 2.5% of the total nationwide. The behavior of GDP at the state level according to economic activities are: 4.1% is represented by primary activities, 33.3% for secondary activities and 62.6%

for tertiary activities, so it is important in this chapter to consider The companies represented are those that have a greater contribution to the GDP that represent 95.9%.

4 Methodology

The methodology proposed for this study is based on a quantitative approach, a descriptive-correlational scope and a non-experimental, transectional, descriptive-correlational research design as stated (Hernández et al., 2014). A quantitative approach is considered because it seeks to analyze and interpret how Mexican companies in the State of Hidalgo perform activities of production system, considering that this system is based on the extraction of resources, rudimentary processing in the use of raw materials, the use by consumers and the pollution that occurs based on the current economies, and therefore there is no sustainability (Priede & Hilliard, 2019). Therefore, it's interesting to analyze how these companies currently work, since it's necessary for companies to have a change in the traditional paradigm of production system and to be able to incorporate themselves into a circular economy, in order to have a transition to green and digital.

The research design was non-experimental, transectional, of a descriptive and correlational type, because the production system that companies currently have is observed, in its natural context to be able to analyze them later, in addition to investigating how the variables on ISO26000 are being practiced to know whether companies work sustainable, likewise to evidence the practices they implement with the organizations digitalization, the study was performed at a single moment with companies from the State of Hidalgo (Hernández et al., 2014). In order to estimate a statistically representative sample, a confidence level of 95% was determined for the sample size, a statistical error of 5% and the expected proportions were estimated at 50% of which a minimum sample of 545 micro was obtained of small companies from the State of Hidalgo specifically from the municipality of Tulancingo. Estructured according to Table 1.

Table 1 Structure of the micro and small companies of the sample surveyed

Economic sector	% concentration Economic units	Number of companies surveyed
Trade	55.10%	300
Services	35.10%	192
Manufacturing Industries	9.80%	53

Source: Author's elaboration.based INEGI (2020)

4.1 Conceptualization and Operationalization of Variables

The variables analyzed for this study were ISO26000 features and principles to underlined the way in which companies are sustainable and transition into a circular economy, another of the variables analyzed are digital technologies in industry 4.0, which will allow to establish the green and digital transition. The variables analyzed are registered in a research questionnaire proposed by (Aguilar et al., 2020). This systemic analysis variables were assessed with a five-level Likert scale, from totally agree to totally disagree. Table 2 shows the conceptualization and operationalization of variables.

4.2 Statistical Methods Used

For the analysis of the variables in this investigation, descriptive statistic was used with a distribution of frequency and means, likewise, multivariate analysis was used to analyze whether companies have a significant incidence in the levels of average compliance with environmental performance with respect to sustainability according to ISO 26000 variables, the Kruskal-Wallis test was applied. Likewise, the statistics of principal components were applied to analyze digital technologies that companies apply. To conduct both descriptive and inferential statistics, the Statistical Package for Social Science (SPSS) was used.

Table 2 Conceptualization and operationalization of variables

Variables	Conceptualization	Operationalization
Circular economy: Issues and principles ISO 26000	The ISO 26000 Standard proposes organizations to minimize their environmental impacts and guarantee that their consumption of resources is sustainable (Ardagoña & Isea, 2011)	14-item test applied to the study subject population
Green digital technology	Smart production models can increase competitiveness with a smaller environmental footprint, since companies are using digital tools to map and reduce it, in order to assess their impact on climate change and modify their production processes (Economic Commission for Latin America and the Caribbean [ECLAC], 2022)	14-item test applied to the study subject population

Source: Author's elaboration.based on Aguilar et al. (2020)

5 Results

5.1 Descriptive Analysis

To verify the reliability of the research instrument applied to Mexican organizations in the State of Hidalgo, the Cronbach's Alpha test was performed, which is presented in Table 3, with a value of 0.906, which turns out to be reliable.

We proceeded to analyze the results in a descriptive way to study the current state of how Mexican companies in the State of Hidalgo work to achieve sustainability and environmental care, because it is necessary for companies to have a change in the traditional system paradigm of production processes and manage to incorporate into a circular economy, that is, they have to move into to a green and digital transition. A table of means and standard deviation was designed to determine compliance with environmental performance, as shown in Table 4.

Table 4 shows how the Mexican companies of Hidalgo perform involving the circular economy in regards of environmental care, the main findings related to the item *"in my company we try to prevent pollution and waste production that impacts the environment, and we try to recycle and care for the environment"* with an average of 4.2073, another relevant aspect, is that companies are *willing to take social responsibility for the impact of the activities carried out* with an average of 4.2110, Likewise, another important fact was that companies *encourage employees to behave in a socially responsible way* with an average of 4.2128, another important finding was that *companies try to abide by everything that the rules and laws require them* with an average of 4.16, These are relevant aspects that can be considered for a circular economy; however, companies must change their paradigm since, although it is true, companies under this study recognize these actions because they are oriented to totally agree in Likert scale, however in practice they have to change the paradigm in order to migrate towards a circular economy in practice that allows promoting an improvement in sustainability.

Another technique that was used for non-parametric studies is the tree technique that sought to analyze the incidence that companies related to the circular economy if they use it in the decision making process and then transition to a green and digital circular economy. According to Martínez and Fernández (2008a, 2008b), the tree technique is a graphic representation that shows the progressive breakdown of the factors or means that can contribute to a certain effect or objective, it is the segregation of the elements that make up the object of study. The representation of the three technique can be seen in Fig. 1.

As seen in Fig. 1, companies surveyed for the case of the circular economy are only evaluated through some dimensions such as environmental care, which is the

Table 3 Alfa de Cronbach

Alfa de Cronbach	No. of elements
0.906	14

Source: SPSS V.22 (2022)

Table 4 Descending classification of the social responsibility items, ISO 26000 matters

Item Number	Item description	Average	Standard deviation
5	In my company we are responsible with our clients: We take care of their confidentiality, we attend their complaints, we communicate without information deceit which interests them regarding prices, costs, terms of service, contracts and we offer safe products	4.4771	0.77645
4	In my company we avoid getting involved in malpractices such as bribery, corruption, piracy or unfair competition and we do not allow it in our clients and suppliers either	4.2569	1.03259
9	In my company we have rules that promote ethical behavior	4.2349	0.99256
3	In my company we seek the Well-being of workers, listening to their needs, offering them a safe environment, access to health services and a balanced personal life	4.2202	1.04633
7	In my company, we are willing to take social responsibility for the impact of our activities	4.2110	1.03499
13	In my company we encourage employees to behave socially responsible	4.2128	1.01759
2	In my company we try to prevent pollution and waste production that impact the environment, and we try to recycle and take care of the environment	4.2073	1.00511
14	In my company we do concrete actions that are socially responsible	4.1835	1.04111
11	In my company we try to abide by everything that the rules and laws require us to	4.1651	1.05991
8	When required, in my company we are willing to publicize our activities, particularly those that could impact society	4.0697	1.10093
10	At my company we are interested in those who may be affected by our operations.	4.0367	1.12726
1	In my company we have policies and procedures to respect human rights and to avoid complicity in acts that violate civil, political, economic, social, cultural and labor rights and to avoid discrimination.	3.9890	1.26312
6	The company promotes community development activities with people from the community where it is located. We are concerned about not causing and even solving local social problems. We prefer local over foreign suppliers.	3.6422	1.45725
14	In my company we try to respect behavioral international standards.	3.6037	1.48175

Source: Author's elaboration

dependent variable that is interpreted as node 0 where 86% of the companies are biasedly *in agreement* that they are environmental friendly from their perspective, 14% comment that they do not take care for the environment, a higher percentage relatively with the number of companies surveyed, companies do not implement the elements of the circular economy strictly, they only use some aspects to determine if they are environmental friendly, immediately we can observe that the dependent

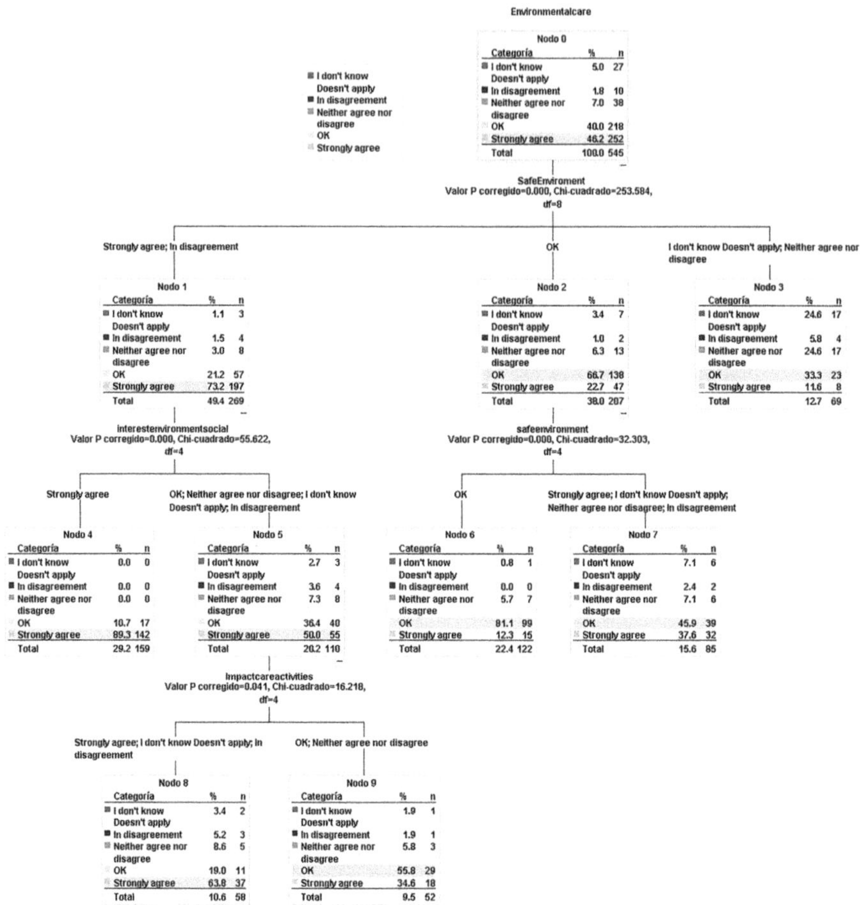

Fig. 1 Tree model of the representation of the circular economy for environmental care. *Source*: SPSS V.22 (2022)

variable branches into three nodes, node 1, 2 and 3 belonging to the environmental variable, surely this is the main predictor variable, we can see that the chi-square of node 1 is higher than the one correspondent to node number two, the fact that we are interested in knowing if companies really care about the environment for a circular economy, so it can be seen in node 1 that of 49.4% of the companies studied, 73% strongly agree that there is a safe environment.

Node number 1 branches again with node 4 and node 5 and they belong to the variable of interest one social environment, including the item *in my company we encourage employees to behave socially responsible*, we observe that in node 4 there is greater interest with the social environment and being socially responsible with 80.3%, about node 5 only 50% strongly agreed on their interest with social environment. However, the tree shows that node five branches out from node 8 and node 9, which refers to the care and the impact on activities, with the item *in my company we are willing to respond to society for the impact of our activities*, in node 8 it's observable that 63.8% of the companies are willing to respond to society for the impact generated by their activities towards the environment. Therefore, it is concluded that the nodes in charge of defining environmental care within a circular economy, that is, the variables that influence care for the environment are: Node 0, node 1, node 4, node 5, and node 8.

5.2 *Multivariate Analysis*

An inferential analysis was carried out to determine if the business sector had a significant incidence in the average compliance levels of environmental performance, for which the Kruskal-Wallis statistical test was used, as can be seen in Table 5.

According to Table 5, the items rlated to caring for the environment were grouped with the Kruskal-Wallis test, so they were significantly related to what was disclosed by the companies, it means that there is sufficient evidence to sustain the environmental care through practices but only with the variables safe environment, impact of activities, regulation of regulations and laws, they avoid doing malpractices and also they are interested in the social impact, however the State of Hidalgo is far from having a circular economy with the required features, that's why it's necessary to propose strategies that allow a paradigm shift towards a sustainable economy.

It is considered that current organizations have a sustainable economy, however, they have not transitioned to the circular economy, only a few elements are considered for the common well-being of society, the variables to be considered are: efficiency in the use of raw materials, recycling and recovery, reuse and repair, regenerate, share, optimize, recycle, virtualize, exchange, as some of the categories that could be used in the circular economy.

Table 5 Kruskal-Wallis test statistic $P < 0.05$

	Atmosphere sure	Impact Insurance of activities Standards	And Laws	Actions SR	Avoid Bad practices	Interest with the Social impact
H de Kuskal-Wallis	160.72	97.23	113.56	119.38	98.63	125.65
gl	4	4	4	4	4	4
Sig. asintótica	0.000	0.000	0.000	0.000	0.000	0.000

[a] Test Kruskal Wallis
[b] Grouping variable: Environmental care
Source: SPSS V.22 (2022)

Regarding how companies benefit from digital tools, and to analyze the green in digital transitioning, we performed descriptive-explanatory statistics that consisted of 15 items with a scale of: 5. I know what it is and I use it, 4. I know what it is, I know how to use it, but I don't need it, 3. I know what it is, I don't know how to use it, but I would learn if I need it, 2. I know what it is, I don't know how to use it and I feel that it is difficult to learn it, and 1. I don't know what it is. Statistical data obtained gives the guidelines for the entrepreneurs paradigm shift, due to the following results:

Table 6 refers to *if the facility provides access to internet service instead of using mobile phone,* results show that only 37.8% know what it is and use it.

Table 7 displays the results for *if the company uses a computer, tablet or some electronic device to manage the company* results demonstrate 33.9% know what it is and use it, while 16% do not know or use it and the rest are not interested in using neither.

Table 8 shows the results for the question *if they have a website or social networks to show the catalog of products or services*, only 20% of the surveyed companies answered that they know what it is and use it.

Table 9 relates to the question *if they use electronic devices to make card payments in the business establishment*, and as can be seen, only 16.7% of the micro and small businesses in Hidalgo use an electronic device to make payments.

Table 10 shows the companies that have a system for handling complaints and suggestions online, not by email, so it can be seen that only 9.9% of the companies know what it is and use it. This aspect involves an opportunity to make a paradigm shift to digital

The next item is to analyze whether companies have any specialized software that provides security in terminals, internet portals, points of sale, mobile devices and administrative systems. Table 11 shows that 15.6% know what it is and use it for what they need. In order to accomplish a paradigm shift of Mexican companies in the state of Hidalgo it's imperative that they begin to work with digital tools that can guarantee green in their organization.

Another of the important items to analyze is if Mexican companies in the State of Hidalgo conduct automatic backup in the company's information cloud. The responses are shown in Table 12, only 14.3% know what it is and use it, therefore

Table 6 Access internet traditional

		Frequency	Percentage
Valid	I do not know what it is	91	16.7
	I know what it is, I don't know how to use it and I feel it's hard to learn	109	20.0
	What is it, I don't know how to use it but I would learn easily if I needed it	47	8.6
	I know how to use it but I don't need it	92	16.9
	I know what it is and I use it	206	37.8
	Total	545	100.0

Source: Author's elaboration

Table 7 Use of any digital device

		Frequency	Percentage
Valid	I do not know what it is	87	16.0
	I know what it is, I don't know how to use it and I feel it's hard to learn	87	16.0
	What is it, I don't know how to use it but I would learn easily if I needed it	78	14.3
	I know how to use it but I don't need it	108	19.8
	I know what it is and I use it	185	33.9
	Total	545	100.0

Source: Author's elaboration

Table 8 Internet page social networks

		Frequency	Percentage
Valid	I do not know what it is	22.4	122
	I know what it is, I don't know how to use it and I feel it's hard to learn	17.1	93
	What is it, I don't know how to use it but I would learn easily if I needed it	15.2	83
	I know how to use it but I don't need it	25.3	138
	I know what it is and I use it	20.0	109
	Total	100.0	545

Source: Author's elaboration

that the paradigm shift of entrepreneurs who begin to transcend to digital is an area of opportunity.

6 Conclusions

After analyzing the results amongst 545 Mexican companies in the State of Hidalgo as a results on how companies carry out the circular economy and digitization, the following conclusions are reached, regarding the tree statistics, it's found that the variable safety environment is the best predictor of environmental care obtaining *totally* as a response in five categories: strongly agree, agree, neither agree nor disagree, disagree, and do not or do not apply. The highest probability is in the interest with the social environment with 89.3% and prevails between companies with the safer environment, and the lowest probability of environmental care 33.33% its observed among companies that don't have a safe environment, companies that don't avoid malpractice and that are not concerned about the impact of their activities.

As can be seen in the representation of the tree statistics, there are nine nodes, which indicate for the environmental care variable, node 1, 2 and 3 are added, which refers to the fact that companies sustain a safe environment by seeking the

Table 9 Bank terminal

		Frequency	Percentage
Valid	I do not know what it is	119	21.8
	I know what it is, I don't know how to use it and I feel it's hard to learn	106	19.4
	What is it, I don't know how to use it but I would learn easily if I needed it	85	15.6
	I know how to use it but I don't need it	144	26.4
	I know what it is and I use it	91	16.7
	Total	545	100.0

Source: Author's elaboration

Table 10 Attention internet complaints

.		Frequency	Percentage
Valid	I do not know what it is	152	27.9
	I know what it is, I don't know how to use it and I feel it's hard to learn	87	16.0
	What is it, I don't know how to use it but I would learn easily if I needed it	82	15.0
	I know how to use it but I don't need it	170	31.2
	I know what it is and I use it	54	9.9
	Total	545	100.0

Source: Author's elaboration

well-being of workers, listening to their needs and in general giving them a safe environment that includes access to health services and a balanced personal life. The common agreement is that they work in a safe environment as a dimension of environmental caring. However, it should be highlighted that currently companies are not taking the steps needed to move into the circular economy, they only follow parameters established by ISO 26000 on environmental, reason why a paradigm shift is proposed for Mexican companies in the State of Hidalgo so that they work within circular economy standards to protect and ensure sustainability for companies and society.

Regarding digitalization, it can be concluded that the Mexican companies in the State of Hidalgo, subject of this study, are lacked on digital tools according to the frequencies that were based on the data analysis, so it is necessary to propose a paradigm shift for companies so they start working within a circular economy that transitions to green and digital, this will ensure sustainability of organizations and thereby allow them to have a competitive advantage and be able to take care of the social environment.

It is necessary to consider that the global trend is working towards the circular economy and there is a prevailing need for companies to have that green and digital transition that enables to be sustainable organizations and that in return generates competitive advantages, achieving this transition in organizations in order to accomplish the paradigm shift of companies and attain a sustainable economy.

Table 11 Software portal internet

		Frequency	Percentage
Valid	I do not know what it is	149	27.3
	I know what it is, I don't know how to use it and I feel it's hard to learn	9	17.1
	What is it, I don't know how to use it but I would learn easily if I needed it	77	14.1
	I know how to use it but I don't need it	141	25.9
	I know what it is and I use it	85	15.6
	Total	545	100.0

Source: Author's elaboration

Table 12 Cloud backup

		Frequency	Percentage
Valid	I do not know what it is	181	33.2
	I know what it is, I don't know how to use it and I feel it's hard to learn	76	13.9
	What is it, I don't know how to use it but I would learn easily if I needed it	68	12.5
	I know how to use it but I don't need it	142	26.1
	I know what it is and I use it	78	14.3
	Total	545	100.0

Source: Author's elaboration

Even though the findings presented were that the organizations do not conduct the green and digital circular economy, which is required in the short term so Mexican organizations of the State of Hidalgo can move towards a green and digital circular economy, it will allow a gradual change and improvement of the ecosystem. The paradigm shift for companies is due to the fact that in the performed research, a traditional economy is carried out and that they do not work with digitization, so it is important that organizations have this path towards the transition of the green and digital circular economy. As indicated by Espalialt (2017), cited in Falappa et al., (2019) the circular economy generates great environmental advantages, benefits and added value for companies, therefore, according to the results, the generation of the circular economy is mainly required in Hidalgo companies that will allow achieving environmental objectives and being able to have sustainability.

Although it is true and according to the results, it allows corroborating the theory that, according to Herrán (2012), maintains that it is important to transcend a green and digital economy, understanding green economy as sustainable development through an adequate and correct economy, whose Its objective is to: improve the well-being of the human being and social equity, and reduce environmental risks and scarcity, reduce carbon emissions, use resources efficiently and be socially inclusive, this is how companies in the State of Hidalgo are recommended to transcend to a green and digital economy and according to the results in their entirety do not have good circular economy practices.

Therefore, it is essential to have truly sustainable companies in Hidalgo through a circular economy, since with the accelerated growth at a global level, if they do not transcend to a green and digital economy, they will not be able to be competitive in the market, and it is necessary to mention that before The COVID-19 pandemic has been working with companies to carry out the digital transformation that is important for organizations today.

References

Aguilar, O., Peña, N., & Posada, R. (2020). *Innovation and industry 4.0 in micro and small companies in Latin Americ*. Mc Graw Hill.

Ardagoña, A., & Isea, R. (2011). ISO 26000, a guide for the social responsability of organizations. *Notebooks of the " La Caixa" Chair of Corporate Social Responsibility and Corporate Governance, 11*, 8–10. http://comunicarseweb.com.ar/download.php?tipo=acrobat&view=1&d,ato=1308234290_catedralacaixa_vol11_Final_ tcm5-66526.pdf

Balboa, C., & Domínguez, M. (2014). Circular Economic as a framework for ecodesign: the ECO-3 model. Technical informant. https://www2.uned.es/egi/publicaciones/articulos/Economia_circular_como_marco_para_el_ecodiseno_el_modelo_ECO-3.pdf

Becker, R. (2018). Green Economic and SDG Indicators, UNSD Guatemala. Time for global action, for people and planet. https://www.cepal.org/sites/default/files/courses/files/2018-01_1.1_metodologia-indicadores-economia-verde.pdf

Ceballos, S., Brambila, J., & Pérez, V. (2022). Urban solid waste and circular economy in Pachuca, Hidalgo, Mexico. *University Act., 32*, 1–16.

Cerdá, E., & Khalilova, A. (2016). Circular Economic. Circular Economic, Strategy and Industrial Competitiveness. ISSN 0422-2784, N° 401

Cortés, G. (2020). The circular economy. Key ideas for understanding a new model of economics resources. Sciences Communications Center. https://repositorio.uautonoma.cl/bitstream/handle/20.500.12728/3246/ECONOMIA_CIRCULAR.pdf?sequence=1&isAllowed=y

Deckymn, S. (2018). Circular Flanders: Adaptive policy for a circular economy. In H. Lehmann (Ed.), *Factor X: Challenges, implementation strategies and examples for a sustainable use of natural resources*. Springer.

Delegation of the European Union to Mexico. (March 03, 2022). Presentation of the EUROCLIMA+ Action Plan in Mexico, within the framework of the Europe Team Initiative for a green transition model in Mexico. https://www.eeas.europa.eu/delegations/mexico/presentaci%C3%B3n-del-plan-de-acci%C3%B3n-euroclima-en-m%C3%A9xico-en-the-framework-of-the-initiative_en?s=248

Development Bank of Latin America. (2022). Climate Change in Latin America and the Caribbean. https://www.cambio16.com/cambio-climatico-podria-elevar-la-pobreza-en-58-millones-en-america-latina-y-el-caribe/

Economic Commission for Latin America and the Caribbean [ECLAC]. (2022). Digital technologies for a new future. United Nations. https://www.cepal.org/es/publicaciones/46816-tecnologias-digitales-un-nuevo-futuro

Ellen MacArthur Foundation. (2019). Completing the picture: how the circular economy helps tackle climate change. Cowes, 26 de septiembre

Espalialt, C. (2017). *Circular economy and sustainability. New approaches to creation of value*. CreateSpace. https://wolfypablo.com/documentacion/documentos/2017-10/710%20Economia_circular_y_sostenibilidad.pdf

Esposito, M., Tse, T., & Soufani, K. (2018). The circular economy: An opportunity for renewal, growth, and stability. *Thunderbird International Business Review, 60*(5), 725–728. https://doi.org/10.1002/tie.21912
Estevez, R. (2016). What it is the Green Economy? Eco-intelligence. https://www.ecointeligencia.com/2016/01/economia-verde/
European Commission. (02 of 03 of 2022). Towards a green, digital and resilient economy: our European growth model. https://ec.europa.eu/commission/presscorner/detail/en/ip_22_1467
Falappa, M., Lamy, M., & Vázquez, M. (2019). From a linear to a circular economy in the 21st century. Undergraduate thesis, National University of Cuyo. https://ediunc.bdigital.uncu.edu.ar/objetos_digital es/14316/falappa-fce.pdf
Geissdoerfer, M., Savaget, P., Bocken, N., & Hultink, E. (2017). The circular economy – A new sustainability paradigm? *Journal of Cleaner Production, 143*, 757–768. Elsevier Ltd. https://doi.org/10.1016/j.jclepro.2016.12.048
Hernández, S., Fernández, C., & Baptista, L. (2014). *Investigation Methodology*. Mc Graw Hill.
Herrán, C. (2012). The Path to a green economy. México https://flacsoandes.edu.ec/web/imagesFTP/1340319587.documento_fes_09156.pdf
ISO 26000. (2010). Norma Internacional. Guía de Responsabilidad Social. https://americalatinagenera.org/wp-content/uploads/2014/09/U4ISO26000.pdf
Jaca, C., Ormazábal, M., Prieto, V., Santos, J., & Viles, E. (2018). *Economía circular. Guía para PYMES*. Ediciones Universidad de Navarra, Eunsa.
Jiménez, D. (2020). What is Sustainable Development? Corporate social responsibility. https://www.researchgate.net/publication/324532686_What_is_Sustainable_Development
Martínez, R., & Fernández, A. (2008a). Árbol de Problemas y áreas de Intervención. México-CEPAL 2008.
Martínez, R., & Fernández, A. (2008b). Methodologies and Instruments for the Formulation, Evaluation and Monitoring of Social Programs. Problem Trees and Areas of Interventions. United Narion, 60 ECLAL.
Ministry of Economíc. (2016). https://www.gob.mx/cms/uploads/attachment/file/175903/hidalgo_2016_1116.pdf
Ministry of Environment and Natural Resources. (January 10, 2017). Urban Solid Waste (MSW). https://www.gob.mx/semarnat/acciones-yprogramas/residuos-solidos-urbanos-r
Morales, E. (2016). Hidalgo Technological Observatory: The circular economy response to the problem of garbage. https://otech.uaeh.edu.mx/noti/index.php/articulo-otech/la-economia-circular-respuesta-al-problema-de-la-basura/
Municipal Development Plan. (2016). http://planestataldedesarrollo.hidalgo.gob.mx/pdf/PMD/2016-2020/PMD_Tulancingo2020-2024.pdf
National Institute of Statistics and Geography [INEGI]. (2014). Statistical and Geographic Yearbook of Hidalgo 2020. Recovered the 06 de Junio de 2020, de https://www.datatur.sectur.gob.mx/ITxEF_Docs/HGO_ANUARIO_PDF.pdf
National Institute of Statistics and Geography [INEGI]. (2020a). National Census of Municipal Governments and Territorial Demarcations of Mexico City 2011-2019. Basic tabulated. https://www.inegi.org.mx/temas/residuos/default.html#Tabulados
National Institute of Statistics and Geography [INEGI]. (2020b). Statistical and Geographic Yearbook of Hidalgo 2020. Recovered the 06 de Junio de 2020, de https://www.datatur.sectur.gob.mx/ITxEF_Docs/HGO_ANUARIO_PDF.pdf
Pérez, A. (2016). *Management control and dashboard: From diagnosis to action*. Signo Vital Ediciones.
Pietro-Sandoval, V., Jaca, C., & Ormazabal, M. (2017). Circular economy: Relationship with the evolution of the concept of sustainability and strategies for its implementation. *Memorandum research Engineering*, (15), 2301–1106.
Priede, T., & Hilliard, I. (2019). The circular economy in the food industry. *Dossier Economists Whiteout Borders., 35*, 36–40. https://abacus.universidadeuropea.com/bitstream/han-

dle/11268/9281/Dossieres%20EsF%2035%20RSC%20en%20la%20industria%20alimentaria.pdf?sequence=2

Raya, L. (2020). Tecnología Verde. *Revista digital de Acta.* Cedro editorial. https://www.acta.es/medios/articulos/ciencias_y_tecnologia/073001.pdf

Saucedo, J. A., Mendoza, A., & Alvarado, M. (2019). Collection of solid waste in municipal areas: Urban logistics. *Sustainability, 11*(19), 5442. https://doi.org/10.3390/su11195442

Senate of the Mexican Republic. (2021). Approval of the General Circular Economy Law https://comunicacionsocial.senado.gob.mx/informacion/comunicados/1208-aprueba-el-senado-ley-general-de-economia-circular

Solórzano, G. (2018). Circular economy and future prospect. In P. Tello, D. Campani, & R. Sarafian (Eds.), *Integral Management of Urban Solid Waste.* Inter-American Associations of Sanitary and Enviromental Engineerig. (AIDIS).

Zaman, A. (2010). Comparative study of municipal solid waste treatment technologies using life cycle assessment method. *International journal of Environmental Science and Technology, 7*, 225–234, Berlín, Springer, marzo.

How Does the Circular Economy Address the Sustainable Development Goals? Evidence from the Brazilian Legal Amazon

Luciana Aparecida Barbieri da Rosa, Marcos Cohen, Waleska Yone Yamakawa Zavatti Campos, Maria Carolina Martins Rodrigues, Maria Jose de Sousa, and Lucas Veiga Avila

Abstract The study main objective is to verify whether the circular business model helps in the implementation of SDGs in the Amazon and which objectives foster the circular economy. The Legal Amazon company adapted the models of (Yadav et al. (Journal of Cleaner Production 277: 124186, 2020); Rossi et al. (Journal of Cleaner Production 247: 119137, 2020), Da Vermunt et al. (Journal of Cleaner Production 222: 891–902, 2019)) to analyse the strategies to be used in implementing the CE principles, and the UN model to comply with the SDGs by generating more sustainable operations. The results of the analysis allow us to analyse that the analysed company has EC practices that contribute to the achievement of the sustainable development goals, highlighting mainly SDG 2—Zero hunger, SDG15—Terrestrial life and SDG17—Revitalize the global partnership for sustainable development, for which they are found some practices by the interviewee. In addition, it should be noted that the limitations of the study are related to the single case approach, making it impossible to extrapolate the findings to the Rondônia scenario. On the other hand, the results of this exploratory research will allow for the formulation of a more detailed conceptual model and some hypotheses that will be tested in a future quantitative study. In this study the sample will include not only companies located in the state of Rondônia, but also in other legal states of the Amazon region.

L. A. B. da Rosa (✉) · M. Cohen · Waleska Yone Yamakawa Zavatti Campos
PUCRio, Rio de Janeiro, Brazil
e-mail: mcohen@iag.puc-rio.br; waleskazavatti@alumni.usp.br

M. C. M. Rodrigues
Universidade do Algarve (CinTurs), Faro, Portugal

M. J. de Sousa
ISCTE-IUL, Lisbon, Portugal

L. V. Avila
Universidade Federal de Santa Maria, Rio Grande do Sul, Brazil
e-mail: lucas.avila@ufsm.br

C. Rego et al. (eds.), *Entrepreneurship, Technological Change and Circular Economy for a Green Transition*, Studies on Entrepreneurship, Structural Change and Industrial Dynamics, https://doi.org/10.1007/978-3-031-48079-9_5

Keywords Circular economy · Sustainable development goals · Sustainability · Amazon

1 Introduction

The environment suffers several impacts caused by pollution from industrial activities (Ogunmakinde et al., 2022). In addition, numerous natural resources used in industries are finite and susceptible to depletion due to continuous extraction. According to the paradigm of linear economy, industries, in addition to producing waste, also absorb a significant amount of resources (Hossain et al., 2020; Ogunmakinde et al., 2022).

In this scenario, overcoming a LE towards a CE can be the way out to achieve environmental sustainability in various contexts and industries, whose focus on the UN's SDGs can be facilitator of this process of paradigmatic change (Fatimah et al., 2020; Lukin et al., 2022).

Sustainability is a broad term that involves three dimensions: social, economic and environmental (Elkington, 1998). The circular economy presents sustainable foundations of reuse, recycling and reuse, being closely related to sustainability, especially in its environmental dimension (Azizuddin et al., 2021). Thus, the SDGs can narrow the theoretical and practical gaps, guiding how CE can help in shaping a better society. So, the 17 sustainable development goals work together as a compass that guides governments, companies, and civil society organizations toward the sustainability trip.

Given the importance of sustainable practices in the Brazilian Amazon region from a social and environmental economic point of view (Silva et al., 2019), this chapter analyse the adoption of CE practices effects of SDGs in the Brazilian Amazon. The question that the article intends to answer is: Do circular business models assist in the implementation of SDGs in the Brazilian Amazon region? If so, what SDGs are fostered in the circular economy in the locus of study?

By answering these questions this article seeks to contribute in a practical way to the extent that circular economy business models can be a facilitator of SDGs in the Amazon region. This is especially important because the Amazon is considered the filter of the world (Oliveira, 1991), having 3.3 million kilometers forested. In addition, contribute to the theme of sustainability by indicating the ways in which the CE can favor the development of a more just society, of a preserved environment without disregarding economic aspects, in other words, how CE is capable of assisting sustainable development from the perspective of triple bottom line (Elkington, 1998), especially in developing countries.

After the introduction, we have the theoretical framework, responsible for presenting the concepts of CE and its connection with the SDGs. The third section addresses the methodological applied to the study and then we have the section where results are presented, followed by the final considerations ending with the references of the citations used.

2 Theoretical Framework

2.1 *Circular Economy and Sustainable Development Goals*

Based on growing concerns about the natural resource usage and the improper disposal of waste, environmental management issues occupy a large part of the agenda of sustainable actions of companies, communities and large organizations interested in sustainable development. With population growth and new consumption habits, and the consequent growth of the industrial sector, waste generation and greenhouse gas emissions also increase (Reich et al., 2023; Ye et al., 2023).

In this scenario, the critical points are the generation and inappropriate disposal of waste in organizations and the excessive consumption of resources and materials, since the present unsustainable organizational model is threatened by the limited availability of natural resources, causing instability and uncertainty in the prices of companies (Rodriguez-Anton et al., 2019). Different authors point out the difficulty of matching current consumption and production levels and availability of natural resources for future generations (Schroeder et al., 2019; Virmani et al., 2022).

In this sense, the circular economy proposes a sustainable transition, moving away from the approaches of the linear economy (Zhu et al., 2022). In other words, there is a contrast between the linear economy model, based on the indiscriminate use of resources and waste generation, and the circular economy, based on a regenerative system with a view to reducing waste and emissions (Reich et al., 2023; Zhu et al., 2022).

Regarding World Wide Fund For Nature (WWF-World Wide Fund for Nature, 2012), in 2015 humanity uses about 1.5 Planets to provide the resources it consumes and absorb the waste it generates, which shows that planet Earth takes up to a year and a half to deal with the resources used and generated in a year. In addition to the damage to the environment, this has impacts on the way in which business is developed (Antunes et al., 2022; Genovese et al., 2017; Leitão, 2015). From the point of view of the final consumer, it can be observed that the consumption behavior is not only related to the products, but also to the process of production of goods and services (Morren & Grinstein, 2016; Susanty et al., 2021).

With the competition of organizations for resources and the concentration of supply, society and industry depend on costly imports from some supplier countries, so supply chain management must also be thought of in favor of local sustainability (Kumar Shetty & Subrahmanya Bhat, 2022). Research highlights that negative effects on the supply and price of materials, along with population growth, will further reduce the creation of externalities in the real cost of raw material (Kumar Shetty & Subrahmanya Bhat, 2022; Taborga et al., 2018).

Therefore, it is crucial to adopt a new strategy for how nations develop economically and how businesses are run, being central the environmental, social and economic rationality (Awan & Sroufe, 2022). Effective material management brings benefits in environmental protection and cost reduction through the optimization of consumption and production (Awan & Sroufe, 2022; Dougherty & Hardy, 1996).

Based on this concern, many companies and the government began to think about changing the forms of natural resource usage and the disposal of waste, which gave rise to the concept of circular economy (Reich et al., 2023).

By redesigning new business models and products, the circular economy has established itself as an alternative sustainability model that aims to keep the flow of materials and goods with the highest value and utility (Lukin et al., 2022; Prieto-Sandoval et al., 2018). The circular economy seeks a more holistic approach, turning waste into new resources, using renewable energy, and minimizing or eliminating toxic elements (EMAF, 2015).

The circular economy concept, which is more recent than sustainability, demonstrates different reasons and aims (Awan & Sroufe, 2022). While the circular economy has a clear scope to close the loop and eliminate waste, sustainability was justified by the environmental movement after the publication of the Report. Brundtland (Klewitz & Hansen, 2014). In the field of sustainability, green innovation not only involves radical changes, but also encompasses the accumulation of knowledge for the sake of improving processes and products (da Rosa et al., 2022).

The definition of circular economy has evolved over the years, being refined by different ideas and is not associated with a single author (EMAF, 2015). In the late 1970s, many scholars attempted to develop a paradigm for circular economy through theoretical-methodological relationships involving systems engineering, physics, and industrial economics (EMAF, 2015; Yuan et al., 2006). At the time, this concept predicted the collapse of the linear economic model of transformation, extraction, and outflow of products (García-Barragán et al., 2019).

In the mid-1970s, architect and industrial analyst Walter Stahel drafted a report for the European Commission, titled "Jobs for Tomorrow: The Potential for Substituting Manpower for Energy," co-authored by Geneviève Reday-Mulvey (Stahel & Reday-Mulvey, 1981), which offers a vision of an economy in cycles, where it is possible to transform its energy impact into: waste and resource savings, unemployment and waste management and reduction, into work (Bocken et al., 2016; Goyal et al., 2018). Stahel is also known for co-founding the term "cradle to cradle", which works with a closed-loop approach to production processes (Braungart & McDonough, 2009; Goyal et al., 2018).

The "cradle to cradle" concept continued to be developed by chemist Michael Braungart in partnership with architect William MacDonough, who together published the book.: "Cradle to cadle: Remaking the way we make things" (Braungart & McDonough, 2009). The book is based on a design-driven philosophy for effectiveness, which considers the materials of an industrial process the nutrients needed for future processes (Braungart & McDonough, 2009; Goyal et al., 2018; Leitão, 2015). The composition of products can be developed by continuous recovery and reuse, in order to eliminate waste, recovering the values of these materials (EMAF, 2015).

Stahel contributed to the development of the field of industrial sustainability in the 1980s, so much so that in 1982 he published an article entitled "Product Life Factors" that addressed product life cycles and the "performance economy," now known as the circular economy. Stahel sees the importance of selling services, not products, and seeks intelligent solutions for systems (Stahel, 2016).

However, the concept was in disuse until the 1990s. From there, the concept received more and more attention and environmental issues became more necessary. When Germany established the Law on Closed-Cycle Waste Management and Toxic Substances in 1996, it served as a pioneer in the implementation of the circular economy idea (Chen et al., 2020; Su et al., 2013). A law promoting a circular economy was passed in Japan in 2000, and the Japanese government promoted the creation of pilot projects to move society toward the circular economy (Ghisellini et al., 2016). China is now introducing the EC as a new development model with the aim of making China's economic system more sustainable (Su et al., 2013). A common feature of the German and Eastern circular economy is the desire to prevent environmental degradation and promote the conservation of scarce resources through effective management (Su et al., 2013).

The Ellen MacArthur Foundation, a non-profit organization, was established in 2010. Its mission is to encourage the adoption of the circular economy as a means of reducing the use of hazardous chemicals that harm the environment and advancing the use of renewable energy (EMAF, 2015). The institution also emphasizes waste disposal through the improved design of materials, systems, and business goods. In fact, the Ellen MacArthur Foundation released the first of a series of papers in 2012, which marked a significant global emergence of the circular economy theme (EMAF, 2015).

This scenario allowed the theme of the circular economy to prioritize the search to develop products and processes designed for the more rational use of natural resources, facilitating their recovery, through reform, recycling, remanufacturing or reuse (EMAF, 2015). This way of designing products for rational use is called a "restorative" (or "regenerative") economy, in which it seeks to reincorporate materials into biological or productive cycles, aiming at their renewability (EMAF, 2015; Rossi et al., 2020).

The European Union in 2015 adopted the circular economy (CE) as a model to enable the achievement of the targets set by 2050 (Ghisellini et al., 2016; Kirchherr et al., 2018; Vermunt et al., 2019). Faced with this new strategic vision, the circular economy causes fundamental structural changes in the industrial symbiosis, in which the residue of one operation can be input for another operation, and so on, which means using a resource for as long as possible (Awan & Sroufe, 2022; Lim et al., 2022).

The circular economy model consists of influencing each and every size and type of business, which is in activity, through cleaner production approaches, eco-design, by industrial symbiosis with logistical and functional partnerships that have an effect on regional development or even in organizations that allow the systemic redesign of cities, in the so-called concept of sustainable-cities (Antunes et al., 2022; Babbitt et al., 2018).

Through recycling, reuse, and recovery, the EC permits the return of materials to the production process, acting as fresh secondary raw materials that compete with those that are currently on the market (Nikolaou et al., 2021).

As a result, the closed cycle presupposes the minimization of the need for new energies, materials, emissions, waste and the decrease of environmental pressures linked to the extraction of resources (Niero & Olsen, 2016).

In order to create a waste-free world with a responsible attitude toward resources, the circular economy integrates a number of zero-waste ideas, including resource efficiency, closed loop systems, recycling, and reuse (Ghisellini et al., 2016; Sarkar et al., 2022; Su et al., 2013). Given this context, the circular economy presents itself as an industrial model of sustainability, which aims to reshape the linear production chain, for a new circular concept of circulation production of resources (Ogunmakinde et al., 2022; Yadav et al., 2016, 2020).

In summary, the circular economy, which is distinguished between biological and technological cycles, attempts to sustain components, products, and resources with increased value. Therefore, waste is now considered new resources, restricting waste and reducing dependence on sources of raw materials (EMAF, 2015).

The circular economy, which is based on the reuse, reduction, recovery, and recycling of energy and materials, is a major model for promoting the decoupling between growth in resource consumption and economic growth (Reich et al., 2023). In light of the fact that the circular economy is seen as the operationalization of sustainable development on a global scale, the notion of circular economy is of utmost significance for the many industries (Ghisellini et al., 2016; Murray et al., 2017; Corrado & Sala, 2018).

It is noted that, in the long run, the circular economy can reduce costs with risks in volatility, reduce costs in the raw material and establish more resilient economic systems. Stahel (2016) describes an estimated cost reduction of up to $500 million in material savings and an increase in the workforce by about 4% overall.

The SDGs refer to seventeen goals of the 2030 Agenda, created in 2015 (United Nations, 2015). The SDGs are: 1—No poverty, 2—Zero hunger, 3—Good health and well-being, 4—Quality education, 5—Gender equality, 6—Clean water and sanitation, 7—Affordable and clean energy, 8—Decent work and economic growth, 9—Industry, innovation and infrastructure, 10—Reduced inequalities, 11—Sustainable cities and communities, 12—responsible consumption and production, 13—Climate action, 14—Life below water, 15—Life on land, 16—Peace, justice and strong institutions, 17—Partnerships for the goals. (United Nations, 2015).

Sustainability in the Amazon can be fostered through an industry approach that combines the circular economy and SDGs, to reduce waste, improve resource efficiency, favor social development and environmental protection (Ogunmakinde et al., 2022).

3 Methodology

This exploratory single case study was carried out between January and March 2022 in a company based in Nova União/RO through a presential interview an on-site visit and observation of the production processes seeking to prove the P1 proposition. The company is in the Brazilian Amazon, selected according to convenience, accessibility, and representativeness in the sector.

To analyses the strategies used by companies in the Legal Amazon in adoption of the CE, the models of Vermunt et al. (2019), Rossi et al. (2020), Yadav et al. (2020), in relation to fulfilment of the SDGs (United Nations, 2015). Figure 1 depicts dimensions and categories of the conceptual model of the research.

The interview was recorded with the proper authorization of the respondent and later transcribed, seeking a better analysis of its content. The interviewee was the director of the company, in which its foundation was in 2017. It was chosen to nominate in a fictitious way as Jamari to preserve his anonymity (Table 1).

Used the content analysis method in the Bardin (2016, p. 47) defines as "a set of communication analysis techniques that uses systematic and objective procedures for describing the content of messages in order to make inferences or deduction of knowledge regarding the causes and antecedents of messages and/or the consequences of those messages."

The recorded data was transcribed based on a script of topics and grouped using NVivo software. NVivo is a data analysis software that assists qualitative research, and in this way, it was possible to organize, categorize and evidence the results found, allowing to respond the problematic of the search.

Based on the dimensions and categories of analysis, we sought to analyses the EC Strategies and their relationship fulfilment of the SDGs in Jamari company belonging to the Brazilian Amazon. The results will be presented and discussed below.

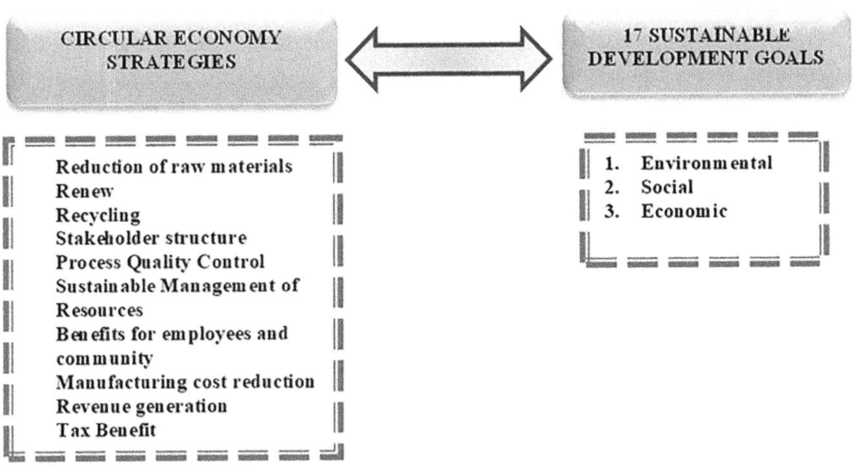

Fig. 1 Dimensions and Categories of Research Conceptual Model. *Source*: Own elaboration

Table 1 Structure of the qualitative interview protocol

Section	Title	Questions
I	Respondent profile	05
II	Characterization of the company	08
III	Circular economy strategies	21
IV	Sustainable development goals	17

Source: Own elaboration

4 Results

The company based in the Northern Region of Brazil was named, in this study, Jamari. The choice for the name was guided by the river that crosses the State of Rondônia. The data were described from three categories of analysis: characterization of companies, CE strategies and SDGs.

4.1 Characterization of the Company

The company Jamari began its activities in 2013, in the municipality of Nova União and has at its core conserve the Amazon rainforest, growing cocoa and impacting the environment as little as possible. The manager highlights the importance of the Amazon biome that covers different species of cocoa in terms of genetics, flavor and quality of the soil making the fruit have an excellent quality and tied to the tripod of sustainability.

The company has inclusion actions through its SCM to guide community to the EC. It is important to highlight that sustainable cultivation restores areas that were deforested, contributing to the improvement of life of the community and especially of the Amazon biome.

The production of cocoa generates income for the family and at the same time contributes to the conservation of forests and the ecosystem, predominating the agroforestry system, impacting the forest as little as possible. He points out that: "(…) this is our challenge to produce cocoa by promoting the adoption of sustainable practices."

The company, since 2017, has stood out with distinguished awards for the quality of the cocoa it produces. "We're going to do it differently because the market is open to quality." In 2017, the company was awarded its almond in first place at the Rondônia Rural Show and in second place at the ninth Chocolate Festival, in Ilhéus-BA, in 2018. More recently, in 2021, it obtained the first place in the Quality and Sustainability contest of Rondônia (Concacau). Table 2 presents a synthesis of the main characteristics analysed.

The main evidence highlighted by the interviewee regarding CS strategies are described next.

4.2 *Strategies of CE*

The CE causes fundamental structural changes in industrial symbiosis, in which waste from one operation can be reused in another, and so on, which means utilizing a resource for as long as possible (Awan & Sroufe, 2022).

In this sense, the company Jamari has in its process practices that develop solutions to reduce and improve the management of waste generated in its processes. In this context, the Jamari interviewee explains that the company:

> ...has always sought to innovate in this sense of bringing biodegradable packaging to our packaging. However, in our country, whenever we talk about natural or biodegradable, it is said that it is an inaccessible cost. So, some packaging, are already biodegradable. However, as we still have plastic packaging, so today we are still part of a chain, where environmental compensation is worked.

When it comes to the motivation of employees, collaborators, and external partners in the development of strategies aimed at sustainability and circular economy, the company seeks to form these partnerships through research, incentives, training, and qualification of employees. He emphasizes that "betting really without knowledge is able to mobilize any sector. And in the case of circular economies, people should be able to see that there is potential for everyone to benefit together". As a result, efficient resource management offers advantages in terms of environmental protection and cost reduction through decreased consumption and production.

In this scenario, Nascimento et al. (2022) state that micro, small, medium enterprises have significant challenges in the adoption EC, as they have limited resources and must deal with the constant technological transformation and globalization.

Based on these inspirations, numerous studies have been conducted on a worldwide on the fundamental importance of adoption the CE (Lim et al., 2022; Sarkar et al., 2022), highlighting how these practices assist in the integration of production chains that go beyond waste management and aim to maximize resource efficiency, as well as contribute to a triangular balance (Tukker, 2015; Lukin et al., 2022).

Table 2 Characterization of the analysed company

Organizational characteristics	Enterprise Jamari
Year of foundation	2013
Number of employees	4
Location	Nova União (RO)
Product marketed	Cacao
Sector of activity	Food
Company size	Small
Partnerships with other companies	Yes
Partnerships institutions or research center	Yes
Circular economy practice	Yes

Source: Data from the Research and Own Elaboration

The theme of CE relates to the efficient use of resources aimed at new business models that aim to optimize processes and reuse raw materials, aiming at the use of renewable inputs. Regarding the environmental dimension says that "seeks to reducing the use of raw materials, opting for innovative processes that consume less". In addition, it "uses waste management to make processes more efficient and uses technologies to achieve this strategy."

It is known that Circular Economic in the use of resources allocated in new business models that aim to optimize processes and reuse raw materials. When it comes to environmental considerations, the company strives to reduce the number of raw materials. In addition, it has waste management practices to make processes more effective and the use of technologies aimed at sustainability bias (Ye et al., 2023).

These explanations are in line with the practices carried out by the company Jamari stresses the importance of treating cocoa residues, that is, the "casqueiro", which is what is left of the cocoa, because "with the breaks we make a lot and then, in the process of decomposition there are fungi, which will decompose and for these fungi not to become evil, in the crop limestone or lime is used and thrown on top of it, then it decomposes. From the moment it decomposes we put it in the cocoa plantation".

According to de Jabbour et al. (2019) CE is a presupposition for new looks at ongoing modes of production. In this sense, Morseletto et al. (2017) denote in their study that in order to make CE effective in the company, strategies aimed at optimizing resources must first be developed.

Thus, it is highlighted that the CE a possible solution to minimize inequalities resulting from current production methods (Reich et al., 2023). In this regard, Azizuddin et al. (2021) highlight in their study that developing strategies aimed at optimizing resources is the first step in making EC effective in an organization.

From these practices, it can be shown that circular economy ally in the contribution of the sustainability tripe to the extent that it impacts on the conservation and reuse of resources, combats environmental degradation, promotes the growth of the local economy of the communities that use it (Babbitt et al., 2018).

The supporting evidence explained by the interviewee in the context of the SDG's is presented below.

4.3 Sustainable Development Goals

Since 2015, the 2030 Agenda has been the subject of discussion and reflection in several international meetings, providing multilateral forums. As a result, the 17 SDGs and their 169 goals and 232 indicators are driving public reflection and action in relation to medium and long-term government plans up to 2030, spanning several governments (Rodriguez-Anton et al., 2019). In this light, the core objectives of the SDGs are to overcome extreme poverty and hunger, to have quality education throughout life to protect the environment and to promote more peaceful societies, aiming for a more sustainable and inclusive world.

Brazil is a developing country that belongs to the BRIC group of countries and has a high capacity of resources and industrialization. It is notorious to highlight that it has in its territory the largest biodiversity in the world (about 12%) and, especially in the northern region, the Amazon Forest, needing support for sustainable development actions (OCDE, 2005).

So the connection of themes EC and SDG are under development. Rodriguez-Anton et al. (2019) point out that Circular Economy may be a possible solution regarding demand for raw materials, price volatility as well as the global population growth.

The strands of today's economy are still based on an open linear system of extraction, transformation, distribution, consumption, and disposal (Murray et al., 2017). In this sense, it is known that industrialization is one of the fundamental causes of environmental degradation that leads to obsolete productive models. Burger et al. (2019) state that adoption of a production system based on sustainability principles, or circular model, is necessary in this perspective to linked to the premises of the 2030 agenda.

The results, evidence examined in the case conducted by the Jamari company are presented. The discussion is presented according to the models of Vermunt et al. (2019), Rossi et al. (2020), Yadav et al. (2020), and the model SDGs of the with the United Nations (2015), which allowed an analysis of EC strategies and their relationship with the according of the SDGs.

In Table 3, results found in the study

5 Conclusions

The chapter was to find whether the circular business helps in the implementation of SDGs in the Amazon and which objectives foster the CE. In the regard, the phenomenon was investigated by qualitative methods through a study in a company producing chocolate from Rondônia called fictitious as Jamari company.

With that the data found that the analysed company has EC practices that helper in reach of the SDGs, highlighting mainly SDG 02, SDG 15, and SDG 17, for which they are found some practices by the interviewee.

By observing the repercussions of these relations, the findings provide a connected view, a symbiosis between sustainability and CE in contexts of management of the UN's SDGs in the Legal Amazon. In addition, the results offer a deeper understanding of the role of EC management and how it can improve economic, social, and environmental performance in the Amazon.

When we consider the effects of these relationships, the findings provide a connected view that combines sustainability and CE in the management of the SDGs. In addition, the results provide in-depth analysis EC management and how it can improve the environmental, social, and economic performance of the region in the State of Rondônia, in the Legal Amazon.

Table 3 Summary of evidence

		Analysis Categories	Jamari
Strategies for the Circular Economy	Environmental Dimension	Reduction	Yes
		Renew	Yes
		Recycling	No
		Stakeholder structure	Partial
		Process Quality Control	Yes
	Social Dimension	Sustainable Management of Resources	Yes
		Benefits for employees and the community	Yes
	Economic Dimension	Manufacturing cost reduction	Partial
		Revenue generation	Yes
		Tax Benefit	No
Sustainable Development Goals		1 - No poverty	No
		2 - Zero hunger	Yes
		3 - Good health and well-being	No
		4 - Quality education	No
		5 - Gender equality	No
		6 - Clean water and sanitation	No
		7 - Affordable and clean energy	No
		8 - Decent work and economic growth	No
		9 - Industry, innovation and infrastructure	No
		10 - Reduced inequalities	No
		11 - Sustainable cities and communities	No
		12 - responsible consumption and production	No
		13 - Climate action	No
		14 - Life below water	No
		15 - Life on land	Yes
		16 - Peace, justice and strong institutions	No
		17 - Partnerships for the goals	Yes

● = Yes (identified by evidence) = No (not identified by evidence = Partial (partially identified)

It is noteworthy that the conclusions can support the strengthening of sustainable practices capable of positively influencing the achievement of the SDGs, especially those that are fundamental reality of communities and organizations.

However, CE strategies help in important questions for the corporate environment analysed. Therefore, the model proposed to analyse the relationship between the CE strategies adopted by Vermunt et al. (2019), Rossi et al. (2020) and Yadav et al. (2020) in relation to achievement of the SDGs used by the United Nations (2015), they allow a technical and objective view of the strategies and practices of the CE that support the achievement of the SDGs.

In addition, it should be noted that the limitations of the study are related to the single case approach, making it impossible to extrapolate the findings to the

Rondonian scenario. On the other hand, the results of this exploratory research will allow for the formulation of a more detailed conceptual model and some hypotheses that will be tested in a future quantitative study. In this study the sample will include not only companies located in the state of Rondônia, but also in other legal states of the Amazon region.

References

Antunes, J. C. C., Eugénio, T., & Branco, M. C. (2022). Circular economy for cities and sustainable development: The case of the Portuguese City of Leiria. *Sustainability (Switzerland), 14*(3), 1726. https://doi.org/10.3390/su14031726

Awan, U., & Sroufe, R. (2022). Sustainability in the circular economy: Insights and dynamics of designing circular business models. *Applied Sciences (Switzerland), 12*(3), 152. https://doi.org/10.3390/app12031521

Azizuddin, M., Shamsuzzoha, A., & Piya, S. (2021). Influence of circular economy phenomenon to fulfil global sustainable development goal: Perspective from Bangladesh. *Sustainability (Switzerland), 13*(20), 1–16. https://doi.org/10.3390/su132011455

Babbitt, C. W., Gaustad, G., Fisher, A., Chen, W. Q., & Liu, G. (2018). Closing the loop on circular economy research: From theory to practice and back again. *Resources, Conservation and Recycling, 135*(May), 1–2. https://doi.org/10.1016/j.resconrec.2018.04.012

Bardin, L. (2016). *Análise de conteúdo*. Edições 70.

Bocken, N. M. P., de Pauw, I., Bakker, C., & van der Grinten, B. (2016). Product design and business model strategies for a circular economy. *Journal of Industrial and Production Engineering, 33*(5), 308–320. https://doi.org/10.1080/21681015.2016.1172124

Braungart, M., & McDonough, W. (2009). *Cradle to cradle. Remaking the way, we make things*. North Point Press.

Burger, M., Stavropoulos, S., Ramkumar, S., Dufourmont, J., & van Oort, F. (2019). The heterogeneous skill-base of circular economy employment. *Research Policy, 48*(1), 248–261. https://doi.org/10.1016/j.respol.2018.08.015

Chen, T. L., Kim, H., Pan, S. Y., Tseng, P. C., Lin, Y. P., & Chiang, P. C. (2020). Implementation of green chemistry principles in circular economy system towards sustainable development goals: Challenges and perspectives. *Science of the Total Environment, 716*(1), 136998. https://doi.org/10.1016/j.scitotenv.2020.136998

Corrado, S., & Sala, S. (2018). Food waste accounting along global and European food supply chains: State of the art and outlook. *Waste Management, 79*, 120–131. https://doi.org/10.1016/j.wasman.2018.07.032

da Rosa, L. A. B., Gomes, C. M., Campos, W. Y. Y. Z., Rodrigues, C., Godoy, T. P., & Kneipp, J. M. (2022). Influencing factors of the innovation power in the adoption of sustainability strategies. *Sustainability, 14*(12668), 1–15. https://doi.org/10.3390/su141912668

de Jabbour, A. B. L. S., Rojas Luiz, J. V., Rojas Luiz, O., Jabbour, C. J. C., Ndubisi, N. O., de Oliveira, J. H. C., & Junior, F. H. (2019). Circular economy business models and operations management. *Journal of Cleaner Production, 235*, 1525–1539. https://doi.org/10.1016/j.jclepro.2019.06.349

Dougherty, D., & Hardy, C. (1996). Sustained product innovation in large, mature organizations: Overcoming innovation-to-organization problems. *Academy of Management Journal, 39*(5), 1120–1153. https://doi.org/10.2307/256994

Elkington, J. (1998). *Cannibals with forks: The triple-bottom-line of 21st century*. N. S. Publishers, Gabriola Island.

EMAF. (2015). Growth within a circular economy vision for a competitive Europe, London. In *Towards the circular economy*. Ellen MacArthur Foundation. Retrieved from http://www.ncbi.nlm.nih.gov/pubmed/26849997%0A. https://doi.org/10.1111/jne.12374

Fatimah, Y. A., Govindan, K., Murniningsih, R., & Setiawan, A. (2020). Industry 4.0 based sustainable circular economy approach for smart waste management system to achieve sustainable development goals: A case study of Indonesia. *Journal of Cleaner Production, 269*, 122263. https://doi.org/10.1016/j.jclepro.2020.122263

García-Barragán, J. F., Eyckmans, J., & Rousseau, S. (2019). Defining and measuring the circular economy: A mathematical approach. *Ecological Economics, 157*(December 2018), 369–372. https://doi.org/10.1016/j.ecolecon.2018.12.003

Genovese, A., Acquaye, A. A., Figueroa, A., & Koh, S. C. L. (2017). Sustainable supply chain management and the transition towards a circular economy: Evidence and some applications. *Omega (United Kingdom), 66*, 344–357. https://doi.org/10.1016/j.omega.2015.05.015

Ghisellini, P., Cialani, C., & Ulgiati, S. (2016). A review on circular economy: The expected transition to a balanced interplay of environmental and economic systems. *Journal of Cleaner Production, 114*, 11–32. https://doi.org/10.1016/j.jclepro.2015.09.007

Goyal, S., Esposito, M., & Kapoor, A. (2018). Circular economy business models in developing economies: Lessons from India on reduce, recycle, and reuse paradigms. *Thunderbird International Business Review, 60*(5), 729–740. https://doi.org/10.1002/tie.21883

Hossain, M. U., Ng, S. T., Antwi-Afari, P., & Amor, B. (2020). Circular economy and the construction industry: Existing trends, challenges, and prospective framework for sustainable construction. *Renewable and Sustainable Energy Reviews, 130*(May), 109948. https://doi.org/10.1016/j.rser.2020.109948

Kirchherr, J., Piscicelli, L., Bour, R., Kostense-Smit, E., Muller, J., Huibrechtse-Truijens, A., & Hekkert, M. (2018). Barriers to the circular economy: Evidence from the European Union (EU). *Ecological Economics, 150*(December 2017), 264–272. https://doi.org/10.1016/j.ecolecon.2018.04.028

Klewitz, J., & Hansen, E. G. (2014). Sustainability-oriented innovation of SMEs: A systematic review. *Journal of Cleaner Production, 65*, 57–75. https://doi.org/10.1016/j.jclepro.2013.07.017

Kumar Shetty, S., & Subrahmanya Bhat, K. (2022). Green supply chain management practices implementation and sustainability –A review. *Materials Today: Proceedings, 52*, 735–740. https://doi.org/10.1016/j.matpr.2021.10.135

Leitão, A. (2015). Economia circular: uma nova filosofia de gestão para o séc. XXI. *Portuguese Journal of Finance, Management and Accounting, 1*(2), 23.

Lim, J. S., Li, C., Van Fan, Y., & Klemeš, J. J. (2022). How circular economy and green technology can address sustainable development goals? *Journal of Cleaner Production, 333*(December 2021), 2021–2023. https://doi.org/10.1016/j.jclepro.2021.130161

Lukin, E., Krajnović, A., & Bosna, J. (2022). Sustainability strategies and achieving SDGs: A comparative analysis of leading companies in the automotive industry. *Sustainability (Switzerland), 14*(7), 1–15. https://doi.org/10.3390/su14074000

Morren, M., & Grinstein, A. (2016). Explaining environmental behavior across borders: A meta-analysis. *Journal of Environmental Psychology, 47*, 91–106. https://doi.org/10.1016/j.jenvp.2016.05.003

Morseletto, P., Biermann, F., & Pattberg, P. (2017). Governing by targets: Reductio ad unum and evolution of the two-degree climate target. *International Environmental Agreements: Politics, Law, and Economics, 17*(5), 655–676. https://doi.org/10.1007/s10784-016-9336-7

Murray, A., Skene, K., & Haynes, K. (2017). The circular economy: An interdisciplinary exploration of the concept and application in a global context. *Journal of Business Ethics, 140*(3), 369–380. https://doi.org/10.1007/s10551-015-2693-2

Niero, M., & Olsen, S. I. (2016). Circular economy: To be or not to be in a closed product loop? A life cycle assessment of aluminium cans with inclusion of alloying elements. *Resources, Conservation and Recycling, 114*, 18–31. https://doi.org/10.1016/j.resconrec.2016.06.023

Nikolaou, I. E., Jones, N., & Stefanakis, A. (2021). Circular economy and sustainability: The past, the present and the future directions. *Circular Economy and Sustainability, 1*(1), 1–20. https://doi.org/10.1007/s43615-021-00030-3

OCDE. (2005). *Manual de Oslo: Diretrizes para a Coleta e Interpretação de dados sobre Inovação Tecnológica* (p. 184). OCDE, Eurostat e Financiadora de Estudos e Projetos. https://doi.org/10.1787/9789264065659-es

Ogunmakinde, O. E., Egbelakin, T., & Sher, W. (2022). Contributions of the circular economy to the UN sustainable development goals through sustainable construction. *Resources, Conservation and Recycling, 178*(October 2021), 106023. https://doi.org/10.1016/j.resconrec.2021.106023

Oliveira, N. C. (1991). Amazônia, pulmão do mundo? *Conjuntura Econômica, 199.*

Prieto-Sandoval, V., Jaca, C., & Ormazabal, M. (2018). Towards a consensus on the circular economy. *Journal of Cleaner Production, 179*, 605–615. https://doi.org/10.1016/j.jclepro.2017.12.224

Reich, R. H., Vermeyen, V., Alaerts, L., & Van Acker, K. (2023). How to measure a circular economy: A holistic method compiling policy monitors. *Resources, Conservation and Recycling, 188*(August 2022), 106707. https://doi.org/10.1016/j.resconrec.2022.106707

Rodriguez-Anton, J. M., Rubio-Andrada, L., Celemín-Pedroche, M. S., & Alonso-Almeida, M. D. M. (2019). Analysis of the relations between circular economy and sustainable development goals. *International Journal of Sustainable Development and World Ecology, 26*(8), 708–720. https://doi.org/10.1080/13504509.2019.1666754

Rossi, E., Bertassini, A. C., dos Santos Ferreira, C., do Amaral, W. A. N., & Ometto, A. R. (2020). Circular economy indicators for organizations considering sustainability and business models: Plastic, textile, and electro-electronic cases. *Journal of Cleaner Production, 247*, 119137. https://doi.org/10.1016/j.jclepro.2019.119137

Sarkar, B., Debnath, A., Chiu, A. S. F., & Ahmed, W. (2022). Circular economy-driven two-stage supply chain management for nullifying waste. *Journal of Cleaner Production, 339*(June 2021), 130513. https://doi.org/10.1016/j.jclepro.2022.130513

Schroeder, P., Anggraeni, K., & Weber, U. (2019). The relevance of circular economy practices to the sustainable development goals. *Journal of Industrial Ecology, 23*(1), 77–95. https://doi.org/10.1111/jiec.12732

Silva, L. D. J. S., Meneghetti, G. A., Pinheiro, J. O. C., Santos, E. M., & Parintins, D. M. (2019). O Extrativismo Como Elemento De Desenvolvimento E Sustentabilidade Na Amazônia: Um Estudo a Partir Das Comunidades Coletoras De Castanha-Do-Brasil Em Tefé, AM. *Revista Destaques Acadêmicos, 11*(2), 168–187. https://doi.org/10.22410/issn.2176-3070.v11i2a2019.2271

Stahel, W. R. (2016). Circular economy. *Nature, 24*(March), 435–531. https://doi.org/10.36661/2596-142x.2019v1i1.10902

Stahel, W. R., & Reday-Mulvey, G. (1981). *Jobs for tomorrow: The potential for substituting manpower for energy*. Vantage Press.

Su, B., Heshmati, A., Geng, Y., & Yu, X. (2013). A review of the circular economy in China: Moving from rhetoric to implementation. *Journal of Cleaner Production, 42*, 215–227. https://doi.org/10.1016/j.jclepro.2012.11.020

Susanty, A., Puspitasari, N. B., Prastawa, H., Listyawardhani, P., & Tjahjono, B. (2021). Antecedent factors of green purchasing behavior: Learning experiences, social cognitive factors, and green marketing. *Frontiers in Psychology, 12*(December), 1–14. https://doi.org/10.3389/fpsyg.2021.777531

Taborga, C. P., Lusa, A., & Coves, A. M. (2018). A proposal for a green supply chain strategy. *Journal of Industrial Engineering and Management, 11*(3), 445–465. https://doi.org/10.3926/jiem.2518

United Nations. (2015). *Transforming our world: The 2030 agenda for sustainable development.* United Nations General Assembly.

Vermunt, D. A., Negro, S. O., Verweij, P. A., Kuppens, D. V., & Hekkert, M. P. (2019). Exploring barriers to implementing different circular business models. *Journal of Cleaner Production, 222*, 891–902. https://doi.org/10.1016/j.jclepro.2019.03.052

Virmani, N., Saxena, P., & Raut, R. D. (2022). Examining the roadblocks of circular economy adoption in micro, small, and medium enterprises (MSME) through sustainable development goals. *Business Strategy and the Environment, 31*(7), 2908–2930. https://doi.org/10.1002/bse.3054

WWF-World Wide Fund for Nature. (2012). Living planet report. Retrieved from Living planet report website: http://awsassets.panda.org/downloads/lpr_2012_summary_booklet_final.pdf

Yadav, G., Mangla, S. K., Bhattacharya, A., & Luthra, S. (2020). Exploring indicators of circular economy adoption framework through a hybrid decision support approach. *Journal of Cleaner Production, 277*, 124186. https://doi.org/10.1016/j.jclepro.2020.124186

Yadav, M., Kamboj, S., & Rahman, Z. (2016). Customer co-creation through social media: The case of 'crash the Pepsi IPL 2015. *Journal of Direct, Data and Digital Marketing Practice, 17*(4), 258–258. https://doi.org/10.1057/s41263-016-0008-7

Ye, Q., Umer, Q., Zhou, R., Asmi, A., & Asmi, F. (2023). How publications and patents are contributing to the development of municipal solid waste management: Viewing the UN sustainable development goals as ground zero. *Journal of Environmental Management, 325*(PB), 116496. https://doi.org/10.1016/j.jenvman.2022.116496

Yuan, Z., Bi, J., & Moriguichi, Y. (2006). The circular economy: A new development strategy in China. *Journal of Industrial Ecology, 10*(1–2), 4–8. https://doi.org/10.1162/108819806775545321

Zhu, B., Nguyen, M., Sarm Siri, N., & Malik, A. (2022). Towards a transformative model of circular economy for SMEs. *Journal of Business Research, 144*(January), 545–555. https://doi.org/10.1016/j.jbusres.2022.01.093

Lean and Green: 3Ms Meets 5Rs Creating Environmental Value for the Planet

Alvair Silveira Torres Junior and Ronaldo Akiyoshi Nagai

Abstract This chapter proposes operationalizing the Circular Economy in companies through the Lean Thinking management model. For this purpose, the authors demonstrate the conceptual correspondences between the principles of Circular Economy and Lean Thinking based on a common principle of reducing waste and minimizing the use of resources. It also discusses the development of sustainable products whose lean LPPD technique can leverage with the questioning of each design stage from what is valuable for the customer and the planet. Finally, the theoretical discussion is exemplified through a clothing company that applies the concept of lean thinking in its product development and operations with promising results for a green transition.

Keywords Circular economy · Green transition · Lean manufacturing · Green-lean · R principles

1 Introduction

In the last decade, the concepts and objectives of the circular economy and green transition have been increasingly detailed and explored in academic, business, and government discussions at regional and global levels. However, a transition to a circular economy model is impacted by several challenges and critical factors that make the transition arduous. Therefore, in addition to technological development to

A. S. T. Junior (✉) · R. A. Nagai
University of São Paulo – School of Economics, Business and Accounting, São Paulo, Brazil
e-mail: alvair@usp.br; rongai@usp.br

C. Rego et al. (eds.), *Entrepreneurship, Technological Change and Circular Economy for a Green Transition*, Studies on Entrepreneurship, Structural Change and Industrial Dynamics, https://doi.org/10.1007/978-3-031-48079-9_6

a circular economy, changes in philosophies, mindsets, practices, and processes in entrepreneurship are equally important (Cohen, 2006; Söderholm, 2020).

A management concept has been the subject of research and application in academia and organizations to waste reduction since the 1990s: Lean Manufacturing, which originates in the early twentieth century in one of the largest automakers in the world, Japan (Ward, 2007; Womack & Jones, 1997; Womack et al., 1990). However, this concept was aligned with the economic pattern of the time, focusing on waste concerning the efficient use of resources. As a result, lean manufacturing helped countless organizations in the implementation of efficient production chains, allowing the reduction of operational and production costs and, consequently, the achievement of greater profits.

Lean manufacturing is based on the managerial philosophy or mindset called Lean Thinking, centered on identifying and eliminating waste to create value for the customer, along with a specific process for solving problems on the way to minimizing waste. Thus, Lean thinking uses several management tools to critically analyze the production value stream compared to the consumption stream and identify three main categories of misalignment between the production process and demand (known as the 3Ms): Muda (waste originating from the application of extra resources), Muri (the unnecessary overcapacity in the process) and Mura (imbalance, lack of regularity and stability in the process).

Circular economy, on the other hand, has been defined in almost as many ways as there are circular economy researchers and practitioners (Kirchherr et al., 2017). Its popularity among scholars and practitioners occurs because it is viewed as an operationalization for businesses to implement the much-discussed idea of sustainable development (Murray et al., 2017). However, a so tractioned concept employed by various stakeholders can blur its definition. This point has been criticized by stakeholders and practitioners (Lewandowski, 2016; Murray et al., 2017). Thus, in addition to finding and systematizing circular economy, it is equally important to propose frameworks to operationalize the concept from an academic and practical perspective. Otherwise, it may collapse or remain in a deadlock due to permanent conceptual retention (Bocken et al., 2017).

Accordingly, considering that the practice of minimizing waste is aligned with the principles of circular economy, this chapter aims to use the Lean thinking framework in management to eliminate waste, redirect its focus beyond customer value, and achieve value for stakeholders and the planet within the context of the green transition. Therefore, the arrangement is as follows. In the second section, we first describe the concepts of circular economy based on the 5Rs principles: Rethink, Refuse, Reduce, Reuse, and Recycle.

Then, lean management concepts like LPPD (Lean Product and Process Development), SBCE (Set-based Concurrent Engineering), and 3Ms concept (Muri, Muda, Mura) are outlined, followed by our proposal of 'Green-Lean', a combination of 5Rs principles with 3Ms concept. Then, the third section outlines the research methodology, followed by the results and findings in the fourth section. Finally, we conclude with implications for future research in the fifth section.

The following sections will present a clear and objective connection between Lean Manufacturing concepts, which may be familiar to many organizations, and the circular economy's practices. Accordingly, to alert academia, organizations, and society that the transition to a circular economy may be less complex than imagined by these parties.

Ultimately, this chapter intends to contribute to constructing a framework or a practical approach, which we affectionately call 'Green-Lean', so that the parties involved can finally migrate from the conceptual plane to the circular economy practices. The chapter could benefit a wide range of researchers, entrepreneurs, consolidated enterprises, government, and society.

2 Circular Economy and Waste Management

2.1 *Circular Economy*

The roots of the circular economy are found in the work of Boulding (1966). On this occasion, the Author suggests that circular systems in the global economy are essential for humanity to guarantee its existence in the long term. In the 1980s, Pearce and Turner (1989), inspired by Boulding, argued that the traditional, linear economy would no longer be able to guarantee the sustainability of the world and a new vision of the economy, analyzed from a holistic perspective and a system thinking approach, should be proposed. The circular economy is currently one of the most discussed terms among environmental and economic scientists.

One of the current most recognized definitions is proposed by MacArthur et al. (2015), based on a core defining element: the "restorative use" of resources: 'While great strides have been made in improving resource efficiency, any system based on consumption, rather than on the restorative use of resources, entails significant losses along the value chain' (MacArthur et al., 2015).

The concept of a circular economy is based on three principles: designing out waste and pollution, keeping products and materials in use, and regenerating natural systems. The change in the mindset requires a shift in the way we produce and consume goods, as well as the development of new business models that prioritize sustainability. Companies like Patagonia have already embraced circularity with initiatives like take-back programs and closed-loop production systems (Ellen MacArthur Foundation, 2020).

The circular economy is a model that allows rethinking the current society's economic practices and is inspired by the functioning of Nature itself. It is inseparable from the innovation and design of products and systems. It is included in a development framework based on the principle of "closing the life cycle" of products, allowing the reduction in the consumption of raw materials, energy, and water. Promotes the development of new relationships between companies, which simultaneously become consumers and suppliers of materials reincorporated into the production cycle (Ellen MacArthur Foundation, 2012, 2013, 2014).

Thus, the model is designed to be restorative and regenerative rather than linear and wasteful. It is a system that aims to keep resources in use for as long as possible, minimize waste, and promote renewable resources. According to a report by the World Economic Forum, transitioning to a circular economy could add $4.5 trillion to the global economy by 2030. Additionally, it could help reduce greenhouse gas emissions by 39% by 2050 (Lacy & Rutqvist, 2015).

However, the definition of circular economy is vast. Kirchherr et al. (2017) gathered 114 circular economy definitions coded into 17 dimensions. Despite that, these numerous definitions gravitated over a core principle known as the 'R framework'. Initially, the circular economy emerged from the 3R framework: Reduce, Reuse, and Recycle, which can also be found at the core of the 2008 Circular Economy Promotion Law of China (Ghisellini et al., 2016; Ogunmakinde, 2019).

Later, the European Union Waste Framework Directive added 'Recover' as the fourth 'R' (Backes, 2020), also proposed by Sihvonen and Ritola (2015). Scholars and practitioners proposed other frameworks beyond the '4R', like the '6Rs' (Reduce, Reuse, Recycle, Recover, Redesign, and Remanufacturing) first coined by Joshi in proceedings of the IV Global Conference (Joshi et al., 2006). Finally, Potting et al. (2016) proposed the '9R' framework (Refuse, Rethink, Reduce, Reuse, Repair, Refurbish, Remanufacture, Repurpose, Recycle, and Recover).

This chapter, which aims to connect the green transition to the consolidated concepts of lean manufacturing, will adopt the '5R' framework comprising 'Refuse', 'Rethink', 'Reduce', 'Reuse', and 'Recycle'. Therefore, it will mainly focus on *ex-ante* strategies of smarter product use and manufacturing, which is at the core of the Lean Thinking mindset (of reducing all types of waste). Figure 1 discloses the Potting et al. (2016) '9R' framework and the selection of 'Rs' for this chapter's development.

The following subsections will detail the lean manufacturing concepts that will sustain this chapter's 'Green-Lean' framework (or practical approach) subject.

2.2 LPPD: Lean Product and Process Development and SBCE: Set-Based Concurrent Engineering

The concept of "Lean" was initially introduced by Krafcik (1988) and gained widespread recognition through the influential management book titled "The Machine that Changed the World," authored by Womack et al. (1990). Lean methodology strongly emphasizes maximizing efficiency in terms of outputs by minimizing resource utilization. In contrast to traditional mass production practices, Lean aims to achieve substantial reductions across various dimensions, including human labor, factory space, tool investment, and engineering hours dedicated to new product development (Womack et al., 1990).

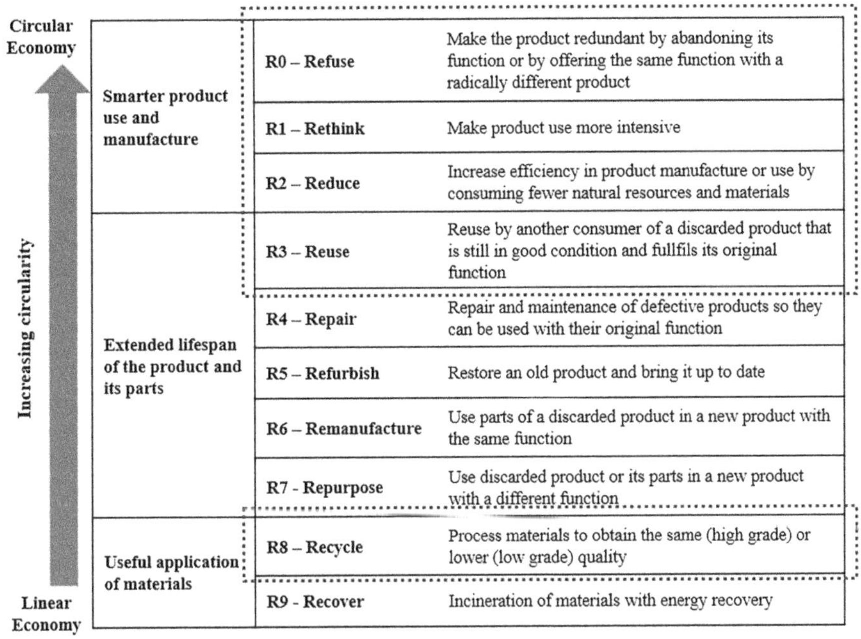

Fig. 1 Circularity strategies and 'Rs' selection for the study. *Source*: adapted from Potting et al. (2016)

Through the MIT International Motor Vehicle Program, Womack's research pointed out that Toyota had created a production system with operational tools such as Kanban and Just-in-time, among others, whose functionality required the development of a qualified workforce to deal with the operational routine both in terms of its stabilization and in the pursuit of continuous improvement. This new use of human resources requires a new approach to communication and leadership at all levels of the organization. (Nagai & Torres Jr., 2021).

Derived from the Toyota lean production system, the Lean Product and Process Development (LPPD) methodology is founded on optimizing value and minimizing waste. Khan et al. (2011) state that the LPPD approach has emerged to cater to the requirements of European manufacturing firms, aiming to surpass lean manufacturing practices by integrating Lean thinking into the product design and development process. (Khan et al., 2011, as cited in Nagai & Torres Jr. (2021).

Some companies have utilized this approach to enhance their manufacturing processes (Baines et al., 2006; Khalil & Stockton, 2010), while only a few have applied Lean Thinking to drive product and process innovation (Al-Ashaab et al., 2016 and Sobek et al., 1999). The model incorporates five key concepts: value focus (VF), knowledge-based environment (KBE), continuous improvement (CI) through Kaizen, chief engineering (CE), and Set-Based Concurrent Engineering (SBCE), with SBCE serving as the guiding framework for the LPPD model. (Nagai & Torres Jr.., 2021).

Set-Based Concurrent Engineering (SBCE) refers to a systematic procedure wherein a multitude of solutions for distinct sub-assemblies and components are concurrently developed, as Ward (2007) highlighted. This concurrent development process progressively narrows down based on the advancement of testing and prototyping activities, thereby establishing a knowledge base that facilitates cohesive decision-making (Al-Ashaab et al., 2016; Sobek et al., 1999). SBCE encompasses a comprehensive framework comprising five fundamental sets of principles: (1) Strategic value research and alignment; (2) Mapping the design space; (3) Simultaneous creation and exploration of multiple concepts; (4) Integration through an intersection; and (5) Ensuring feasibility prior to commitment. (Al-Ashaab et al., 2016; Sobek et al., 1999, as cited in Nagai & Torres Jr.., 2021).

2.3 *Managing Waste in Lean Thinking: Muri, Muda, and Mura (3Ms)*

The concepts of Muri, Mura, and Muda are at the heart of lean manufacturing, a management philosophy that focuses on eliminating waste and continuously improving processes. These three Japanese words refer to different types of waste in manufacturing and other processes. Therefore, Muri, Mura, and Muda are interconnected, and addressing one can lead to improvements in the others (Ohno, 1995; Ward, 2007).

Muri is a term used to describe overburden or excess in a process. It refers to any task that is too difficult, too complex, or beyond the capacity of the people or machines involved in the process. Muri often leads to waste in defects, delays, or rework (Ohno, 1995; Ward, 2007). For example, overloading a production line with work may lead to machine breakdowns, operator fatigue, and increased error rates. To eliminate Muri, processes must be designed to match the people and machines' capacity, and the workload should be evenly distributed.

Mura refers to unevenness in a process, known as waste originating from process variation. It refers to any variation or inconsistency in the process that leads to waste (Ohno, 1995; Ward, 2007). Mura often arises due to poor planning, inadequate training, or insufficient resources. For example, suppose a production line is not designed to handle a particular product. In that case, it may lead to product quality variation or delays in the production process. To eliminate Mura, processes must be standardized, and any deviation must be identified and addressed. Reducing process variation has often attracted extensive attention through initiatives such as Six Sigma, standard work, 5S, and TPM (Bicheno, 2018).

Muda refers to any waste in a process. It includes anything that does not add value to the product or service being provided. For example, Muda is often caused by overproduction, defects, waiting, over-processing, excess inventory, unnecessary motion, or unused talent. To eliminate Muda, processes must be streamlined, and waste must be identified and eliminated wherever possible (Ohno, 1995).

The concepts of Muri, Mura, and Muda are central to the Toyota Production System (TPS), the basis for lean manufacturing. The TPS was developed by Taiichi Ohno and Eiji Toyoda in the 1950s and 1960s, revolutionizing manufacturing in Japan. The goal of this system was to eliminate waste and maximize efficiency in the manufacturing process.

The Toyota Production System is based on several principles, including Just-in-Time (JIT) production, continuous improvement, and respect for people. JIT production involves producing only what is needed when required and in the necessary quantity. The continuous improvement consists in identifying and eliminating waste on an ongoing basis. Finally, respect for people involves treating employees with respect and empowering them to improve the process (Womack, 2002). The concepts of Muri, Mura, and Muda have been applied to a wide range of industries and processes beyond manufacturing, including healthcare (Daultani et al., 2015), software development (Nikiforova & Bicevska, 2020), service industries (Oey & Nofrimurti, 2018) and even in the dining industry (Nagai & Torres Jr.., 2021).

Eliminating Muri, Mura, and Muda requires a systematic approach involving all levels of the organization. It is not enough to address these concepts at the operational level; they must be part of the organization's culture. In addition, it requires a continuous improvement mindset, a willingness to challenge the status quo, and a commitment to providing the necessary resources to achieve lean manufacturing.

In conclusion, the concepts of Muri, Mura, and Muda are essential to lean manufacturing. They represent a waste of resources, unevenness, and overburden, respectively. These concepts must be addressed simultaneously to achieve lean manufacturing. Eliminating Muri, Mura, and Muda requires a systematic approach that involves all levels of the organization and must be part of the organization's culture. A commitment to continuous improvement and a willingness to challenge the status quo are essential for lean manufacturing.

2.4 *Waste Reduction (3Ms Management) and 5Rs Principles: The 'Green-Lean'*

The transition to a green economy, which involves the evolution of the linear economy to a circular economy, has as one of its basic foundations in the 5Rs principles (Rethink, Refuse, Reduce, Reuse, and Recycle). When we align these circular economy principles with lean thinking, we will see that their management techniques for eliminating waste gain scope to positively impact the green transition of entrepreneurship and companies in general.

The 3Ms, for example, are reframed with the circular economy. Muda now is the waste of using natural resources in an unmeasured way, Muri becomes the overload in Nature's restoration capacity, and Mura is the oscillation or instability of consumption caused by agents of market exploitation of mass consumption.

On the other hand, 'Rethink', 'Refuse', and 'Reduce' are closely connected to the design process of products and services, an *ex-ante* activity, which can be contributed to the Lean Thinking principles, Set-Based Concurrent Engineering (SBCE) and Lean Product and Process Development (LPPD). Based on the Lean Thinking concept, the LPPD has its roots in maximizing value while minimizing waste in designing new products and processes (Ward, 2007).

In turn, the SBCE is a process where sets of design solutions for the same objective are developed in parallel (Ward, 2007). The concurrent development process gradually converges as testing and prototyping progress, resulting in the generation of relevant knowledge intricately linked to our proposal to attain optimal solutions for implementing a circular economy. Additionally, parallel development holds the potential to enhance process stability, mitigate process irregularities, and mitigate waste resulting from process variation (Nagai & Torres Jr.., 2021).

In this sense, the objective to achieve maximum resource efficiency through innovative models which sustain the 'Rethink' principle could be associated with the practice of recreating ideas, processes, and dynamics of a product encouraged by the continuous improvement (Kaizen) from Toyota Production System (TPS) concepts (Linder, 2017).

The 'Refuse' and 'Reduce' principles match the core idea of reducing waste from the Muda, Mura, and Muri perspectives. While the former refers to making a product redundant by abandoning its function, offering the same function with a radically different product, or using specific materials or production processes to make the economy more circular (Haupt et al., 2017), the latter comprises using fewer natural resources and the encouragement of dematerialization in the process (Reike et al., 2018). As a consequence, reducing the overload in Nature's restoration capacity (minimizing Muri), providing production and designing process stability avoiding resource waste or idle resources (minimizing Mura), and resulting in the reduction of natural resources use in an unmeasured way (minimizing Muda).

Finally, the 'Reuse' and 'Recycling' principles can receive direct and relevant contributions from the *ex-ante* processes of better and greener design promoted by the LPPD and SBCE, especially from de avoidance of Muda. For example, in the production process, 'Reuse' and 'Recycling' can also benefit from identifying waste of water, energy, and various materials by investigating their root causes and mapping reuse and recycling flows. Also, the structuring of value chains centered on recycling and reuse. Therefore, a first glance, it would not be an exaggeration to suggest that the ideas of Rethink, Refuse, Reduce, Refuse, and Recycle (5Rs) have an intimate relationship with Lean Manufacturing (and 3Rs waste elimination) in its essence (Fig. 2).

Nonetheless, despite the solid and intimate relationship and the promising synergy of these concepts, the issue of eliminating waste in organizations, from the perspective of Lean Manufacturing principles, has focused on the operational and financial competitiveness of organizations to the detriment of an approach that is more related to contributions transition to a green economy. This fact justifies the need to intensify multidisciplinary research, such as the potential to bring concepts already consolidated in academia and organizational practices to discussions

5R Principles		Definition	Lean Thinking Principles	Conceptual Link
Ex-ante	Refuse	Make the product redundant by abandoning its function or by offering the same function with a radically different product	SBCE LPPD Muri Muda Mura	• SBCE and LPPD: design products with longer lifespans, reduction of waste in designing new products and processes, and recreation of an idea through Kaizen (continuous improvement) • MURI: reduce the overload in Nature's restoration capacity; • MURA: development process stability, unevennes avoiding resources waste; • MUDA: avoid the use of natural resources in an unmeasured way, through inventory reduction (lean product flow, pull system, etc)
	Rethink	Make product use more intensive		
	Reduce	Increase efficiency in product manufacture or use by consuming fewer natural resources and materials		
Post Consumption	Reuse	Reuse by another consumer of a discarded product that is still in good condition and fulfills its original function	Muda	• MUDA: reuse and recycle avoid the use of natural resources in an unmeadured way, through overall consumption and demand reduction.
	Recycle	Process materials to obtain the same (high grade) or lower (low grade) quality		

Fig. 2 Relationship between the 5R and Lean Principles. *Source*: Author's elaboration

involving the circular economy. The following sections will try to approach this multidisciplinary research through a case study of a company whose sustainable practices are referenced globally.

3 Methodology

For this research, we adopted a single case study. Despite the crucial historical role played by case studies in research in the management field, many scholars have criticized the method. Critics argue that a case study has limitations regarding its virtual inability to provide a solid basis for generalizing study findings, which is more severe when single case studies are employed (Piekkari & Welch, 2018). Yet, authors like Yin (2009) call attention to the importance of theory building from single case studies, which can enable the creation of theories fitting the base concepts precisely to the details of a particular case. It contrasts with multiple-case studies, where researchers retain only the relationships replicated across most or all cases (Eisenhardt & Graebner, 2007). The single case study utilized in-depth interviews to collect data on the experience of a successful entrepreneur in the apparel industry.

An interview is one of the most common and powerful ways to try to understand our human condition. Consequently, it has become a classic technique for obtaining information in the social sciences, with wide adoption in sociology, communication, anthropology, administration, education, and psychology. Although previously used in journalism, ethnography, psychology, and market and opinion research, its emergence as a methodological theme can be identified in the 1930s in the context of American social work publications, receiving significant contribution in the 1940s in the studies of Carl Rogers in patient-oriented psychotherapy.

However, the in-depth interview is a methodological resource that seeks, based on theories and assumptions defined by the researcher, to collect answers from the subjective experience of a source selected because it holds information that one wants to know. In this way, as in the qualitative research, data are not only collected but also the result of interpretation and reconstruction by the researcher dialoguing intelligently and critically with reality. In this journey of discoveries, the questions make it possible to explore or deepen a subject, describe processes and flows, understand the past, analyze, discuss, and make prospects. They also make it possible to identify problems, micro-interactions, patterns, and details, obtain value judgments and interpretations, characterize the richness of a theme, and explain phenomena of limited scope.

In-depth interviews are a valuable qualitative method used in academic research to gather detailed information about an individual's experiences, perceptions, opinions, and beliefs. Conducting an in-depth interview involves several key steps. The first step is identifying the appropriate participants based on their expertise or experience related to the research question (Boyce & Neale, 2006; Rutakumwa et al., 2020). For this step, we selected the CEO and founder of an apparel company recognized for its sustainable and innovative practices.

Once the participants are identified, the researcher must obtain informed consent to participate in the study. This includes explaining the study's purpose, what the interview entails, and any risks or benefits involved. To fulfill this step, researchers organized a pre-session with the interviewee to outline the study's scope, publication's profile, target audience, and other titles from the same collection. Additionally, documentation like the proposed structure of the chapter was shared beforehand.

As a subsequent step, we prepared the questions in light of the literature presented in the previous sections. We later discussed them with the research team, allowing them to capture the intended phenomenon. The output of this stage comprised an in-depth interview script, which aimed to collect answers to corroborate our proposal to combine the concepts of lean thinking and circular economy toward a green transition in organizations.

Then the secondary data analysis was followed by a two-hour interview with the founder and the session transcription. Finally, the transcription was shared with the researchers for further discussion to exchange ideas and perceptions regarding the entrepreneur's information to find the relationship between the company's reality, lean practices, and the transition to a green economy. Results are presented in the following section.

4 Results

4.1 Sustainable Production, Designing Practices, and Lean Thinking

This research was carried out at Movin, belonging to Infinitu X, a company launched in 2011 to accelerate the transition to a more conscious and technological lifestyle. In addition to Movin, the Infinitu X portfolio comprises Weloop (branded online) and Tropik Cosmetics (cosmetics). The research adopts a qualitative approach, using document analysis and an in-depth interview with the CEO and Founder of Infinitu X and Movin to deepen the analysis. The documents accessed were: the company website, technical reports, company-specific folders, video recordings about the company, photographs, and case studies about the company.

Since 2011, the brands have received several awards, such as Best For The World 2017 and 2018 from B-Corporation, Management Excellence, Most Notable and Sustainable Company from PR Newswire, and Bronze Packaging Award from Brazil Design Awards, to name a few. Movin was launched in the same year (2011) and was the first Brazilian apparel company to obtain the B corporation certification. For the founder, the certificate is relevant because it *recognizes and validates new organizational models that encourage business power to solve social and environmental problems, always in line with the concept of a solidarity economy*, highlighting that *there are 2933 certified companies worldwide in 150 types of industries in 64 countries. Patagonia, a company widely recognized for its sustainable practices, is an important exponent of transitioning to a circular economy in the apparel industry.*

The brand is dedicated to the essential, minimalist, conscious lifestyle, guided by the equation design + technology + ethics. From the founder's point of view, (…) *this tripod expresses our desire to create an environmental and social impact undertaking, combining our experiences and trajectories in technology, design, and prototyping*. Therefore, applying multiple solutions, it aims to offer products based on the principles of (1) sustainable materials, (2) low-impact technological processes, (3) transparency, and (4) traceability.

Regarding sustainable materials and low-impact technological processes, the company adopts recycled PET (polyethylene terephthalate) fabric (or recycled polyester), widely used in sustainable fashion, which is crucial in reducing plastic waste disposed of in the environment. On top of the 'well-known' PET fabric, Movin has initiatives to promote what they call 'smart materials'. For example, some articles employ 'bioleather', a 100% natural, biological, compostable, and vegan fabric. It comprises a web of nanocrystalline cellulose fibers synthesized by bacteria of the *acetobacter* genre during cultivation.

The production of the biotissue in the laboratory is sustainable, with simple, non-polluting, and renewable materials and a reduced water and carbon footprint. Compared to traditional fabric production processes, biofabric preparation uses greener technologies. Moreover, it takes 3–4 weeks to be ready, between growing, harvesting and washing, drying, and processing the material according to end-use. Additionally, these biomaterials can be dyed with natural pigments.

Another example is the development of bamboo fibers. Bamboo (also known as bamboo viscose) is a biodegradable fiber with an extremely light touch. In its planting, it is unnecessary to use fertilizers, as it is a very fast-growing plant. There is also no need to use pesticides, as bamboo is naturally resistant to pests. In addition, bamboo fiber uses little water in its irrigation process and absorbs more carbon dioxide from the air than cotton and wood. From a functional perspective, products made from bamboo fibers have an antibacterial function, inhibit odors, and have a quicker drying process.

According to the founder, *sourcing sustainable materials in the market is demanding. First, because the current scale is, from a financial perspective, unfeasible for most suppliers. Second, it is still a challenge to change the mindset of entrepreneurs in the market. For this reason, we act as an investor and promoter of the sustainable supply chain, supporting potential suppliers to convert their portfolio to a 'greener line-up gradually'.*

As for transparency practices, Movin discloses data like the material composition, yield, source (location of the suppliers), price, cost breakdown (including by production process—Fig. 3), and, most unusually, the company profile supplying its materials. As shown in Fig. 4, Movin discloses, for every product in their portfolio, the location of the supplier, the number of associates, the person in charge, and the associate's wage level. Movin believes that (…) promoting a sustainable business must promote good practices throughout the chain. Therefore, suppliers' associates must have decent compensation in line with market standards.

Concerning traceability, Movin makes efforts to conduct the audit, whenever possible in loco, to verify and guarantee sustainable practices along the chain, tracking the source of raw material, the development of products, suppliers' compliance with labor and environmental laws, and companies' certifications. This tracking can be done on multiple levels, ranging from individual products to entire lots, depending on the verified variable and the value adds driven by the monitoring. *Visiting our suppliers and partners is an excellent opportunity to see the phenomenon, be in contact with our partners' associates, know the satisfaction level of people, and especially to feel if the partner is truly aligned with our principles.*

Regarding the connection with lean thinking, the principles adopted and practiced by Movin—sustainable materials, low-impact technological processes, transparency, and traceability—are directly related (Fig. 5). Out of the 5Rs principles, 'Reduce', 'Reuse', and 'Recycle' can be directly associated with the company's practices. For example, using smart materials (PET recycled fabric, biotissue, and bamboo fiver) reduces water, carbon, and pesticide footprint while reusing and recycling materials, consequently reducing Muda. Also, the traceability of suppliers and the constant monitoring contributes to establishing a solid and fluid process, avoiding Mura. Finally, transparency and traceability practices are fundamental to providing reliability to customers and partners, being the basis of this system.

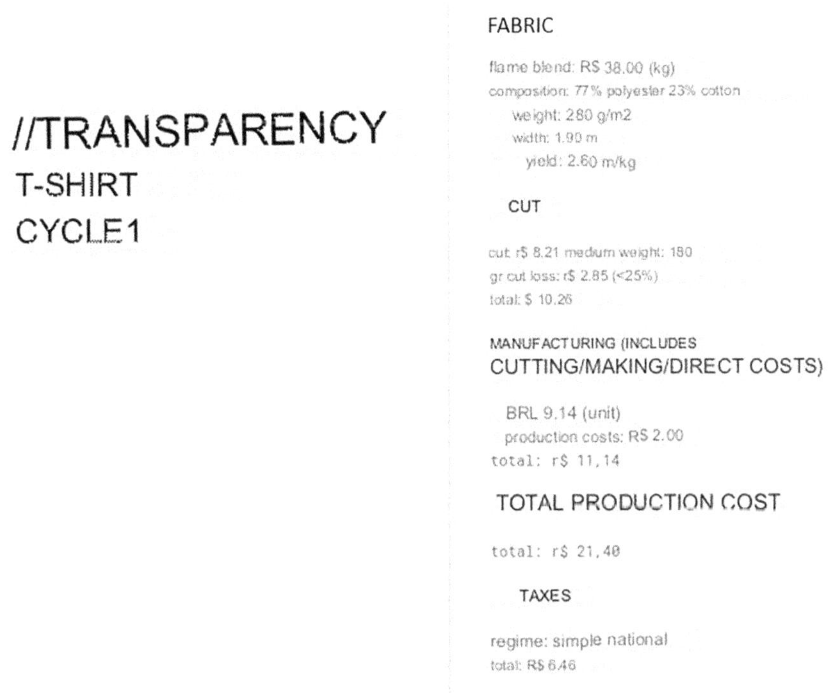

Fig. 3 Price and cost breakdown for 'Cycle 1 – T-Shirt'. *Source*: Movin, available at https://startmovin.com/en/pages/transparencia-camiseta-cycle1

Fig. 4 Supplier's location, associates, and wage level data. *Source*: Movin, available at https://startmovin.com/en/pages/transparencia-camiseta-cycle1

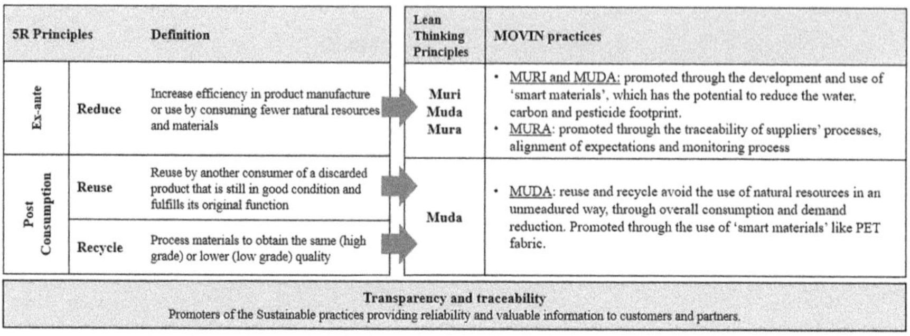

Fig. 5 Connecting 'Rs' and lean thinking principles with Movin practices. *Source*: Author's elaboration

4.2 MOVIN ZRO® Project and Lean Thinking

Furthermore, committed to continuous innovation, the company launched MOVIN ZRO®, a project in which the presale and on-demand format is applied to eliminate stocks and bio-tissue research through a partnership with a research institute (BIOTECAM). Finally, to strengthen the green economy initiatives, in 2020, the company launched the Spinoff WeLoop, which focuses on creating the future wardrobe connected to the principles of circular economy and the Internet of Things (IoT).

The MOVIN ZRO® project is one of the founder's great hopes for transitioning to a green economy, as it breaks paradigms in this sector. For example, the current linear model of the fashion industry, which accounts for 10% of human carbon dioxide (CO_2) emissions, generates nearly 92 million tons of waste (Dahlbo et al., 2017). In other words, about 30% of the clothes produced each season are never even sold (Niinimäki et al., 2020; Runnel et al., 2017).

According to the founder, (…) *the MOVIN ZRO® project is one of a kind. Considering that, according to our estimates and sector statistics, around 30% of the volume produced by the segment is discarded. It becomes, literally, garbage. The collections of the main brands are produced in a 'volume-based, push system', charging higher values for pieces at launch and offering aggressive discounts on "leftover collections". This logic does not adhere to the circular and sustainable economy, as there is minimal incentive to reward customers loyal to your brand and aligned with your principles. On the contrary, in terms of prices for the customer, we are rewarding the overproduction.*

MOVIN ZRO® project is divided into four steps: (1) Co-creation; (2) Step 01—presales; (3) Step 02—production; and (4) Step 03—smart delivery. Co-creation step regards active hearing and communication with the Movin community. The company interacts with customers through social media and other channels and, based on data analysis, co-creates the product design to be launched. Then, the approved design moves to Step 01—presales. In this step, each approved design in the previous step has a minimum and maximum target. The minimum target seeks the breakeven point to launch the product, while the maximum target is established to *value quality, not quantity*. If the campaign does not reach the minimum goal, the design/product is frozen. In the case of success, reaching the minimum goal, Step 02—production starts.

During Step 02, only what was purchased in the presale will be produced, avoiding stock excess. Thus, production status will be updated on the product page so the Movin community can follow and understand the various manufacturing steps. In addition, it aligns with the company's commitment to transparency and product traceability during all stages. Finally, during Step 03—smart delivery, the production is consolidated into a single shipment to the Movin distribution center. Through this consolidating sales, the company can reduce impacts on product logistics, promote stability in the process and avoid unevenness. Once the shipments are organized, products are delivered directly to customers, using more efficient transport to reduce CO_2 gas emissions.

In short, if in the traditional model (where inventories are produced) excess unsold inventory, which represents around 30% of all production, is included in the final price, in practice, the consumer is paying for the excess: (…) *as in the MOVIN ZRO® project we only produce what is sold, extra costs are cut and more competitive prices are possible. Brands that work with inventories end up liquidating excess production, rewarding only those who buy during the promotion, that is, customers who value the brand the least and are only looking for the price factor. We invert this logic and give bonuses to the first presale buyers, valuing partners who align with the company's values.* The correlation between the steps of the MOVIN ZRO® project and lean thinking is shown in Fig. 6 below:

It is also important to highlight that the above process can drastically reduce the lead time from product conception to final customer delivery. According to the founder, a typical apparel design have an average of seven 'prototyping-validation' processes, which involves the design process, proofing, adjustments, and redesign. These *(…) 'comings and goings' themselves are considered inefficient in the process, a cause of Muda because we had to spend more resources, materials, time, and*

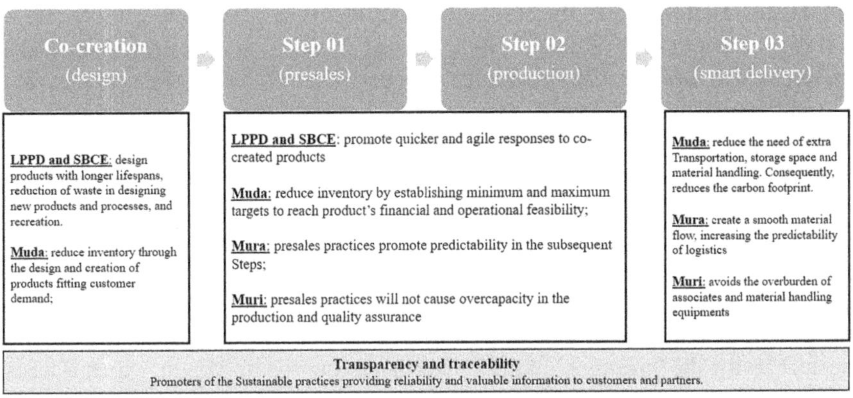

Fig. 6 Correlation of MOVIN ZRO® project with lean thinking. *Source*: Author's elaboration

people to reach the ideal product. The adoption of lean thinking, specially LPPD and SBCE, in the process, combined with our database, can potentially reduce the lead time from 30 to 3 days.

It was mentioned that the average inventory waste in the fashion industry is around 30% of the total volume produced. In the case of Movin, on the contrary, the inventory's loss level is close to 5%. For this reason, the founder calls our attention to the fact that Movin is not, in reality, a fashion company. *When we label a 'fashion company', we have the image of a fast-fashion business model in which green economy and sustainable practices are relegated to a secondary plan. Rather than a 'fashion company', I believe that Movin should be considered a 'clothing company' as this sounds more functional, practical, and responsible, reflecting the real purpose of the business model.*

During the interview, the founder mentioned plans to increase the 'aftersales traceability'. The 'aftersales traceability' can extend products' lifespan and promote the 'Rethink' principle by employing augmented reality (AR) and integrating the product with smart home appliances. In addition, these technologies can help inform the customers about best practices for, for example, washing, rinsing, drying, and storing the garment. Finally, the aforementioned embedded technologies can be integrated with second-hand marketplaces, promoting the 5Rs 'Reuse' principle.

From the single case study, the examination of the company's documents, and in-depth interviews, it was possible to establish a direct relationship between the well-known concepts of Lean Manufacturing and the Circular Economy. In addition, Movin initiatives to promote 'smart materials' seem to minimize the water, carbon, and pesticide footprint and environmental damage.

The case study aimed to connect the green transition to the consolidated concepts of lean manufacturing, adopting the '5R' framework ('Refuse', 'Rethink', 'Reduce', 'Reuse', and 'Recycle'). In addition, it focused on proposing ex-ante strategies of smarter product use and manufacturing, which is at the core of the Lean Thinking mindset, which is based on waste reduction (of all types and sources).

As a result, the following 'Green-Lean' framework can be proposed (Fig. 7):

The use of environmental-friendly materials like PET fabric and bamboo fibers has the potential to reduce the consumption of inputs in agriculture activity. These initiatives are framed in 'Refuse', 'Rethink' and 'Reduce' *ex-ante* principles, which are preferable from the environmental impact point of view. Transparency and traceability, in turn, proved to be essential practices to provide reliability and strengthen ties between customers, partners, and the company.

Even more promising, the MOVIN ZRO® project encapsulates all lean manufacturing concepts brought to this research. For example, the co-creation process and applying LPPD and SBCE concepts are fundamental to designing more sustainable products, potentializing the principles of 'Refuse', 'Rethink', and 'Reduce'. Additionally, the co-creation step allowed the company to reduce the prototyping and designing process. As a result, it reduces intermediate inventories and material waste. Finally, the 'inversion' of the value chain proposed by Movin, which is based on rewarding the customers participating in the co-creation and presales steps, giving these customers higher discount rates instead of adopting the market practice of

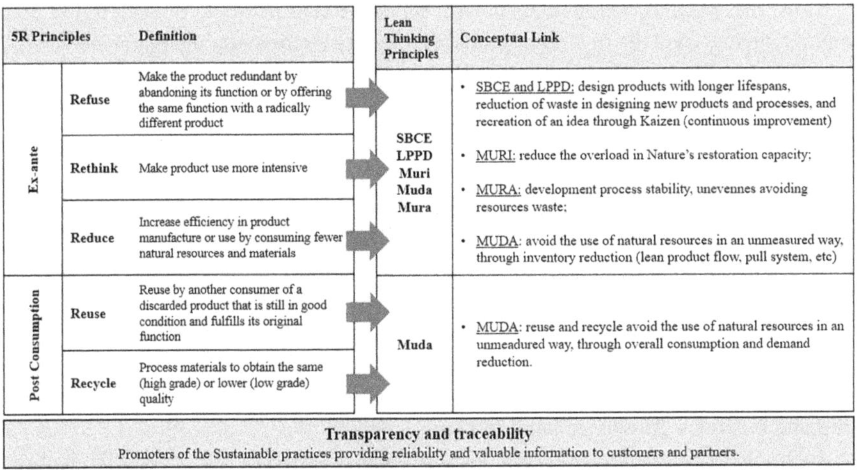

Fig. 7 Green-Lean framework proposal. *Source*: Author's elaboration

offering discounts on overstock products, seemed to be a key variable to reduce the inventory loss level to 5% (while the industry average is 30%).

In sum, the 'Green-Lean' framework, lean management concepts like LPPD (Lean Product and Process Development), SBCE (Set-based Concurrent Engineering), and 3Ms concept (Muri, Muda, Mura), combined with the 5Rs principles can be a reliable framework for companies pretending to migrate from the conceptual plane to the circular economy practices.

5 Conclusions

This chapter has established a compelling connection between Lean Manufacturing and Circular Economy concepts, offering a promising avenue for organizations to navigate the complex landscape of the Green Transition. By incorporating principles from Lean Manufacturing and Circular Economy, the proposed 'Green-Lean' framework presents an innovative and practical approach that integrates entrepreneurship, technological change, and circular economy principles into sustainable business practices.

Green-lean framework is an innovative approach because it allows the entrepreneur to develop new businesses, endowed with economic and socio-environmental viability in the green transition. This viability occurs through the construction of a common concept for the operationalization of new businesses: the elimination of waste (3Ms meets 5Rs). Waste in the Green-Lean approach is recognized as an element that does not create value for the customer and society, nor for nature seen as the protagonist.

Thus, the waste of value in costs with products kept in stock, considered in lean manufacturing, evolves in Green-lean to also consider as waste the materials of sold products that do not return to compose new products. Eliminating this second waste also reduces costs on the company in a new circular supply chain under development and creates environmental value in a new economic cycle as we tried to demonstrate in the emblematic case.

Entrepreneurship is vital in driving the Green Transition by fostering innovation, identifying opportunities, and creating sustainable business models. The Green-Lean framework recognizes the entrepreneurial spirit as a catalyst for change, encouraging organizations to embrace a mindset that seeks out eco-friendly solutions, explores new markets, and capitalizes on emerging technologies. By embedding entrepreneurship within Lean Manufacturing and Circular Economy principles, organizations can cultivate a culture of sustainability where environmental stewardship and business growth go hand in hand.

Technological change emerges as a pivotal enabler of the Green Transition, offering innovative solutions to address environmental challenges. The Green-Lean framework acknowledges the transformative power of technology, encouraging organizations to embrace digitalization, automation, and clean technologies. As a result, organizations can streamline processes, optimize resource utilization, and reduce environmental footprints by leveraging technological advancements.

Moreover, technological innovation opens doors to novel circular economy practices, such as product life extension, resource recovery, and remanufacturing. Integrating these technological advancements within the Green-Lean framework empowers organizations to drive sustainable growth, enhance operational efficiencies, and unlock new market opportunities.

Within the context of the Circular Economy, the Green-Lean framework serves as a practical guide to aligning business practices with the principles of resource conservation, waste reduction, and closed-loop systems. By embracing circular economy principles, organizations can transition from linear, extractive models to regenerative and sustainable ones. The Green-Lean framework provides a comprehensive approach that fosters circularity through Lean Manufacturing methodologies, emphasizing waste elimination, continuous improvement, and value creation. This integration enables organizations to optimize resource utilization, minimize waste generation, and design products with extended lifecycles and recyclability.

The Green Transition, characterized by pursuing sustainable and environmentally responsible practices, necessitates a holistic approach encompassing entrepreneurship, technological change, and circular economy principles. The Green-Lean framework represents a pragmatic response to these intertwined themes, offering organizations a practical roadmap to navigate the complexities of the Green Transition. By adopting the Green-Lean framework, organizations can foster entrepreneurial mindsets, leverage technological advancements, embrace circular economy practices, and achieve a harmonious integration of economic prosperity, environmental stewardship, and social well-being.

In conclusion, the Green-Lean framework is a timely and relevant approach for organizations seeking to drive the Green Transition. By recognizing the interconnectedness of entrepreneurship, technological change, circular economy, and sustainability, organizations can position themselves as leaders in the quest for a greener and more sustainable future. Furthermore, embracing the principles and practices of the Green-Lean framework empowers organizations to seize the opportunities presented by the Green Transition, creating value for stakeholders while safeguarding the planet for future generations.

Research limitations involve the adoption of a single case study. Single case studies are typically limited in examining causal relationships between variables. This is because there is typically no control group or manipulation of variables, which makes it challenging to conclude causality. Another limitation of single case studies is the potential for observer bias. Researchers intimately involved in a single case may have preconceived notions or expectations about the outcome, which could influence their interpretation of the data. Therefore, future studies could address these limitations using multiple cases and companies with different sizes and locations. Additionally, propose enhancing the 'Green-Lean' framework aiming at quantitative approaches.

References

Al-Ashaab, A., et al. (2016). Development and application of lean product development performance measurement tool. *International Journal of Computer Integrated Manufacturing, 29*(3), 342–354.

Backes, C. (2020). The waste framework directive and the circular economy. In *Research handbook on EU environmental law* (pp. 328–343). Edward Elgar Publishing.

Baines, T., Lightfoot, H., Willians, G., & Greenough, R. (2006). State-of-the-art in lean design engineering: A literature review on white-collar lean. *Proceedings of the Institution of Mechanical Engineers, Part B: Journal of Engineering Manufacture, 220*(9), 1539–1547.

Bicheno, J. (November, 2018). Towards reducing queues: Muri, Mura, muda. In Proceedings of the European lean educator conference, The University of Buckingham (pp. 13–16).

Bocken, N. M., Olivetti, E. A., Cullen, J. M., Potting, J., & Lifset, R. (2017). Taking the circularity to the next level: A special issue on the circular economy. *Journal of Industrial Ecology, 21*(3), 476–482.

Boulding, K. E. (1966). The economics of the coming spaceship earth. In H. Jarrett (Ed.), *Environmental quality issues in a growing economy* (pp. 3–14). Johns Hopkins University Press.

Boyce, C., & Neale, P. (2006). *Conducting in-depth interviews: A guide for designing and conducting in-depth interviews for evaluation input* (Vol. 2). Pathfinder international.

Cohen, B. (2006). Sustainable valley entrepreneurial ecosystems. *Business Strategy and the Environment, 15*(1), 1–14.

Dahlbo, H., Aalto, K., Eskelinen, H., & Salmenperä, H. (2017). Increasing textile circulation—Consequences and requirements. *Sustainable Production and Consumption, 9*, 44–57.

Daultani, Y., Chaudhuri, A., & Kumar, S. (2015). A decade of lean in healthcare: Current state and future directions. *Global Business Review, 16*(6), 1082–1099.

Eisenhardt, K. M., & Graebner, M. E. (2007). Theory building from cases: Opportunities and challenges. *Academy of Management Journal, 50*(1), 25–32.

Ellen MacArthur Foundation. (2012). *Towards the circular economy 1: Economic and business rationale for an accelerated transition*. Ellen MacArthur Foundation.

Ellen MacArthur Foundation. (2013). *Towards the circular economy 2: Opportunities for the consumer goods sector*. Ellen MacArthur Foundation.

Ellen MacArthur Foundation. (2014). *Towards the circular economy: Accelerating the scale-up across global supply chains*. World Economic Forum.

Ellen MacArthur Foundation. (2020). What is the circular economy? Ellen MacArthur Foundation. https://www.ellenmacarthurfoundation.org/circular-economy/concept

Ghisellini, P., Cialani, C., & Ulgiati, S. (2016). A review on circular economy: The expected transition to a balanced interplay of environmental and economic systems. *Journal of Cleaner Production, 114*, 11–32.

Haupt, M., Vadenbo, C., Zeltner, C., & Hellweg, S. (2017). Influence of input-scrap quality on the environmental impact of secondary steel production. *Journal of Industrial Ecology, 21*(2), 391–401.

Joshi, K., Venkatachalam, A., & Jawahir, I. S. (2006). A new methodology for transforming 3R concept into 6R concept for improved product sustainability. In *IV global conference on sustainable product development and life cycle engineering* (pp. 3–6).

Khalil, R., & Stockton, D. (2010). Predicting the effects of cycle time variability on the efficiency of electronics assembly mixed-model, zero buffer flow processing lines. *International Journal of Computer Integrated Manufacturing, 23*(12), 1149–1157.

Khan, M., et al. (2011). Set-based concurrent engineering process within the LeanPPD environment. In *Advanced concurrent engineering* (pp. 433–440). Springer.

Kirchherr, J., Reike, D., & Hekkert, M. (2017). Conceptualizing the circular economy: An analysis of 114 definitions. *Resources, Conservation and Recycling, 127*, 221–232.

Krafcik, J. F. (1988). Triumph of the lean production system. *Sloan Management Review, 30*, 41–52.

Lacy, P., & Rutqvist, J. (2015). *Waste to wealth: The circular economy advantage* (Vol. 91). Palgrave Macmillan.

Lewandowski, M. (2016). Designing the business models for circular economy—Towards the conceptual framework. *Sustainability, 8*(1), 43.

Linder, M. (2017). Ripe for disruption: Reimagining the role of green chemistry in a circular economy. *Green Chemistry Letters and Reviews, 10*(4), 428–435.

MacArthur, E., Zumwinkel, K., & Stuchtey, M. R. (2015). *Growth within: A circular economy vision for a competitive Europe*. Ellen MacArthur Foundation.

Murray, A., Skene, K., & Haynes, K. (2017). The circular economy: An interdisciplinary exploration of the concept and application in a global context. *Journal of Business Ethics, 140*, 369–380.

Nagai, R. A. & Torres Jr., Alvair S. (2021). Lean product and process development and set-based concurrent engineering in the dining industry: The experience of an American-Asian fusion restaurant. *IJAR–International Journal of Action Research, 16*(3), 202–224.

Niinimäki, K., Peters, G., Dahlbo, H., Perry, P., Rissanen, T., & Gwilt, A. (2020). The environmental price of fast fashion. *Nature Reviews Earth & Environment, 1*(4), 189–200. https://doi.org/10.1038/s43017-020-0039-9

Nikiforova, A., & Bicevska, Z. (2020). Application of lean principles to improve business processes: A case study in latvian it company. arXiv preprint arXiv:2007.04696.

Oey, E., & Nofrimurti, M. (2018). Lean implementation in traditional distributor warehouse-A case study in an FMCG company in Indonesia. *International Journal of Process Management and Benchmarking, 8*(1), 1–15.

Ogunmakinde, O. E. (2019). A review of circular economy development models in China, Germany and Japan. *Recycling, 4*(3), 27.

Ohno, T. (1995). *Toyota production system, beyond large-scale production*. Productivity Press.

Pearce, D. W., & Turner, R. K. (1989). *Economics of natural resources and the environment*. Johns Hopkins University Press.

Piekkari, R., & Welch, C. (2018). The case study in management research: Beyond the positivist legacy of Eisenhardt and Yin. In *The SAGE handbook of qualitative business and management research methods* (pp. 345–358). SAGE.

Potting, J., Hekkert, M. P., Worrell, E., & Hanemaaijer, A. (2016). *Circular economy: Measuring innovation in the product chain*. PBL Netherlands Environmental Assessment Agency.

Reike, D., Vermeulen, W. J., & Witjes, S. (2018). The circular economy: New or refurbished as CE 3.0?—Exploring controversies in the conceptualization of the circular economy through a focus on history and resource value retention options. *Resources, Conservation and Recycling, 135*, 246–264.

Runnel, A., Raihan, K., Castle, N., Oja, D., & Bhuiya, H. (2017). The undiscovered business potential of production leftovers within global fashion supply chains: Creating a digitally enhanced circular economy insight from research among fabric and garment factories of China and Bangladesh. Tallin, Estonia.

Rutakumwa, R., Mugisha, J. O., Bernays, S., Kabunga, E., Tumwekwase, G., Mbonye, M., & Seeley, J. (2020). Conducting in-depth interviews with and without voice recorders: A comparative analysis. *Qualitative Research, 20*(5), 565–581.

Sihvonen, S., & Ritola, T. (2015). Conceptualizing ReX for aggregating end-of-life strategies in product development. *Procedia CIRP, 29*, 639–644.

Sobek, D. K., Ward, A. C., & Liker, J. K. (1999). Toyota's principles of set-based concurrent engineering. *Sloan Management Review*, 67–84.

Söderholm, P. (2020). The green economy transition: The challenges of technological change for sustainability. *Sustainable Earth, 3*(1), 1–11.

Ward, A. C. (2007). *Lean product and process development*. Lean Enterprise Institute.

Womack, J. P. (2002). Lean thinking: Where have we been and where are we going? *Manufacturing Engineering, 129*(3), L2–L6.

Womack, J. P., & Jones, D. T. (1997). Lean thinking—Banish waste and create wealth in your corporation. *Journal of the Operational Research Society, 48*(11), 1148–1148.

Womack, J. P., Jones, D. T., & Ross, D. (1990). *The machine that changed the world*. Rawson Associates.

Yin, R. K. (2009). *Case studies: Design and methods* (4th ed.). Sage Publications.

Part II
Agriculture and Agribusiness

An Overview of Circular Business Models in Agribusiness

Barbara Lamolinara, Mário Sérgio Teixeira, Cristina Galamba Marreiros, Vítor Hugo dos Santos Ferreira, and Amaury Pérez-Martínez

Abstract Circular economy is changing the way businesses operate and is pushing companies and organizations from a linear to a closed-loop business approach. Circular business models serve as tools to help managers in this transition. The agribusiness sector has a significant environmental impact but a considerable potential for the implementation of circular practices. However, circular business models have been little discussed in the sector, limiting their application. Thus, this chapter aims to deepen the concept of circular business models in agribusiness to raise awareness and facilitate their adoption. A systematic literature review was conducted, following the PRISMA protocol and using the Bibliometrix software and the app Biblioshiny. As a result, 60 documents were included in the review and bibliometric and content analysed. The interest in the topic is growing, especially in Europe and nations with high agricultural industrialization. The themes emerged and discussed in literature include the challenges, key aspects, and recommendations for circular business models' adoption, the concerns with their sustainability, and their performance measurement. Limitations to current knowledge and future research directions are identified. This chapter provides a valuable contribution to the understanding of circular business models in agribusiness and highlights important aspects useful for entrepreneurs and policymakers.

B. Lamolinara (✉)
Centre for Transdisciplinary Development Studies (CETRAD), University of Trás-os-Montes e Alto Douro, Vila Real, Portugal

Center for Advanced Studies in Management and Economics (CEFAGE), Department of Management, School of Social Sciences, University of Évora, Évora, Portugal

Centre for Rapid and Sustainable Product Development (CDRSP), Polytechnic of Leiria, Rua de Portugal – Zona Industrial, Marinha Grande, Portugal
e-mail: barbara.lamolinara@ipleiria.pt

C. Rego et al. (eds.), *Entrepreneurship, Technological Change and Circular Economy for a Green Transition*, Studies on Entrepreneurship, Structural Change and Industrial Dynamics, https://doi.org/10.1007/978-3-031-48079-9_7

M. S. Teixeira
Centre for Transdisciplinary Development Studies (CETRAD), University of Trás-os-Montes e Alto Douro, Vila Real, Portugal
e-mail: mariosergio@utad.pt

C. G. Marreiros
Center for Advanced Studies in Management and Economics (CEFAGE), Department of Management, School of Social Sciences, University of Évora, Évora, Portugal
e-mail: cristina@uevora.pt

V. H. dos Santos Ferreira
Centre for Rapid and Sustainable Product Development (CDRSP), Polytechnic of Leiria, Rua de Portugal – Zona Industrial, Marinha Grande, Portugal
e-mail: vitor.ferreira@ipleiria.pt

A. Pérez-Martínez
Universidad Estatal Amazónica, Puyo, Pastaza, Ecuador
e-mail: amperez@uea.edu.ec

Keywords Circular business models · Circular economy · Agribusiness · Sustainability

1 Introduction

Nowadays, the business world and its organizations are rethinking and reinventing their way of doing business and creating value (Loorbach & Wijsman, 2013), guided by the sustainable development goals of the United Nations. Circular economy (CE) is gaining ground as the only possible economy to reconcile society's sustainable development, respect for the environment, customer satisfaction with green products, and business profits (Geissdoerfer et al., 2017; Pla-Julián & Guevara, 2019). Therefore, companies are trying to quickly adapt to this new reality and to leave behind the linear economy (Garcés-Ayerbe et al., 2019), trying to avoid being too late in this new market scenario.

To make this change a reality, companies need a tool that allows them to transitioning from a linear economy to a circular economy with their business model. This is where the concept of circular business models (CBMs) comes into play. They are essentially business models for the circular economy (Geissdoerfer et al., 2020). In other words, a circular business model is a business model that innovates itself to integrate sustainable objectives and circular practices (Geissdoerfer et al., 2018a, 2020). That term first appeared in 2006 and then re-emerged in 2013 (Geissdoerfer et al., 2020). It stems from the concept of the business model, the innovation of the business model (Mitchell & Bruckner Coles, 2004), and subsequently the sustainable business model (SBM) (Bocken et al., 2016; Velter et al., 2020). There are many definitions of CBM, however, the main objective of CBM is to reduce the environmental impact and provide better customer value by slowing down, narrowing, intensifying and/or dematerializing resource loops (Geissdoerfer et al., 2020; Lewandowski, 2016; Upadhyay et al., 2019). It is based on closed-loop value chains, industrial ecology and product/service systems, and enables tracking material flows to create value from waste (Donner et al., 2020).

When we think about circular economy, we primarily think of the valorization of waste and the recycling of resources. However, other strategies are part of the CBM and have led, up-to-now, to the recognition of five different types of circular business models: circular supplies/circular design, resource recovery, product life extension, sharing, and product-as-a-service (Donner et al., 2020). Actually, it is important to note that circular business models go beyond simply creating value from waste by closing the loop; they also involve strategies like extending the life of products, reducing waste through eco-design, encouraging reduced consumption, and shifting away from the traditional paradigm of product ownership to a model of sharing products and services. These different CBM strategy types can all be integrated into a single business value chain. Alternatively, it is also possible to adopt only one or some of these circular strategies in a business and still refer to it as adopting CBM.

Currently, the discussion on CBMs is quite popular, but it has not been investigated enough in the agribusiness sector (Singh et al., 2021a). Agribusiness has a significant environmental impact (Donner et al., 2021) but, at the same time, a great potential for organic material and resources recovery, playing a crucial role in sustainable development (D'Amato et al., 2020). However, so far, circular practices in agribusiness have been discussed more at a technological level than at a business level (Ghisetti & Montresor, 2020; Yamoah et al., 2022). This lack of information at the managerial level can hinder the transition to circular economy in companies in the sector and thus slow down their sustainable development (Yamoah et al., 2022). Therefore, there are still many issues to clarify and investigate about CBMs in agribusiness, in order to provide a comprehensive picture that can be useful to the sector's stakeholders.

On the basis of those premisses, this chapter addresses the following research questions: what are the specific characteristics of circular business models in agribusiness? What are the barriers to circular business models adoption in agribusiness? What are the factors that guarantee CBMs success in agribusiness? So far, what are the most discussed topics on CBMs? By answering these questions, this chapter aims to provide readers with a more in-depth framework on CBMs in agribusiness. To achieve this goal, the authors chose to carry out a systematic review of the literature about circular business models in the agribusiness sector. Systematic literature review is a consolidated methodology, and is already widely used in management and social science studies (Snyder, 2019; Tranfield et al., 2003). This methodology guides the collection and analysis of all relevant documents on a topic and provides tools to make an overview of that topic.

The documents to be included in this research were selected following the PRISMA protocol (Page et al., 2021; Tranfield et al., 2003) for the collection and selection of documents in Scopus and Web of Science databases, and using Bibliometrix software (Aria & Cuccurullo, 2017) and its shiny app Biblioshiny (Nasir et al., 2020) as screening aid. Subsequently, the resulting documents were

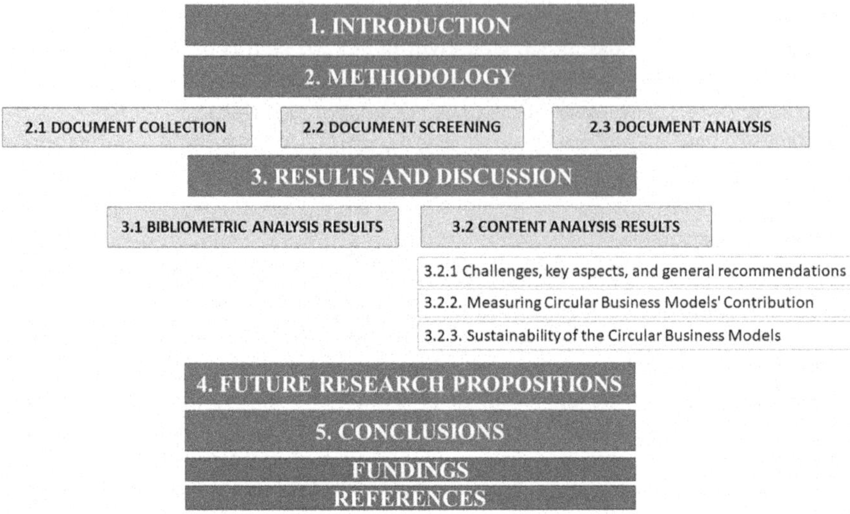

Fig. 1 Structure of the sections and subsections of the chapter

studied with two different and complementary approaches: bibliometric analysis and content analysis. With the results obtained, it is possible to provide a complete overview of the trends, characteristics, information, and topics covered in the documents published dealing with circular business models in agribusiness.

This analysis will be useful for entrepreneurs, managers, and all stakeholders involved in agribusiness, as they will be able to better understand circular business models, recognize their weaknesses and strengths, and reflect on the themes that emerge from literature and proposed for future research. Therefore, this chapter intends to be a clarifying work and, at the same time, a starting point for implementing circular business models and circular transition in an agribusiness. To reach these goals, this chapter is structured in different sections and subsections, depicted on Fig. 1.

2 Methodology

As referred above, a systematic literature review of the published research was implemented in order to have a deeper understanding of CBMs in agribusiness. The systematic review is a methodology to summarize knowledge about a specific topic by collecting literature on that subject (Snyder, 2019). In the present study, the PRISMA protocol (Page et al., 2021) is adopted, since it is a consolidated approach in research about similar topics (e.g., De Carvalho Araújo et al., 2022; Gomes et al., 2022; Hultberg & Pal, 2021; Zhang et al., 2022), to create a dataset of documents to perform the systematic review. PRISMA consists of a 27-item checklist and provides a flow diagram used to guide authors in the systematic literature review and meta-analysis of studies (Snyder, 2019), and to ensure transparency, replicability and objectivity of the study (Tranfield et al., 2003).

Document collection and screening were performed, and the resulting publications were analysed. These processes are described in the following subsections.

2.1 Document Collection

Document collection is the first step of the systematic review. The source of documents must be reliable and recognized by the scientific community. Therefore, both Scopus and Web of Science (WOS) databases were chosen as sources for this review, since they are the largest and most recognized worldwide reference databases platforms for analysing peer-reviewed literature (Singh et al., 2021b). Merging the results of both platforms guarantees a reliable and wide source of documents covering scientific production on CBMs. The selection criteria, Boolean operators and keywords used in both platforms were the same (see Table 1). The keywords used were chosen to select documents focused on CBMs and circular practices in SBMs in the agribusiness sector. To ensure a document collection with a whole perspective on the topic, regarding publications' production over time, document types, and geographical and thematic areas, other refinements except the ones related to the keywords search were not applied.

At first, a keyword combination with the Boolean operator OR was used to find documents related both to sustainable business models and circular business models. Although the review is focusing on CBMs and circular practices, and not on SBMs, the "sustainab* business model*" keyword has been used to avoid the exclusion of studies reporting interesting circular practices and circular waste valorization without naming it as "circular". In some cases, especially when the circular

Table 1 Criteria used for document collection in WOS and Scopus

WOS	SCOPUS
Topic: "**sustainab* business model***" OR "**circular* business model***" OR "**circular* econom* business model***" OR "**CE business model***" OR "**close*-loop business model***" AND Topic: "**agri***" OR "**agro***" OR "**agrar***" OR "**forest***" OR "**farm***" OR "**aquaculture***" OR "**food***"	*TITLE-ABS-KEY: "**sustainab* business model***" OR "**circular* business model***" OR "**circular* econom* business model***" OR CE **business model***" OR "**close*-loop business model***" AND TITLE-ABS-KEY: **agri***" OR "**agro***" OR "**agrar***" OR "**forest***" OR "**farm***" OR "**aquaculture***" OR "**food***"*
All fields	All fields
All types of documents	All types of documents
All years up to present, added "anytime"	All years up to present, added "anytime"
All languages	All languages
Search performed on the 28th of January 2022	Search performed on the 28th of January 2022

Source: adapted from Rodriguez-Soler et al. (2020)

economy debate was still at the beginning, some CBMs and circular practices could have been described and referred to by their authors as SBMs. The asterisk "*" in the search was used to include different word endings, since in literature these concepts are often written as "model" or "models", "sustainable" or "sustainability" business models, "circular" or "circularity" or "circular economy" or "CE" or "close/closed-loop" business models (Ghisellini et al., 2016; Nosratabadi et al., 2019).

The second Boolean operator chosen to refine the search was AND. Since the study aims to focus on the agribusiness sector, keywords for inclusion of agriculture, agri-food, farm, forest, aquaculture, and food industry sub-sectors have been selected. Also in this case, "*" has been used to include all the possible words' endings. It was concluded, through trial search, that the keywords "aquiculture" (and not "aquaculture"), "cradle-to-cradle business model", "wood" or "water" did not add any additional papers related to circular business practices in agribusiness. The search led to a list of 147 documents in Scopus and a list of 122 in WoS. The two lists were exported in BibTeX format to facilitate the merging process described in the following subsection.

2.2 Document Screening

To continue with the bibliometric analysis, a document screening to refine the search was performed. In this step of the bibliographical review, the open-source software Bibliometrix, R-package tool (RStudio version 4.1.2) was used (Aria & Cuccurullo, 2017). Bibliometrix is a science mapping analysis and screening tool useful to help in bibliometric and bibliographical studies (Rodríguez-Soler et al., 2020). Bibliometrix merged the two BibTeX lists of documents, from Scopus and WoS, obtained in the documents' collection process, eliminating 92 duplicates and generating a MS Excel list. All the relevant bibliographical information, such as the title, authors' names and affiliations, abstract, keywords, journal name, year of publication, and the number of citations of the identified documents, was included in the MS Excel spreadsheet, used for further screening by human reading.

The PRISMA diagram in Fig. 2 shows all the steps of the screening performed in this review. Accordingly, excluding and including criteria (Tranfield et al., 2003) for human reading screening have been defined.

The excluding criteria identified and used for the screening of documents were:

E1: duplicate records
E2: books containing chapters already included in the records
E3: proceedings generating an article already included in the records
E4: not retrieved records
E5: out-of-topic records (documents containing the required keywords selected for the review search, but just mentioning them briefly and then focusing on other contexts or sectors; documents focusing on food retail sell strategies, fast food,

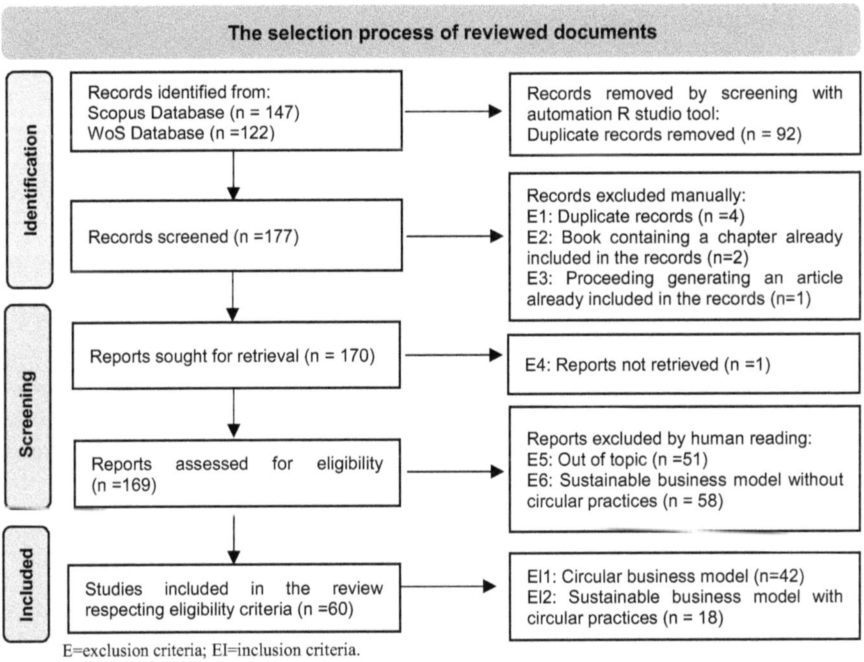

Fig. 2 PRISMA 2020 flow diagram used for the selection and screening process. *Source*: adapted from (Page et al., 2021)

restaurant and franchising coffee shops where the relationship with farmers and agricultural producers was not mentioned; documents about marketing and selling strategies in food retail supply chains; SBM articles referring to waste reduction practices)

E6: Sustainable business model records without circular practices described

The inclusion criteria for the selection of documents were:

EI1: documents focused on CBMs in the agribusiness sector.

EI2: documents focused on SBMs in the agribusiness sector, with a description of circular practices.

The fit of each document that satisfied the eligibility criteria was assessed through careful reading and review of the full-text documents. After the application of the screening steps and the exclusion and inclusion criteria, 60 documents have been identified and included in this systematic literature review—42publications related to CBMs in agribusiness and 18 publications related to SBMs with circular waste valorization practices.

2.3 Document Analysis

At this stage, both bibliometric and content analysis were implemented to have a comprehensive picture of the circular business models discussion in agribusiness. At first, a bibliometric analysis of the resulting 60 documents identified and selected in the previous phase was carried out to have a first understanding of trends, gaps, and peculiarities of the topic investigated. Afterwards, content-based research was conducted to analyse in deep the content of the documents included in this review.

Once identified the documents to include in this review, a final Excel list containing the bibliographical details of the selected documents has been uploaded to Bibliometrix and analysed with its R shiny app Biblioshiny (Nasir et al., 2020) to perform a brief bibliometric analysis and documents characterization. This was followed by a deeper content analysis through human reading. Thus, all publications were read, studied, and analysed to provide insights and discussion about circular business models and circular practices in agribusiness. During content reading, the documents were classified into specific topics and subtopics, as reported in the results and discussion sections.

3 Results and Discussion

This section presents the findings and related discussion of the two different approaches to the systematic review discussed in the methodology section—the bibliometric and the content analysis. The complementarity of the two analyses provides a more comprehensive view of the trends and issues raised on agribusiness CBMs.

3.1 Bibliometric Analysis Results

The selected papers constitute the database of this review and have been descriptively analysed and characterized regarding year and type of publication, country of scientific production, source, and keywords reported (Table 2). As can be seen, the 60 documents in the database have been published since 2017 by 38 different sources and 237 different authors. These results indicate that the topic is recent, and has been taken into consideration by numerous editors interested in the emerging discussion, with a high level of collaboration among authors. Just three documents were written by a single author, and, on average, each document has been written by 4,05 co-authors. In 26.67% of the cases, the documents are authored by co-authors of different countries, indicating that researchers are investigating the topic of CBMs also in an international research collaboration context, exchanging experiences and expertise with other research groups worldwide. Each document has been

Table 2 Characterization of the database

Description	Results
Main information about the data	
Timespan	2017:2022
Sources (Journals, Books, etc.)	38
Documents	60
Document average age (years)	3,02
Average citations per document	12,47
References	4031
Document contents	
Keywords plus (ID)	261
Author's keywords (DE)	267
Authors and Collaboration	
Authors	237
Single-authored document	3
Co-authors per document	4,05
International co-authorships %	26,67

cited, on average, 12,47 times, which indicates a considerable impact of these publications on the scientific literature.

However, considering the most cited references, four articles have been detected as important bases for the development of the theory and concept of the circular business model in agribusiness: the article by Ghisellini et al. (2016), the ones by Bocken et al. (2014, 2016) and the one by Lewandowski (2016). Those four articles have been an important conceptual framework for the documents included in this study database. Instead, Mechthild Donner is the author publishing the most about CBMs in agribusiness, since she authored three documents in the database.

Figure 3 provides a deeper characterization of the documents in the database, reporting the evolution of the number of publications over time and the type and source of publication.

As shown in Fig. 3, there is a growing exponential interest in the topic of CBMs in agribusiness (the year 2022 was not considered since it was not concluded at the time of document's search and collection). The majority of the 60 documents are journal articles, which, nowadays, are the most common format for scientific literature publication. However, it is remarkable that 7 out of 60 documents are reviews, indicating that the academic community felt the need to summarize and organize the emerging discussion about CBMs in agribusiness. "Sustainability" and "Journal of Cleaner Production" are the two journals publishing the most about circular business models and circular practices, and represent suitable journals for authors in the field to submit future works about this issue.

Concerning the countries researching CBMs and circular practices in agribusiness, it is possible to see in Fig. 4 the world map of documents scientific production. The shades of grey represent different productivity rates. Italy is the country with the highest scientific production on the issue, and other European countries, such as

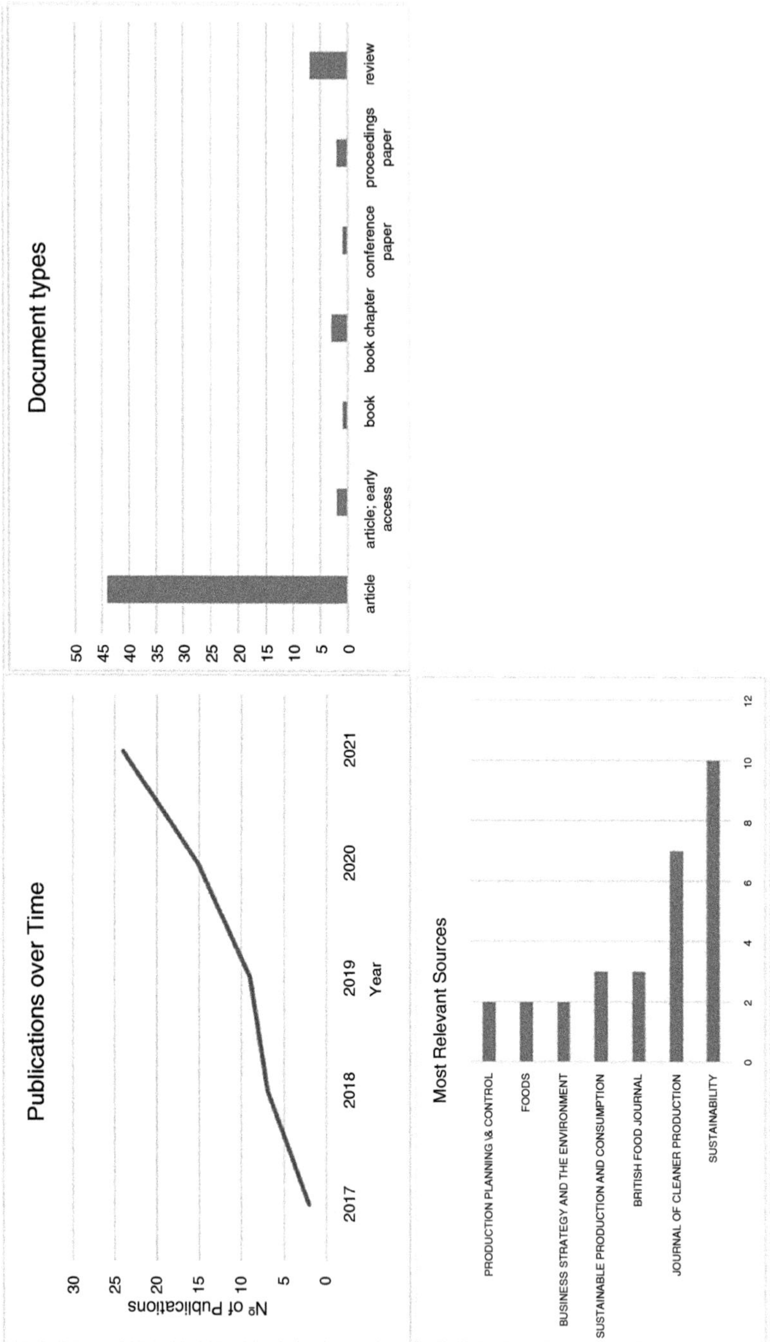

Fig. 3 Analysis of publications over time, document type and source

Fig. 4 World map of the countries' scientific production distribution

Spain and France, also contribute with a relatively high number of scientific publications. European Union programs and projects, such as Horizon 2020, were the main funding for the research published and included in this review. Those European incentives for sustainable development led to investment in research about CE, sustainability, and environmental protection. However, agro-industrialized countries outside Europe, such as the U.S.A., Brazil, China, Chile and Australia, also have a considerable interest in CBMs in agribusiness. With the United Nations 2030 Agenda for Sustainable Development, the concern about CE issues and related research is increasing and spreading, which will probably modify the distribution of the "country scientific production" map in the following years.

For a first glimpse on the main issues and topics addressed in the CBMs literature, before content analysis, the most frequent keywords used in the 60 documents of the database were identified. For this purpose, the 50 most frequent "keywords plus" (words or phrases frequently used in the titles of an article's references and that may appear or not in the title of the article or as author keywords) are taken into consideration to construct the WordCloud showed in Fig. 5. The dimension of the word characters in Fig. 5 is associated with the keyword's frequency. Thus, the bigger the word characters, the more frequent the keyword, the smaller the characters, the less frequent it is. Therefore, we can conclude that, regarding CBM in agribusiness, "innovation" is the most used "keyword plus" in the publications. Indeed, the implementation of CBMs requires an innovation, the transition from a traditional linear business model towards circularity and sustainability (Galvão et al., 2020; Nosratabadi et al., 2019). "Management", "performance", "industry", "design", "waste" and "life-cycle assessment" are widely used words as well.

These results reflect the fact that the CBM is a tool born, discussed, and used in the management area, to improve the sustainability performance of industrial

Fig. 5 WordCloud of the most used Keywords plus

businesses (Geissdoerfer et al., 2020; Marczewska & Kostrzewski, 2020). Moreover, waste valorization, waste reduction and eco-design (Serhan & Yannou-Lebris, 2021) represent essential strategies for CBMs implementation, which performance can be evaluated through life-cycle assessment (Camana et al., 2021).

3.2 Content Analysis Results

In this subsection, a thematic discussion of the various issues addressed in the selected literature is presented. Abstracts and the full text of all publications were read and analysed. The thematic analysis aims to offer a holistic overview of circular business models and circular practices for waste valorization in agribusiness, identifying important theoretical dimensions. All 60 documents were classified and are reported in Table 3 according to the topics discussed and their category or sector in the agribusiness. The following subsections discuss relevant sub-themes that emerged in the content analysis. With the identification of different sub-themes and sub-sectors that emerged in CBM literature, the authors aim to facilitate the reading of the results for interested parties in the various agribusiness sectors.

Table 3 Classification of the database documents into different categories and topics

Circular Business Models (CBMs)		
Category/Sector	Topics discussed	Documents
General recommendations	CBMs key aspects; CBMs typologies in agro-waste; success and risk of CBMs; transition to CE and CBMs; CE implementation of sectorial and cross-cutting challenges	Larsson (2018), Donner et al. (2020), Ghisellini and Ulgiati (2020), Donner et al. (2021), Salvador et al. (2021), Singh et al. (2021a) and Yamoah et al. (2022)
Sustainability measurement	Sustainability assessment of circular practices; performance measurements of CE strategy; CE entrepreneurship education	Cavicchi and Vagnoni (2020), Camana et al. (2021) and Del Vecchio et al. (2021)
Waste to energy/ anaerobic digestion	Manure, fertilizer, anaerobic digestion in CE; Digestate uses and challenges; Economic sustainability of CBM with biogas from animal manure; Sugar-energy CBMs	Yazan et al. (2018), Akerman et al. (2020), Hussain et al. (2020), Oliveira Pavan et al. (2021) and Lamolinara et al. (2022)
Forest	CBMs canvas in forest bio-economy; social capital and CBMs transition in SMEs	Reim et al. (2019), D'Amato et al. (2020) and Alcalde-Calonge et al. (2021)
Food-industry	Consumers' willingness to participate in CBMs; digital platforms for CBMs in food-industry; stakeholders in local food waste circular projects; family business retail CBMs transition; circular cities; retail food and waste loop CBMs; industry 4.0, CE and food supply chains; CBMs in manufacturing and service sector	Nunez-Cacho et al. (2018), Upadhyay et al. (2019), Borrello et al. (2020), Paiho et al. (2020), De Bernardi et al. (2021), Moggi and Dameri (2021), Rincón-Moreno et al. (2019) and Sousa Jabbour et al. (2021)
Agriculture	CBMs in the olive oil industry; coffee CBMs and mushroom production; closed-loop agriculture; CBMs for agriculture; CBMs in the wine industry	Ryabchenko et al. (2017), Winter et al. (2019), Zucchella and Previtali (2019), Nasution et al. (2020), Sehnem et al. (2020), Donner & Radić, (2021), Lagrasta et al. (2021) and Pampuri et al. (2021)
Farming	CBMs in the milk industry; wasted wool valorized in the clothing industry; insect farming; CBMs perception of farmers;	Madau et al. (2020), Nasution et al. (2020), Uvarova et al. (2020), Van PhI et al. (2020), Dagevos and de Lauwere (2021), Martin and Herlaar (2021)
Aquaculture	CBMs case in aquaculture	Zilia et al. (2021)
Agri-tourism	CBMs in tourism	Florido et al. (2019)
Sustainable Business Models (SBMS) with Circular Practices		
Category/Sectors	Topics Discussed	References

(continued)

Table 3 (continued)

Agribusiness sectors	SBMs with circular practices cases in different agro-industrial fields; water reuse in aquaculture/ animal production/ wine production/dairy industry; education in eco-design and eco-innovation; food supply chains and SBMs	Van Genuchten et al. (2017), Brehmer et al. (2018), Hamid and Blanchard (2018), Ribeiro et al. (2018), Cantino et al. (2019), Karlsson et al. (2019), Nosratabadi et al. (2019, 2020), Barbieri and Santos (2020), Fiore et al. (2020), Amoah et al. (2021), Basile et al. (2021), Engelberts et al. (2021), Geldres-Weiss et al. (2021), Kampelmann (2021), Messina et al. (2021), Nayha (2021) and Serhan and Yannou-Lebris (2021)

3.2.1 Challenges, Key Aspects and General Recommendations

Circular business models in agribusiness have been conceived to achieve a perfect dynamic closed-loop resource flow. However, at present, a fully closed-loop system with no minimal leakage is more theoretical than real. According to the study by Larsson (2018), ideal closed-loop businesses at a large scale are not achievable in the short-run, but it is important to start from local small closed-loop ecosystems to arise awareness of consumers and entrepreneurs in favour of circularity. This approach can help to overcome the challenge of scalability, which has been considered a possible barrier to the adoption of circular business model innovation, in the work by Zucchella and Previtali (2019).

Scalability is not the only barrier for CBMs identified in this review. In the selected literature, the lack of a standard circular business model is reported as a barrier to the application of circular economy in agro-industry (Donner et al., 2020; Ghisellini & Ulgiati, 2020; Larsson, 2018). Donner et al. (2020) made a first attempt to create a typology of circular business models for agro-waste valorization, intending to overcome this problem. Their work identified six different types of CBMs in agribusiness, with their own value creation and organizational forms: biogas plant, upcycling entrepreneurship, environmental biorefinery, agricultural cooperative, agropark, and support structure. This means that there is not just a standard type of CBM in agribusiness, but there are slightly different types which are interconnected and interrelated by the same circular purpose, though. Furthermore, if we consider the diversity of the business strategies and value chains within agribusinesses, is understandable that a standard fixed model is difficult to obtain. The agribusiness sector needs CBMs guidelines that can be adapted to the different sub-sectors, and diverse case studies, to allow flexible replicability, as signalised in the study by Zucchella and Previtali (2019).

Although many risks, success factors, barriers, and enablers are common to all business models (traditional linear ones as well), some are specific to circular business models (Donner et al., 2021). When thinking about the critical points for a wealthy CBM, Donner et al. (2021) resume and describe the success factors for the valorization of by-products and agro-waste as follows: flexible in and out logistics, availability of space, regulation on agricultural waste management and subsidies, innovative conversion technologies, involvement of local stakeholders and research organization, joint investment in research and development, acceptance of bio-based production processes, competitive price of bio-based products. In particular, many studies (e.g. Prosperi et al., 2020; Donner et al., 2020, 2021) agree on the fundamental role of governments and companies, which have been identified as the two key players of the circular economy. Therefore, authors agree on the major role of Governments on promote circular economy in waste management through legislation and taxes.

In addition, several studies (D'Amato et al., 2020; Prosperi et al., 2020; Camana et al., 2021; Donner et al., 2021) pointed out that a fundamental enabler for a successful circular business model is the cooperation, collaboration, and commitment of all the stakeholders involved in the circular flow, including suppliers, producers, customers, and the whole circular innovation ecosystem. Indeed, the mentioned studies concluded that, since at the local level the expectations, visions, knowledge, know-how, and investment towards sustainable circular businesses of the main economic agents can converge, local environments and clusters are the most favourable for implementing circular economy business and strategies. The strategic partnerships created with other companies, both with circular and "traditional" models, are fundamental (D'Amato et al., 2020). Interestingly, the work of Donner et al. (2020) explains how the traditional costumer-supplier relationship of the linear economy is substituted by a multi-actor relationship in the circular economy clusters. This interdependency among circular actors rises awareness of need for mutual responsibility and quality control. This fact is pointed out by Donner et al. (2020), who also identify responsibility sharing among all the players involved in the circular ecosystem as a key advantage of the circular business model ecosystem. In these models, businesses stakeholders' responsibilities are often overlapping and, consequently, the quality of a player's outputs is crucial for the quality of another player's inputs in the circular flow. To coordinate the numerous stakeholders, Moggi and Dameri (2021) proposed and established a model of participative governance, to improve engagement, sense of belonging and reciprocal trust among the circular ecosystem, for example by involving participants with courses, events, meetings, and fundraising.

Considering the previous premises about local clusters and the circular ecosystem, it is understandable how geographical proximity among players is crucial for the success of CBMs. This conclusion is also defended by Zucchella and Previtali (2019), which stress the advantage of geographical proximity for the reduction of transport costs and environmental impact of economic activities. However, the content analysis revealed an unsuccessful case of CBMs in a local environment—the case of the study made in Spain by Pereira et al. (2018) about CBM for pasteurized milk vending machines in local communities. That CBM aimed to directly connect farmers and consumers. When comparing this sustainable short supply chain with the supermarket and retail chain, the former failed, from a socioeconomic perspective, because of the farmers' lack of processing and marketing capacities, the difficulties in networking and collaboration with other stakeholders and the need for consumer's awareness and acceptance of the limited range of dairy products offered (Pereira et al., 2018). Thus, having a CBM with a lower environmental impact is not always synonymous with a successful business.

Regarding actors coordination on CBMs, the study by Zucchella and Previtali (2019) defined the figure of the "orchestrator", who represents a player able to create a business model ecosystem with different actors and networks. The orchestrator has the role to manage and lead the complementarity between all members of the network ecosystem, coordinating small and large companies and organizations. He or she is usually not a third-party figure, but part of the ecosystem. This role is fundamental for promoting CBMs and promote circular practices, since at present, general corporate inertia towards CE transition has been reported in the literature (Yamoah et al., 2022). Especially, managers and business leaders' values and beliefs can promote or inhibit stakeholders' engagement in the transition to the CBM. In this case, there is not necessarily a lack of awareness, but there could be a misalignment between the values and beliefs of executives and the circularity goals and values of the organization (Yamoah et al., 2022).

Other aspects can influence the inertia towards the CBM transition. One of those aspects is the achievement of economic self-sufficiency over time (Moggi & Dameri, 2021). Although it can take decades for circular business models to become widely used, their cost-effectiveness will be crucial to build its volume and competitiveness (Larsson, 2018), influencing entrepreneurs' decisions. Therefore, it is strongly recommended to analyse the viability of circular businesses in the short and long term, together with the evaluation of possible risks, before implementing circular practices. Additionally, the technological transition towards CE usually requires significant investment, which can discourage enterprises. Therefore, when a company decides to innovate towards circular principles, it is essential to assess the compatibility of the needed technological innovation with the existing production process and with the resources to be used (D'Amato et al., 2020), to avoid a negative sustainability impact caused by technological incompatibility and the consequent loss of investment.

Another aspect which might cause entrepreneurs' scepticism to drop off the traditional linear business strategy, is the seasonal availability of the products in the agribusiness sector, discussed in the work by Salvador et al. (2021). Seasonal availability of resources forces companies to rethink their product portfolio and diversify production, to make the circular flow continue running. This also requires a consortia investment to adapt to the new business strategy. Yet, different studies (Del Vecchio et al., 2021; Yamoah et al., 2022) recommended CE education programs for entrepreneurship as an incentive to open up to the rethinking of value creation processes, collaborative approaches, participation in the innovation ecosystem and soft-skill acquisition in the CE context. The level of education of managers is positively related to the level of sophistication of the adopted CE practices (Sehnem et al., 2020). Thus, it is expected that education will increase knowledge of CBMs among SME entrepreneurs and will mitigate the resistance to their adoption.

3.2.2 Measuring Circular Business Models' Contribution

The United Nations' sustainable development goals are motivating companies to adopt a new management culture towards sustainability. Measuring outputs and performance through tools and systems is essential to understand if the sustainability actions adopted are working correctly (Donner et al., 2020). In linear businesses, the outputs are more easily measured and sustainability indicators already exist (Kruger et al., 2022). However, measuring indicators in circular businesses may be more complex. In fact, the need for measuring and assessing the contribution of circular business models to sustainability has emerged in this review. Currently, some attempts to measure the sustainability of CBMs have been reported in the literature (Cavicchi & Vagnoni, 2020; Camana et al., 2021; Singh et al., 2021a). However, Yamoah et al. (2022) refer the lack of a standard system of indicators for CE performance.

To overcome this barrier, existing tools such as industrial ecology life, life-cycle assessment, and flow analysis may represent possible solutions. In particular, the life-cycle assessment approach has been used in two different studies (Camana et al., 2021; Singh et al., 2021a), in which it resulted, up to now, as the most suitable tool to assess the effect of CE practices (Camana et al., 2021). Recently, the Ellen MacArthur Foundation created a free business tool called Circulytics (Valls-Val et al., 2022), which supports companies in the transition toward circularity by measuring circularity achievement in their operations. This new tool can help companies and organizations to measure circularity performance, support decision-making, highlight strengths and weaknesses, provide transparency and open up new opportunities for sustainable management. Therefore, the use of this new tool has the potential to become one of the standard procedures to measure the performance of a CBM.

3.2.3 Sustainability of the Circular Business Models

The sustainable business model and the circular business model are two different tools (Bocken & Ritala, 2021; Donner & de Vries, 2021; Goni et al., 2021), although they have similarities and connections (Donner et al., 2020; Galvão et al., 2020; Geissdoerfer et al., 2020). The boundaries between the two are not yet well-defined, and is still difficult in the academia and entrepreneurship environment to distinguish between them (Geissdoerfer et al., 2018b). Because of this, still, undefined boundary, various documents on this review reported circular practices in a SBM context. In some cases, these circular practices are not sufficient to talk about CBM, in other cases, authors refer to a potential CBM just as a SBM. These uncertain definitions might be explained by different reasons—for example due to the lack of awareness of the researchers about the emerging concept of CBM or due to the conceptualization of CBM as a subcategory of SBM. In this case, it is safer for authors to refer to a business model as a SBM when they are not sure if the circular practices described can justify its classification into a CBM. Indeed, the work by Bocken et al. (2014) recognized the CBM as one of the archetypes of the SBM. More specifically, the CBMs have been defined as SBMs which incorporate circular practices and circular value chains to find solutions for the circular economy (Geissdoerfer et al., 2018a).

Both sustainable and circular business models emerged from business models innovation (Mitchell & Bruckner Coles, 2004). In other words, when a new business model is created to respond to sustainability needs and aims, it could be considered a sustainable business model (Nosratabadi et al., 2019) and a sustainability tool. In the same way, when a sustainable business model is innovated with circular practices and is implemented with a circular flow, it can be referred to as a circular business model (Galvão et al., 2020). However, Geissdoerfer et al. (2018a) were the first to introduce the idea that a CBM, even being a subcategory of SBM, may not be sustainable. This might happen because of the potential trade-off between circular implementation and sustainability. For example, in some cases, the working conditions quality of employees may decrease when turning to circular or the circular solutions adopted may be less energetically efficient than new linear technologies (Geissdoerfer et al., 2017).

The relationship and boundaries between SBMs and CBMs, as represented in Fig. 6, was also analysed in this systematic literature review. Content analysis confirmed that, in an ideal situation, the CBM is always sustainable, and it is a subcategory of a SBM that implemented circular practices in a closed-loop value chain. However, in real situations, there are cases inwhich some CBMs may not be sustainable. This is true especially at the economic level and in a short-term timespan (Larsson, 2018; Moggi & Dameri, 2021), when companies implement new circular practices in a previous linear business. The business model of companies undergoing technological innovations and circular production practices implementation can result in an economically unsustainable CBM in the first years due to the

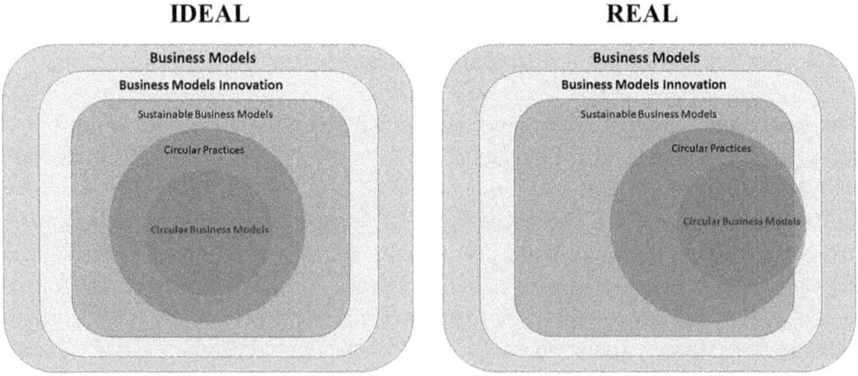

Fig. 6 The relationship between Sustainable Business Models and Circular Business Models

considerable investment required. However, governmental incentives and subsidies for CE transition and innovation may help companies overcome the high cost of CBM implementation (Donner et al., 2021; Lamolinara et al., 2022). It is expectable that in the long-run, after the return on the investment, the CBM would be completely sustainable and self-efficient.

4 Future Research Propositions

From the content analysis of the documents, some themes were identified as still not deeply discussed but which may play an important role in the adoption and success of CBMs. Those issues are interesting suggestions for future research and discussion, and may provide a different point of view on circular transition.

One crucial aspect to be addressed, to continue with the discussion in Sect. 3.2.3, is the economical sustainability of CBMs in agribusiness, in particular at the initial stages, differentiating cases of linear business turning circular versus new companies born already as circular.

Moreover, as mentioned in the work of Salvador et al. (2021), product-service-system, take-back-systems (including reverse logistics), management of seasonal availability of resources, social impacts and rebound effects of CBMs have been scarcely addressed in the literature. More specifically, the adoption of circular practices can have a considerable social impact since it may influence the balance between rural and urban relationships and development, and it may change job offers and conditions (Salvador et al., 2021). The rebound effect has been discussed also in the work by Camana et al. (2021), which points out the possibility of an environmental, social and economic rebound effect of CE practices. For example, some circular practices may increase eutrophication or energy consumption for recycling treatment processes, while others may have a positive sustainable impact in the long-run but not in the short-term (Camana et al., 2021). Thus, circular practices at the beginning of their implementation do not always guarantee a reduction of environmental impact. Additionally, with waste becoming a commodity, and with marketing strategies of lowering the price of recycled products, so they are competitive in the market, there is the risk of a rebound effect of triggering consumption, which contradicts the circular pillar of consumption reduction (Salvador et al., 2021). Thus, it would be interesting to deeply investigate the social impacts and the rebound effects of CBMs adoption.

Another interesting issue that emerged in the review is greenwashing. The work of Yamoah et al. (2022) gave a realistic glimpse into how the commitment to CBM transition, in some cases, could be a form of circular greenwashing. This is an emerging phenomenon to respond to consumers' environmental awareness and green choices' preference. In addition, the study by Borrello et al. (2020) assessed that with an economic advantage for consumer involvement in circular practices, the CBMs are more likely to be successful. Thus, future research should pay attention to the intentions, antecedents, and determinants for companies and consumers to go circular.

Moreover, looking at the future, food and retail industries are an integrating part of the circular city (Paiho et al., 2020), where food waste prevention, reduction, and recycling are part of CE solutions and innovative CBM opportunities. In circular cities, digital platforms and apps for the food industry propose a "win-win-win strategy" that can benefit retailers, the consumers, and the planet (De Bernardi et al., 2021). In fact, these tools have an educational function, promote food waste reduction and recycling by consumers and allow communication among the parties, creating a direct relation. These business-to-customer relations are critical for the success of circular practices, thus they should be deeper investigated.

Lastly, the synergy between CE and Industry 4.0 in the food industry and food waste reduction is recognized as a success crucial point (Sousa Jabbour et al., 2021) for companies, and therefore the relationship between CBMs and Industry 4.0 in agribusiness can be addressed by future research.

5 Conclusion

This chapter presented a systematic review of 60 documents on circular business models (CBMs) in agribusiness, providing insights into their trends and issues related to entrepreneurship and green transition. The review applied two different and complementary approaches, bibliometric and content analysis. The bibliometric analysis revealed that CBMs in agribusiness is a recent topic that has gained exponential attention in the last few years, as well as circular economy, with a considerable impact on the scientific community and a high collaboration between authors. The review highlighted the countries with the highest scientific production in the field and the most relevant sources of publication, with Italy leading the group of the most productive countries and with "Sustainability" and "Journal of Cleaner Production" as the two journals publishing the most about CBMs and circular practices. The most frequent keywords used in the reviewed publications are "innovation", "management", "performance", "industry", "design", "waste", and "life-cycle assessment", suggesting a strong relation between CBMs, management area, industrial technology, and sustainability.

The content analysis identified important theoretical dimensions and discussed relevant sub-themes of the main topic, such as challenges, key aspects, and general recommendations for CBMs, their sustainability and performance measurement.

Authors recognize a connection between CBMs, green transition and technological change. The circular business model in agribusiness has the characteristic of being based on organic processes and products that have great potential for transformation and valorization into different products. Thus, companies investing in technological change and green transformation can improve their circular economy strategies by maximizing their resource recovery and valorization processes.

Moreover, the content analysis explains interesting aspects of the relation between CBMs and entrepreneurship. It is recommended for managers and entrepreneurs to understand the technical nature and potential of their resources, of their final product and of their waste, to identify crucial points to implement circular processes and favour green transition in their company. At the same time, managers should understand market needs and consumer's acceptance for green and recycled products.

The crucial role of governments and companies in promoting circular economy practices were highlighted, as well as the importance of cooperation and commitment among all stakeholders involved in the circular ecosystem. The local environment was pointed out as the most favourable for the success of CBMs. The lack of knowledge about the topic, the lack of a standard circular business model, scalability issues, the investment in technology transition and the legislation were identified as barriers to circular business models implementation. Life-cycle assessment and the "Circulytics" tool rose as suitable methods to assess circular performance in companies. Moreover, the study concluded that even if the CBM is a subcategory of the SBM, there are still cases in which it can be unsustainable, especially at an economic level.

To conclude, future research questions to be addressed were suggested, such as the relationship between CBMs and Industry 4.0 in agribusiness, the importance of greenwashing for the commitment to CBM transition, the intentions, and determinants of bringing companies to circularity and the rebound effect and the social impact of CBMs' adoption. While the study has some limitations, such as being based on a specific combination of databases and keywords, and on scientific literature, instead of documents provided by companies, it provides a valuable contribution to the understanding of CBMs challenges and recommendations in agribusiness and highlights important aspects useful for entrepreneurs and policymakers to succeed in circular economy transition.FundingThis research is supported by national funds through the FCT (Portuguese Foundation for Science and Technology)/ MCTES (PIDDAC) under the projects UIDB/04011/2020, UIDB/04007/2020, UIDB/04044/2020, UIDP/04044/2020, Associate Laboratory ARISE LA/P/0112/2020 and PAMI - ROTEIRO/0328/2013 (N° 022158). Funding was additionally provided by the FCT under the research contract PhD grant with 2021.07012.BD code to Barbara Lamolinara (supported by the European Union "Fundo Social Europeu (FSE)", "Por_Norte" Program and MCTES/Portuguese Republic).

References

Akerman, M., Humalisto, N., & Pitzen, S. (2020). Material politics in the circular economy: The complicated journey from manure surplus to resource. *Geoforum, 116*, 73–80. https://doi.org/10.1016/j.geoforum.2020.07.013

Alcalde-Calonge, A., Ruiz-Palomino, P., & Sáez-Martínez, F. J. (2021). Sustainability in the Forest bioeconomy: How social capital may favour the development of circular economy business models. In *Social capital: Issues, challenges and perspectives*. Nova Science Publishers,. https://www.scopus.com/inward/record.uri?eid=2-s2.0-85112433420&partnerID=40&md5=a2252281bcee64ae017e2dd7b320ba64

Amoah, P., Gebrezgabher, S., & Drechsel, P. (2021). Safe and sustainable business models for water reuse in aquaculture in developing countries. *Resource Recovery and Reuse, 2021*(20), 1–36. https://doi.org/10.5337/2021.212

Aria, M., & Cuccurullo, C. (2017). Bibliometrix: An R-tool for comprehensive science mapping analysis. *Journal of Informetrics, 11*(4), 959–975. https://doi.org/10.1016/J.JOI.2017.08.007

Barbieri, R., & Santos, D. F. L. (2020). Sustainable business models and eco-innovation: A life cycle assessment. *Journal of Cleaner Production, 266*. https://doi.org/10.1016/j.jclepro.2020.121954

Basile, V., Capobianco, N., & Vona, R. (2021). The usefulness of sustainable business models: Analysis from oil and gas industry. *Corporate Social Responsibility and Environmental Management, 28*(6), 1801–1821. https://doi.org/10.1002/CSR.2153

Bocken, N., & Ritala, P. (2021). Six ways to build circular business models. *Journal of Business Strategy.* https://doi.org/10.1108/JBS-11-2020-0258

Bocken, N. M. P., Short, S. W., Rana, P., & Evans, S. (2014). A literature and practice review to develop sustainable business model archetypes. *Journal of Cleaner Production, 65*, 42–56. https://doi.org/10.1016/j.jclepro.2013.11.039

Bocken, N. M. P., Weissbrod, I., & Tennant, M. (2016). Business model experimentation for sustainability. In R. Setchi, R. J. Howlett, Y. Liu, & P. Theobald (Eds.), *Sustainable design and manufacturing 2016* (Vol. 52, pp. 297–306). Springer-Verlag. https://doi.org/10.1007/978-3-319-32098-4_26

Borrello, M., Pascucci, S., Caracciolo, F., Lombardi, A., & Cembalo, L. (2020). Consumers are willing to participate in circular business models: A practice theory perspective to food provisioning. *Journal of Cleaner Production, 259*, 121013. https://doi.org/10.1016/J.JCLEPRO.2020.121013

Brehmer, M., Podoynitsyna, K., & Langerak, F. (2018). Sustainable business models as boundary-spanning systems of value transfers. *Journal of Cleaner Production, 172*, 4514–4531. https://doi.org/10.1016/j.jclepro.2017.11.083

Camana, D., Manzardo, A., Toniolo, S., Gallo, F., & Scipioni, A. (2021). Assessing environmental sustainability of local waste management policies in Italy from a circular economy perspective. An overview of existing tools. *Sustainable Production and Consumption, 27*, 613–629. https://doi.org/10.1016/j.spc.2021.01.029

Cantino, V., Giacosa, E., & Cortese, D. (2019). A sustainable perspective in wine production for common-good management: The case of Fontanafredda biological "reserve". *British Food Journal, 121*(2), 259–274. https://doi.org/10.1108/BFJ-06-2018-0351

Cavicchi, C., & Vagnoni, E. (2020). Sustainable business models in hybrids: A conceptual framework for community pharmacies' business owners. *Sustainability (Switzerland), 12*(19), 8125. https://doi.org/10.3390/su12198125

D'Amato, D., Veijonaho, S., & Toppinen, A. (2020). Towards sustainability? Forest-based circular bioeconomy business models in Finnish SMEs. *Forest Policy and Economics, 110*, 101848. https://doi.org/10.1016/j.forpol.2018.12.004

Dagevos, H., & de Lauwere, C. (2021). Circular business models and circular agriculture: Perceptions and practices of dutch farmers. *Sustainability (Switzerland), 13*(3), 1–15. https://doi.org/10.3390/su13031282

De Bernardi, P., Bertello, A., & Forliano, C. (2021). Digital platforms for circular business model innovation: A case-study to tackle food waste. In *Business model innovation: New Frontiers and perspectives*. Taylor and Francis. https://www.scopus.com/inward/record.uri?eid=2-s2.0-85106654117&partnerID=40&md5=cca272f78a4ce2ba97cd0e59d7f136db

De Carvalho Araújo, C. K., Bigarelli Ferreira, M., Salvador, R., de Carvalho Araújo, C. K. C., Camargo, B. S., de Carvalho Araújo Camargo, S. K., et al. (2022). Life cycle assessment as

a guide for designing circular business models in the wood panel industry: A critical review. *Journal of Cleaner Production, 355*, 131729. https://doi.org/10.1016/J.JCLEPRO.2022.131729
Del Vecchio, P., Secundo, G., Mele, G., & Passiante, G. (2021). Sustainable entrepreneurship education for circular economy: Emerging perspectives in Europe. *International Journal of Entrepreneurial Behaviour and Research, 27*(8), 2096–2124. https://doi.org/10.1108/IJEBR-03-2021-0210
Donner, M., & de Vries, H. (2021). How to innovate business models for a circular bio-economy? *Business Strategy and the Environment, 30*, 1932–1947. https://doi.org/10.1002/bse.2725
Donner, M., Gohier, R., & de Vries, H. (2020). A new circular business model typology for creating value from agro-waste. *Science of the Total Environment, 716*, 137065. https://doi.org/10.1016/j.scitotenv.2020.137065
Donner, M., & Radić, I. (2021). Innovative circular business models in the olive oil sector for sustainable mediterranean agrifood systems. *Sustainability (Switzerland), 13*(5), 1–23. https://doi.org/10.3390/su13052588
Donner, M., Verniquet, A., Broeze, J., Kayser, K., & De Vries, H. (2021). Critical success and risk factors for circular business models valorising agricultural waste and by-products. *Resources, Conservation and Recycling, 165*, 105236. https://doi.org/10.1016/j.resconrec.2020.105236
Engelberts, L., van Rheede, A., Kievit, H., & Nijhof, A. (2021). Appreciating multiple realities in the transformation towards a sustainable dairy sector: An explorative study from the inside-out perspective. *Agronomy-Basel, 11*(11), 2116. https://doi.org/10.3390/agronomy11112116
Fiore, M., Galati, A., Gołębiewski, J., & Drejerska, N. (2020). Stakeholders' involvement in establishing sustainable business models: The case of polish dairy cooperatives. *British Food Journal, 122*(5), 1671–1691. https://doi.org/10.1108/BFJ-04-2019-0263
Florido, C., Jacob, M., & Payeras, M. (2019). How to carry out the transition towards a more circular tourist activity in the hotel sector. The role of innovation. *Administrative Sciences, 9*(2), 47. https://doi.org/10.3390/admsci9020047
Galvão, G. D. A., Homrich, A. S., Geissdoerfer, M., Evans, S., Ferrer, P. S. S., & Carvalho, M. M. (2020). Towards a value stream perspective of circular business models. *Resources, Conservation and Recycling, 162*, 105060. https://doi.org/10.1016/j.resconrec.2020.105060
Garcés-Ayerbe, C., Rivera-Torres, P., Suárez-Perales, I., & La Hiz, D. I. L. D. (2019). Is it possible to change from a linear to a circular economy? An overview of opportunities and barriers for European small and medium-sized Enterprise companies. *International Journal of Environmental Research and Public Health, 16*, 851. https://doi.org/10.3390/IJERPH16050851
Geissdoerfer, M., Morioka, S. N., de Carvalho, M. M., & Evans, S. (2018a). Business models and supply chains for the circular economy. *Journal of Cleaner Production, 190*, 712–721. https://doi.org/10.1016/j.jclepro.2018.04.159
Geissdoerfer, M., Pieroni, M. P. P., Pigosso, D. C. A., & Soufani, K. (2020). Circular business models: A review. *Journal of Cleaner Production, 277*, 123741. https://doi.org/10.1016/j.jclepro.2020.123741
Geissdoerfer, M., Vladimirova, D., & Evans, S. (2018b). Sustainable business model innovation: A review. *Journal of Cleaner Production, 198*, 401–416. https://doi.org/10.1016/j.jclepro.2018.06.240
Geissdoerfer, M., Savaget, P., Bocken, N. M. P., & Hultink, E. J. (2017). The circular economy – A new sustainability paradigm? In *Journal of cleaner production* (Vol. 143, pp. 757–768). Elsevier Ltd.. https://doi.org/10.1016/j.jclepro.2016.12.048
Geldres-Weiss, V. V., Gambetta, N., Massa, N. P., & Geldres-Weiss, S. L. (2021). Materiality matrix use in aligning and determining a firm's sustainable business model archetype and triple bottom line impact on stakeholders. *Sustainability (Switzerland), 13*(3), 1–23. https://doi.org/10.3390/su13031065
Ghisellini, P., Cialani, C., & Ulgiati, S. (2016). A review on circular economy: The expected transition to a balanced interplay of environmental and economic systems. *Journal of Cleaner Production, 114*, 11–32. https://doi.org/10.1016/J.JCLEPRO.2015.09.007

Ghisellini, P., & Ulgiati, S. (2020). Circular economy transition in Italy. Achievements, perspectives and constraints. *Journal of Cleaner Production, 243*, 118360. https://doi.org/10.1016/j.jclepro.2019.118360

Ghisetti, C., & Montresor, S. (2020). On the adoption of circular economy practices by small and medium-size enterprises (SMEs): Does "financing-as-usual" still matter? *Journal of Evolutionary Economics, 30*(2), 559–586. https://doi.org/10.1007/s00191-019-00651-w

Gomes, G. M., Moreira, N., & Ometto, A. R. (2022). Role of consumer mindsets, behaviour, and influencing factors in circular consumption systems: A systematic review. *Sustainable Production and Consumption, 32*, 1–14. https://doi.org/10.1016/J.SPC.2022.04.005

Goni, F. A., Gholamzadeh Chofreh, A., Estaki Orakani, Z., Klemeš, J. J., Davoudi, M., & Mardani, A. (2021). Sustainable business model: A review and framework development. *Clean Technologies and Environmental Policy, 23*(3), 889–897. https://doi.org/10.1007/s10098-020-01886-z

Hamid, R. G., & Blanchard, R. E. (2018). An assessment of biogas as a domestic energy source in rural Kenya: Developing a sustainable business model. *Renewable Energy, 121*, 368–376. https://doi.org/10.1016/j.renene.2018.01.032

Hultberg, E., & Pal, R. (2021). Lessons on business model scalability for circular economy in the fashion retail value chain: Towards a conceptual model. *Sustainable Production and Consumption, 28*, 686–698. https://doi.org/10.1016/j.spc.2021.06.033

Hussain, Z., Mishra, J., & Vanacore, E. (2020). Waste to energy and circular economy: The case of anaerobic digestion. *Journal of Enterprise Information Management, 33*(4), 817–838. https://doi.org/10.1108/JEIM-02-2019-0049

Kampelmann, S. (2021). Knock on wood: Business models for urban wood could overcome financing and governance challenges faced by nature-based solutions. *Urban Forestry and Urban Greening, 62*, 127108. https://doi.org/10.1016/j.ufug.2021.127108

Karlsson, N. P. E., Hoveskog, M., Halila, F., & Mattsson, M. (2019). Business modelling in farm-based biogas production: Towards network-level business models and stakeholder business cases for sustainability. *Sustainability Science, 14*(4), 1071–1090. https://doi.org/10.1007/s11625-018-0584-z

Kruger, S. D., Zanin, A., Durán, O., & Afonso, P. (2022). Performance Measurement Model for Sustainability Assessment of the Swine Supply Chain. Sustainability (Switzerland), 14(16). https://doi.org/10.3390/SU14169926

Lagrasta, F. P., Pontrandolfo, P., & Scozzi, B. (2021). Circular economy business models for the Tanzanian coffee sector: A teaching case study. *Sustainability (Switzerland), 13*(24), 13931. https://doi.org/10.3390/su132413931

Lamolinara, B., Pérez-Martínez, A., Guardado-Yordi, E., Guillén Fiallos, C., Diéguez-Santana, K., & Ruiz-Mercado, G. J. (2022). Anaerobic digestate management, environmental impacts, and techno-economic challenges. *Waste Management, 140*, 14–30. https://doi.org/10.1016/J.WASMAN.2021.12.035

Larsson, M. (2018). Circular business models: Developing a sustainable future. In *Circular business models: Developing a sustainable future*. Springer. https://doi.org/10.1007/978-3-319-71791-3

Lewandowski, M. (2016). Designing the business models for circular economy-towards the conceptual framework. *Sustainability (Switzerland), 8*(1), 1–28. https://doi.org/10.3390/su8010043

Loorbach, D., & Wijsman, K. (2013). Business transition management: Exploring a new role for business in sustainability transitions. *Journal of Cleaner Production, 45*, 20–28. https://doi.org/10.1016/J.JCLEPRO.2012.11.002

Madau, F. A., Arru, B., Furesi, R., & Pulina, P. (2020). Insect farming for feed and food production from a circular business model perspective. *Sustainability (Switzerland), 12*(13). https://doi.org/10.3390/su12135418

Marczewska, M., & Kostrzewski, M. (2020). Sustainable business models: A bibliometric performance analysis. *Energies, 13*(22), 6062. https://doi.org/10.3390/en13226062

Martin, M., & Herlaar, S. (2021). Environmental and social performance of valorizing waste wool for sweater production. *Sustainable Production and Consumption, 25*, 425–438. https://doi.org/10.1016/j.spc.2020.11.023

Messina, C. M., Arena, R., Manuguerra, S., Renda, G., Laudicella, V. A., Ficano, G., et al. (2021). Farmed Gilthead Sea bream (Sparus aurata) by-products valorization: Viscera oil omega-3 enrichment by Short-path distillation and in vitro bioactivity evaluation. *Marine Drugs, 19*(3), 160. https://doi.org/10.3390/md19030160

Mitchell, D. W., & Bruckner Coles, C. (2004). Business model innovation breakthrough moves. *Journal of Business Strategy, 25*(1), 16–26. https://doi.org/10.1108/02756660410515976

Moggi, S., & Dameri, R. P. (2021). Circular business model evolution: Stakeholder matters for a self-sufficient ecosystem. *Business Strategy and the Environment, 30*(6), 2830–2842. https://doi.org/10.1002/bse.2716

Nasir, A., Shaukat, K., Hameed, I. A., Luo, S., Alam, T. M., & Iqbal, F. (2020). A bibliometric analysis of Corona pandemic in social Sciences: A review of influential aspects and conceptual structure. *IEEE Access, 8*, 133377–133402. https://doi.org/10.1109/ACCESS.2020.3008733

Nasution, A. H., Aula, M., & Ardiantono, D. S. (2020). Circular economy business model design. *International Journal of Integrated Supply Management, 13*(2–3), 159–177. https://doi.org/10.1504/IJISM.2020.107848

Nayha, A. (2021). Backcasting for desirable futures in Finnish forest-based firms. *Foresight, 23*(1), 50–72. https://doi.org/10.1108/FS-01-2020-0005

Nosratabadi, S., Mosavi, A., & Lakner, Z. (2020). Food supply chain and business model innovation. *Foods, 9*(2), 132. https://doi.org/10.3390/foods9020132

Nosratabadi, S., Mosavi, A., Shamshirband, S., Zavadskas, E. K., Rakotonirainy, A., & Chau, K. W. (2019). Sustainable business models: A review. *Sustainability (Switzerland), 11*(6), 100272. https://doi.org/10.3390/su11061663

Nunez-Cacho, P., Molina-Moreno, V., Corpas-Iglesias, F. A., & Cortes-Garcia, F. J. (2018). Family businesses transitioning to a circular economy model: The case of "Mercadona". *Sustainability, 10*(2). https://doi.org/10.3390/su10020538

Oliveira Pavan, M. d. C., Ramos, D. S., Soares, M. Y., & Carvalho, M. M. (2021). Circular business models for bioelectricity: A value perspective for sugar-energy sector in Brazil. *Journal of Cleaner Production, 311*, 127615. https://doi.org/10.1016/j.jclepro.2021.127615

Page, M. J., McKenzie, J. E., Bossuyt, P. M., Boutron, I., Hoffmann, T. C., Mulrow, C. D., et al. (2021). The PRISMA 2020 statement: An updated guideline for reporting systematic reviews. *BMJ, 29*, 372. https://doi.org/10.1136/BMJ.N71

Paiho, S., Mäki, E., Wessberg, N., Paavola, M., Tuominen, P., Antikainen, M., et al. (2020). Towards circular cities—Conceptualizing core aspects. *Sustainable Cities and Society, 59*, 102143. https://doi.org/10.1016/j.scs.2020.102143

Pampuri, A., Casson, A., Alamprese, C., Di Mattia, C. D., Piscopo, A., Difonzo, G., et al. (2021). Environmental impact of food preparations enriched with phenolic extracts from olive oil mill waste. *Foods, 10*(5), 980. https://doi.org/10.3390/foods10050980

Pereira, A., Villanueva-Rey, P., Vence, X., Teresa Moreira, M., & Feijoo, G. (2018). Fresh milk supply through vending machines: Consumption patterns and associated environmental impacts. *Sustainable Production and Consumption, 15*, 119–130. https://doi.org/10.1016/j.spc.2018.05.003

Pla-Julián, I., & Guevara, S. (2019). Is circular economy the key to transitioning towards sustainable development? Challenges from the perspective of care ethics. *Futures, 105*, 67–77. https://doi.org/10.1016/J.FUTURES.2018.09.001

Prosperi, M., Sisto, R., Lopolito, A., & Materia, V. C. (2020). Local Entrepreneurs' Involvement in Strategy Building to Facilitate Agro-Food Waste Valorisation within an Agro-Food Technological District: A SWOT-SOR Approach. Sustainability, 12(11), 4523. https://doi.org/10.3390/su12114523

Reim, W., Parida, V., & Sjödin, D. R. (2019). Circular business models for the bio-economy: A review and new directions for future research. *Sustainability (Switzerland), 11*(9), 2558. https://doi.org/10.3390/su11092558

Ribeiro, I., Sobral, P., Peças, P., & Henriques, E. (2018). A sustainable business model to fight food waste. *Journal of Cleaner Production, 177*, 262–275. https://doi.org/10.1016/j.jclepro.2017.12.200

Rincón-Moreno, J., Franco-García, M.-L., Carpio-Aguilar, J. C., & Hernández-Sarabia, M. (2019). Share, optimise, closed-loop for food waste (SOL4FoodWaste): The case of Walmart-Mexico. *Greening of Industry Networks Studies, 6*, 165–190. https://doi.org/10.1007/978-3-319-92931-6_9

Rodríguez-Soler, R., Uribe-Toril, J., & De Pablo Valenciano, J. (2020). Worldwide trends in the scientific production on rural depopulation, a bibliometric analysis using bibliometrix R-tool. *Land Use Policy, 97*, 104787. https://doi.org/10.1016/J.LANDUSEPOL.2020.104787

Ryabchenko, O., Golub, G., Turčeková, N., Adamičková, I., & Zapototskyi, S. (2017). Sustainable business modeling of circular agriculture production: Case study of circular bioeconomy. *Journal of Security and Sustainability Issues, 7*(2), 301–309. https://doi.org/10.9770/jssi.2017.7.2(10)

Salvador, R., Puglieri, F. N., Halog, A., Andrade, F. G. D., Piekarski, C. M., & De Francisco, A. C. (2021). Key aspects for designing business models for a circular bioeconomy. *Journal of Cleaner Production, 278*, 124341. https://doi.org/10.1016/j.jclepro.2020.124341

Sehnem, S., Ndubisi, N. O., Preschlak, D., Bernardy, R. J., & Santos Junior, S. (2020). Circular economy in the wine chain production: Maturity, challenges, and lessons from an emerging economy perspective. *Production Planning and Control, 31*(11–12), 1014–1034. https://doi.org/10.1080/09537287.2019.1695914

Serhan, H., & Yannou-Lebris, G. (2021). The engineering of food with sustainable development goals: Policies, curriculums, business models, and practices. *International Journal of Sustainable Engineering, 14*(1), 12–25. https://doi.org/10.1080/19397038.2020.1722765

Singh, S., Babbitt, C., Gaustad, G., Eckelman, M. J., Gregory, J., Ryen, E., et al. (2021a). Thematic exploration of sectoral and cross-cutting challenges to circular economy implementation. *Clean Technologies and Environmental Policy, 23*(3), 915–936. https://doi.org/10.1007/s10098-020-02016-5

Singh, V. K., Singh, P., Karmakar, M., Leta, J., & Mayr, P. (2021b). The journal coverage of web of science, Scopus and dimensions: A comparative analysis. *Scientometrics, 126*(6), 5113–5142. https://doi.org/10.1007/S11192-021-03948-5/FIGURES/5

Snyder, H. (2019). Literature review as a research methodology: An overview and guidelines. *Journal of Business Research, 104*, 333–339. https://doi.org/10.1016/J.JBUSRES.2019.07.039

Sousa Jabbour, A. B. L., Frascareli, F. C. D. O., Santibanez Gonzalez, E. D. R., & Chiappetta Jabbour, C. J. (2021). Are food supply chains taking advantage of the circular economy? A research agenda on tackling food waste based on industry 4.0 technologies. *Production Planning and Control, 34*, 1–17. https://doi.org/10.1080/09537287.2021.1980903

Tranfield, D., Denyer, D., & Smart, P. (2003). Towards a methodology for developing evidence-informed management knowledge by means of systematic review. *British Journal of Management, 14*(3), 207–222. https://doi.org/10.1111/1467-8551.00375

Upadhyay, A., Akter, S., Adams, L., Kumar, V., & Varma, N. (2019). Investigating "circular business models" in the manufacturing and service sectors. *Journal of Manufacturing Technology Management, 30*(3), 590–606. https://doi.org/10.1108/JMTM-02-2018-0063/FULL/XML

Uvarova, I., Atstaja, D., Grinbergs, U., Petersons, J., Gegere-Zetterstroma, A., & Kraze, S. (2020). Transition to the circular economy and new circular business models - an in-depth study of the whey recycling. *IOP Conference Series: Earth and Environmental Science, 578*(1), 012019. https://doi.org/10.1088/1755-1315/578/1/012019

Valls-Val, K., Ibáñez-Forés, V., & Bovea, M. D. (2022). How can organisations measure their level of circularity? A review of available tools. *Journal of Cleaner Production, 354*, 131679. https://doi.org/10.1016/J.JCLEPRO.2022.131679

Van Genuchten, E. J. S., Mulder, I. J., & Schaaf, N. (2017). Strategies for food longevity. In C. Bakker & R. Mugge (Eds.), *Product lifetimes and the environment (plate)* (pp. 139–143). IOS PRESS. https://doi.org/10.3233/978-1-61499-820-4-139

Van PhI, C. P., Walraven, M., Bézagu, M., Lefranc, M., & Ray, C. (2020). Industrial symbiosis in insect production— A sustainable eco-efficient and circular business model. *Sustainability (Switzerland), 12*(24), 1–14. https://doi.org/10.3390/su122410333

Velter, M. G. E., Bitzer, V., Bocken, N. M. P., & Kemp, R. (2020). Sustainable business model innovation: The role of boundary work for multi-stakeholder alignment. *Journal of Cleaner Production, 247*, 119497. https://doi.org/10.1016/j.jclepro.2019.119497

Winter, A., Pedro, E., Slasko, J., Battaglini, J., Faelker, M., Kivipelto, R., Duarte, A. J., Malheiro, B., Ribeiro, C., Justo, J., Silva, M. F., Ferreira, P., & Guedes, P. (2019). Waste to fungi an EPS@ISEP 2019 project. In M. A. C. Gonzalez, F. J. R. Sedano, C. F. Llamas, & F. J. GarciaPenalvo (Eds.), *TEEM'19: Seventh international conference on technological ecosystems for enhancing Multiculturality* (pp. 115–122). Assoc Computing Machinery. https://doi.org/10.1145/3362789.3362921

Yamoah, F. A., Sivarajah, U., Mahroof, K., & Peña, I. G. (2022). Demystifying corporate inertia towards transition to circular economy: A management frame of reference. *International Journal of Production Economics, 244*, 108388. https://doi.org/10.1016/j.ijpe.2021.108388

Yazan, D. M., Cafagna, D., Fraccascia, L., Mes, M., Pontrandolfo, P., & Zijm, H. (2018). Economic sustainability of biogas production from animal manure: A regional circular economy model. *Management Research Review, 41*(5), 605–624. https://doi.org/10.1108/MRR-02-2018-0053

Zhang, Q., Dhir, A., & Kaur, P. (2022). Circular economy and the food sector: A systematic literature review. *Sustainable Production and Consumption, 32*, 655–668. https://doi.org/10.1016/J.SPC.2022.05.010

Zilia, F., Bacenetti, J., Sugni, M., Matarazzo, A., & Orsi, L. (2021). From waste to product: Circular economy applications from sea urchin. *Sustainability, 13*(10), 5427. https://doi.org/10.3390/su13105427

Zucchella, A., & Previtali, P. (2019). Circular business models for sustainable development: A "waste is food" restorative ecosystem. *Business Strategy and the Environment, 28*(2), 274–285. https://doi.org/10.1002/bse.2216

The Circular Economy as a Creator of Value in Agribusiness 4.0: Proposals for a Life

Andrea Mara Pimenta-Alonso and Elena González-Rodrigo

Abstract The circular economy has come with momentum to settle in the countries as a new model of economic growth. Thus, sustainable development and the objectives set forth by the United Nations for its achievement and expansion highlight the concern and preoccupation of society for these issues. For this reason, this chapter deals with four specific aspects of the agricultural industry sector, which aim to improve our lives and the sustainability of our planet. These aspects refer to the use of biomass residues and their contribution to the food sector, the generation of biofuel and electricity from biogas for agricultural use, as well as the reuse of treated wastewater for irrigation. Finally, a decrease in cotton production and the use of regenerated cotton by the textile industry will be considered. Based on the theoretical framework, four specific proposals are established to achieve the aggregate global growth requirement posed by Sustainable Development Goal 8, compared to the reduction in the use of CO_2, together with the reduction in the use of resources (Goals 6, 7, 9, 12 and 15). These proposals cover aspects such as the reuse of organic and biodegradable wastes, the reuse of water through different irrigation methods and its application based on its components, and the reuse of cotton from existing clothing. All of this provides economic agents, administration, companies, and households with opportunities for reflection and implementation of a more sustainable economic and social model.

Keywords Circular economy · Sustainable development · Agribusiness · Well-being · Textile industry

A. M. Pimenta-Alonso
ESIC University, Pozuelo de Alarcón, Spain
e-mail: andreamara.pimenta@esic.university

E. González-Rodrigo (✉)
ESIC University; ESIC Business and Marketing School, Pozuelo de Alarcón, Spain
e-mail: elena.gonzalez@esic.university

C. Rego et al. (eds.), *Entrepreneurship, Technological Change and Circular Economy for a Green Transition*, Studies on Entrepreneurship, Structural Change and Industrial Dynamics, https://doi.org/10.1007/978-3-031-48079-9_8

1 Introduction

The Circular Economy (CE) is a new economic system that includes both innovation in business models and changes in consumer behaviour. Economic agents put into practice the reuse of materials and products with the aim of extending their useful life and finding new uses for them (Lewandowski, 2016).

Concern and care for the environment play a key role in this alternative to neoclassical economics that seeks to respond to the scarcity of resources, whether these are endogenous or exogenous to economies, through the sustainable management of their resources, changes in social behaviour, business initiatives and marketing models (Hysa et al., 2020). Its main purpose is to find economic well-being, accompanied by respect for the environment and the impact they can have on the next generations (Kirchherr et al., 2017).

Sustainable development, on the other hand, focuses on environmental, economic and social benefits as a whole (Lehtonen, 2004). Despite this, both ideas share points in common, such as integration into development, the importance of cooperation, the need for regulations and incentives to put them into practice, and innovation in the business model, which have the power to cause a technological change in the industry (Geissdoerfer et al., 2017).

The search for proposals for technological change for the growth of food production in an environmentally sustainable and respectful way, goes through the abandonment of the closed cycle of extract–produce–discard, and by the adoption of waste reuse technology, reaching towards a model that aims to retain the value of resources, products, parts and materials. Several organizations of international importance, such as the World Business Council for Sustainable Development or the Direção Geral das Atividades Econômicas from Portugal, consider relevant and highlight this new strategic concept of CE, which by creating a system that allows the useful life of resources and materials to be extended, causes optimal reuse through transformation and recycling.

However, financial institutions continue to show more interest in projects considered more polluting than green initiatives, since the risks associated with these new technologies are greater and their rate of return is relatively low, which highlights the need for incentives for green finance (Sachs et al., 2019; Van der Waal & Thijssens, 2020).

Sometimes it is also possible to find conflicts between the Sustainable Development Goals (SDG) themselves (United Nations, 2015) and the structure of the countries that adopt them (Eisenmenger et al., 2020; Goralski & Tan, 2020; Shi et al., 2019). In this sense, we can mention cultural conditions, since they affect the sensitivity of individuals regarding the care and maintenance of the environment (De Guimarães et al., 2020). On the other hand, regulations for the protection and care of the environment, incentives for administrations and the promotion of sustainable consumption promoted by organizations and associations should drive the economies so that they can get going (Santika et al., 2019). These, in their entirety, are some of the measures that allow economic growth to occur whilst meeting the objectives of Sustainable Development (Salvia et al., 2019; Di Baldassarre et al., 2019).

In this regard, studies can be found that propose an adaptation of some aspects to facilitate compliance. However, the requirement of aggregate global growth (Goal 8) of the SDG, compared to the reduction in the use of CO_2, together with the reduction in the use of resources (Goals 6, 7, 9, 12, and 15) seem to be in opposition (Hickel, 2019; Eisenmenger et al., 2020). In any case, it is notorious that, for a green transition of production and current life models to occur, political, socioeconomic, and cultural changes must first take place (Schot & Steinmueller, 2018; Pichler et al., 2017; Brand, 2016).

Taking the above into account, this study aims to explore and describe the most innovative proposals in the field of agribusiness, taking as a reference the most widespread and economically viable practices, either in the European Union, which has been a pioneer in the implementation of CE practices in the agricultural sector (Barros et al., 2020), or other regions that provide ideas that favour the environment and a sustainable economy.

More specifically, in this chapter emphasis will be placed on four aspects that affect agro-industry, promoting discussion of its effectiveness and viability to contribute to improving our lives: the use of biomass residues and their contribution to the food field; the generation of biofuel and electricity from biogas for agricultural use; the reuse of treated wastewater for irrigation; and the decrease in cotton production and the use of regenerated cotton by the textile industry. Next, the advantages and specific contributions of each of these proposals will be shown, in order to finally establish a series of conclusions and future lines of study.

2 Use of Biomass Residues in Agribusiness

The use of waste from agro-industrial activity may be the point that stands out in the performance of Brazilian agribusiness on the international scene within the field of Bioeconomy (Gouvêa et al., 2023). According to Vaz Júnior (2020), whilst producing biomass, this industry also generates a large amount of waste that requires more advanced technology to reduce its impact on the environment and public health. However, this waste can give rise to opportunities for its use as a renewable industrial raw material, following the precepts of green chemistry, Bioeconomy and CE, in conjunction with the Sustainable Development Goals defined by the United Nations (United Nations, 2015).

In this sense, agribusiness can represent a fundamental link in the integration of renewable raw materials in the production processes. Kinnunen et al. (2022), when referring to the food industries, point out that this sector is responsible for generating tons of waste, causing serious environmental problems with the production of organic waste. For these authors, the Circular Bioeconomy of organic waste and the food industry has been receiving increasing attention since this waste can represent usable resources for the synthesis of useful products.

In addition to this waste, it is also possible to take advantage of animal waste, manure, urban and industrial waste to fertilize some crops, which implies significant economic savings when using this waste and not using chemical compounds that currently represent a good percentage of the costs of agricultural production. We should not forget that the main advantage is to promote "environmental care" and the reduce global pollution, which is why it is very attractive to use it on large crops. Another point to take into account is that its use can facilitate the cleaning of mountains and the field in general, avoiding unwanted situations, such as fires.

In order to encourage these practices, the European Union tries to adopt policies that affect governments to try to transform the economies of its member states (Tsoukalis, 2022). This means implementing a profound restructuring, both of society and of the productive structures of the countries, which implies a higher cost in the short and medium term, with results that often slow down productivity (Marino & Pariso, 2020). In return, it is expected to achieve environmentally sustainable as well as economically efficient regions (Aiginger & Rodrik, 2020).

For this, the concepts of Bioeconomy and CE seek an alignment and a transformation of the basic ideas from which they were born, to arrive at what the latest technological propositions call Circular Bioeconomy (Comissão Europeia, 2015). For this organism, bio-based waste can be used for the production of new products and as an alternative to products and energies based on fossil fuels, presenting advantages related to its renewability, biodegradability or composting possibilities based on renewable natural resources, such as the production of productive chains made up of biomass.

To exemplify which wastes are economically viable to be reused, we mention the use of dry residues from the extraction of cassava starch as a substitute for maize in the feed for lactating cows in periods of high maize prices. Even though the use of this residue can cause a reduction in the intake of dry matter and nutrients, as well as a reduction in milk production, with impacts on its profitability, its use continues to be economically efficient (Fernandes et al., 2015).

The use of these wastes in animal feed makes them an economical and nutritionally acceptable alternative raw material. Through their transformation and enrichment, using silage technology, they constitute a more sustainable and cleaner production option, reducing environmental impacts. Latin American countries such as Chile, Cuba, Brazil, Panama, Mexico and Colombia, rich in a diversity of by-products and agro-industrial residues, have managed to incorporate these residues into the animal diet. In this same sense, in China, the use of this waste under processes that reduce its rotting and decomposition to obtain material rich in energy, proteins and vitamins are known as ECO-FEED, and it is applied successfully in the feeding of laying hens and growth and fattening pigs (Ramírez et al., 2017).

Going further, we can think about the intelligent use of food loss and waste that begins in the field and goes through its transformation and consumption, such as waste from restaurants and other food outlets. According to the report on the food waste index (UNEP, 2021), some 931 million tons of food—17% of the total food available to consumers in 2019—ended up in the rubbish bins of households, retailers, restaurants and other food services, which is equivalent, globally, to 121 kg of wasted food per capita year.

The acceptance of this type of product in the market is a key aspect of the agricultural sector. The CE has highlighted the importance of care for the environment, involving a significant part of consumers in the care and respect for ecological and sustainable products. In this regard, it is worth noting the consumption of WTV foods that prevents products from being "lost" in the supply chain (Coderoni & Perito, 2020).

Although in developed countries there is a growing trend towards the consumption of healthy organic products, the truth is that they are more expensive. For this reason, and because of life habits, focused on personal fulfilment through work done outside the home, the consumption of processed foods in homes is high (it is estimated that two billion people are overweight). This affects the environment through energy consumption (30% of the planet's total consumption[1]) and the generation of waste from the packaged products we purchase.

If we add one-third of the food that ends up in the trash before being consumed, due to failures in the distribution chain or poor planning of daily household menus, this non-recoverable economic cost represents around one billion dollars. This not only supposes irresponsible consumption, but also overloads the fertility of the land, livestock, the marine environment and, in general, natural resources to generate food.

In 2020, according to the FAO, the global average cost and affordability data for a healthy diet (CoAHD) was \$3.54 per person per day. If we do the maths, a person will spend \$1292 annually to maintain a healthy diet. According to data from the World Bank, in 2021, Afghanistan, Syria, and Mozambique, amongst a total of approximately 33 countries, had a GDP per capita below this value, and, in addition, this income had to satisfy all personal needs, not just food.

Regarding developing countries, the food industry is very widespread, yet it allows the use of components that are banned in the first world because of their high health and hygiene restrictions and pollutant emissions. In these countries, the dietary habits are energy-dense, but it is often not a balanced diet.

In underdeveloped countries, the production system is based on monoculture and hunting, without existing health regulations. Their diet is not very varied, and they have energy deficiencies.

3 The Generation of Biofuel and Electricity from Biogas for Agricultural Use

With the growing demand for energy and the geopolitical complications to obtain the necessary quantities to supply industries and private use, the production and use of renewable energies seem increasingly established options. Whilst it is true that these energies are not yet widely used, most of those that are normally used are produced from existing fossil resources and raw materials, anaerobic digestion

[1] https://www.un.org/sustainabledevelopment/es/sustainable-consumption-production/.

(AD) can improve the production and quality of biogas which provides multiple uses (Achinas et al., 2017).

In this matter, the European Union is the world leader in biomethane production, introducing economic, environmental and climatic benefits, both as fuel for vehicles, including those in the agricultural sector, and for the natural gas network (Scarlat et al., 2018).

As an example of concern and progress in the use of this type of energy, the Biogas Roadmap is an instrument that is used to articulate what is established in the Spanish National Integrated Energy and Climate Plan (Ministry for the Ecological Transition and the Demographic Challenge, 2020). This document confirms that the biogas value chain presents a high degree of technological maturity, so the activation of the market can be immediate (Ministerio para la Transición Ecológica y el Reto Demográfico, 2022). Of course, the direct use of biogas in locations close to its production should be prioritized, favouring the appearance of synergies with related industries, as well as its use in transport, when electrification is not the most economically and environmentally efficient option.

In order to promote this development in the short term, it is considered necessary to value the renewable origin of biogas through the creation of a system of guarantees of origin, which allows consumers and companies to differentiate their added value compared to gases of fossil origin and ensure the corresponding sustainability standards (Ministry for the Ecological Transition and the Demographic Challenge, 2022).

The production of biogas by anaerobic decomposition is a useful way to treat biodegradable waste since it produces a valuable fuel, in addition to generating an important by-product called digestate as effluent. This material, liquid or solid, contains nitrogen, phosphorus, potassium, calcium and other elements that are highly recommended as fertilizer for the soil (OAS, 2023). Likewise, biogas can be used to produce electricity through turbines or gas-generating plants, or to generate heat in ovens, stoves, dryers, boilers or other gas combustion systems, duly adapted for this purpose. On the other hand, biomethane produced from biogas can gradually displace natural gas of fossil origin in its usual uses, since it is fully interchangeable, especially in those applications with high energy consumption or difficult to electrify, such as heavy transport or industries requiring intensive use of thermal energy.

In addition to its ecological value, the removal of tons of organic substrate from the environment and the energy value of the cubic meter of biogas also contribute to justify its use, with 60% methane and 40% CO2, equivalent to 0.7 litres of gasoline; 2.4 kW-hours of electricity; 0.6 m^3 of natural gas or 1.3 kg of wood. In terms of biogas productivity, a case study carried out by the FAO in conjunction with the UNDP, amongst other organizations, shows that the manure of ten cattle can generate the equivalent of four 15-kilo cylinders of liquefied gas per month, and fertilize 300 m^2 at a rate of 160 nitrogen units/ha (FAO, 2011).

Biomass-producing sources to give rise to biogas, called organic substrate, are very versatile: sludge from sewage treatment plants (STP); urban solid waste (MSW) from landfills; manure, slurry and waste from livestock activity; the remains of agricultural or industrial activity, and energy plantations.

4 Reuse of Treated Wastewater for Irrigation

Wastewater can be understood as the flow of water used and discharged from residences, companies, industries, commercial activities and institutions and which is directed to treatment stations. These wastewaters are later categorized and defined according to their sources of origin (Sousa & Uceda-Maza, 2017).

The World Health Organization (WHO, 2013) considers that the use of wastewater in agriculture is a procedure capable of combining recycled water and nutrients but recommends that it has to be done safely to maximize the uses for public health and environmental benefits.

Evidence of the need to take advantage of this type of resource is given by the World Bank's demographic growth projections, which indicate that by 2050 the planet's population will exceed ten billion inhabitants and that said population will have to have food and fibre to meet their basic needs. It is estimated that by this same year, agricultural production should increase by approximately 70%. To achieve this level of food production, water is a fundamental input and plays an important role in food security (World Bank Group Water Global, n.d.).

In the approach of the CE for the sustainability of water, it is essential to highlight that the water sector can be one of the main protagonists in a transformation of linear economic systems to more circular models (Barros et al., 2020). In addition, in the field of water, the concept of CE materializes in the return of water use repeatedly, just as it occurs in the natural cycle (Melgarejo-Moreno et al., 2019).

On the other hand, it is necessary to take into consideration that every reuse process has its negative side, and in the case of water, the main agents that cause possible contamination must be removed. It is true that there are beneficial elements in terms of the possible potential of this element when used in agriculture, but more research is still needed on the possible presence of soil contaminants that have not yet been analyzed (Fallahi et al., 2021).

The multiple uses of wastewater offer many opportunities to produce sustainable and more stable access to water (Caselles-Osorio et al., 2018), as well as to generate valuable products, such as agricultural nutrients, soil amendments and fuel (Hua et al., 2022). Currently, irrigated agriculture represents 20% of the total cultivated area and contributes 40% of the total food production worldwide. It is, on average, at least twice as productive per unit of land as rainfed agriculture, allowing for greater production intensification and crop diversification (World Bank Group Water Global, n.d.).

Some cities in southern Brazil are already adopting water reuse, knowing that the management of water resources, in addition to technical issues, also involves social factors, institutions, and administrative procedures (Hua et al., 2022). This practice is also widespread in regions that suffer from water shortage resources, for example Japan, the Mediterranean basin (Israel, Tunisia, Cyprus, Spain), the United States (mainly the state of California), Asia in general and the Persian Gulf.

However, it must be remembered that there is a need to adapt the use of treated wastewater for agricultural irrigation. The main factors that determine its adequacy are its salinity, the presence of heavy metals and pathogens that can cause adverse effects in humans, plants and soils (Ofori et al., 2021). This adaptation involves carrying out treatments (primary or secondary) so that the water reaches the quality level required for new use. Consequently, in addition to all the aspects included in the design of a traditional irrigation system, it is necessary to define well the aspects associated with the collection and treatment of water, in order to implement an irrigation system that uses reclaimed water.

The characteristics of the water used for irrigation will also influence the definition of the water distribution mechanisms. Dispersal mechanisms by flooding or surface runoff can lead to the discharge of irrigation water in inappropriate areas. In addition, spraying techniques cause the dispersal of wastewater droplets, which can pose problems for farmers and nearby residential or publicly accessible areas. This type of technique is even prohibited in some regions for foods that are eaten raw, since the reclaimed water comes into direct contact with the product. The aspects related to the choice of irrigation methods are summarized in Table 1.

In this line, several studies show that consumers present food neophobia (FN) and aversion to foods processed in new ways, that is, food technology neophobia (FTN), if they are not clear about the process used (Coderoni & Perito, 2020). To inform the consumer in relation to food production with the use of reclaimed water, the FAO proposes a classification of crops based on the level of risk for consumers and farmers.

The classification of crops based on the level of risk for consumers and farmers if reclaimed water is used for irrigation is as follow: (i) Weak risk: industrial crops not intended for human consumption (e.g. cotton and sisal), crops treated with heat or dried before their destination for human consumption (e.g. cereals, oilseeds and sugar beets), fruits and vegetables grown exclusively for canning or other treatment that effectively destroys pathogenic germs, fodder crops and other animal feed crops that are harvested and sun-dried before consumption; (ii) Medium risk: pastures and herbaceous forage crops, crops for human consumption that are not in direct contact with wastewater, if it does not collect anything from the soil and that sprinkler irrigation is not used (e.g. arboriculture and vineyards), crops for human

Table 1 Advantages of the uses of different types of irrigation

Irrigation type	Advantages	Measures to adopt in its use
Flood	Cheap	It is necessary to protect people in direct contact with the product
	No need for slopes	
Surface runoff	Cheap	It is necessary to protect people in direct contact with the product
Aspersion	Average yield	Filtration to avoid clogging the mechanisms
	No need for slopes	Set the distance for your placement
Micro irrigation	High performance	Filtration to avoid clogging the mechanisms
	High health protection	

Source: Own elaboration

consumption after cooking (from: potatoes, aubergines and beets), crops for human consumption whose skin is not eaten (e.g. citrus and bananas), all crops not identified as "High Risk" if overhead irrigation is used; (iii) High risk: all foods that are eaten raw or grown in close contact with wastewater effluents (e.g. lettuce and carrot), sprinkler irrigation, regardless of crop type, within 100 m of residential areas or places with public access (Lazarova & Bahri, 2008).

To end up with this fear in relation to food production, it is necessary to guarantee that the use of reclaimed water is safe, in order to stimulate the reuse of water on a global scale and increase the confidence of the population in this practice. Therefore, the production and supply of reclaimed water for agricultural irrigation should be allowed only according to a permit issued by the competent authorities.

However, there remains the need to mention the drawbacks of water recycling, since we are talking about expensive water treatment and regeneration techniques, in addition to the need for very rigorous legislation in relation to the quality of the water obtained to avoid possible health problems. The change of mentality on the part of the consumer must begin with an intense campaign to raise awareness of the need for a change in the way of producing food, in order to produce more, but in a sustainable and respectful way with the resources that nature offers us.

5 Reduction in Cotton Production and Use of Regenerated Cotton by the Textile Industry

Following the line of elimination of waste produced by the industry and by consumption in general and focusing on the CE, we can also mention the textile industry and its value chain. In this sense, cotton, as the virgin fibre that it is, uses a large number of agricultural resources for its production. It is true that the cotton production chain is very important for the world economy and its production cannot be interrupted without promoting an alternative economy in exchange.

But if we return to the data from the World Bank mentioned in proposal three, in which the concern for the growth of the world's population makes us rethink our current uses of the land and how we are going to manage to feed the ten billion of inhabitants expected to be reached in the next 30 years, it comes to mind using more natural resources for food production and less for agricultural products that are not intended for food.

In this sense, there is a growing concern with the quality/fertility of agricultural land used for food production and its possible scarcity in the future. At the moment, FAO studies (FAOSTAD, 2020) suggest that this will not happen globally, although in some regions and areas there is already a serious shortage, which may worsen. By the end of the twentieth century, in regions such as the Near East and North Africa, 87 percent of the land suitable for agriculture was already under cultivation, whilst in South Asia that figure was as high as 94 percent. What is happening in several countries in these areas is that the land balance is negative, that is, more land is cultivated than is suitable for rainfed crops, either due to its drought condition or its orography. In FAO projections for the year 2030, more than 80 percent of arable area expansion is expected to occur in sub-Saharan Africa and Latin America.

Although there is still surplus land in these regions, if fertilizer use is not increased to compensate, this could result in poor soil and stagnant or diminished yields. In South Asia, the situation will be even more complicated—with 98 percent already in cultivation, over 80 percent of the increase in production will have to come from yield increases as only 5–6 percent can come from the expansion of the arable surface. The forecast is that cropping intensity will increase in all developing regions. This will occur by shortening fallow periods and increasing multiple harvests, which will be possible in part thanks to the growth in irrigated areas (FAOSTAD, 2020).

Regarding the use given to these farmlands, which, according to the World Bank, reached 4738 million hectares (World Bank, 2020) in 2020, the area devoted to cotton cultivation occupies an area of 31.8 million hectares (FAOSTAD, 2020) and is located in seasonally dry subtropical and tropical areas. The main producing countries are India, China, the United States, Brazil and Pakistan. Together, these countries account for more than three-quarters of world production (OECD, 2020).

Regarding concern for the environment, cotton production is far from the ideal of sustainability that we defend in the CE. How to handle the dilemma between environmental sustainability and sustainability of less favoured economies is what leads us to seek new solutions and business opportunities. One of the hypotheses to handle this situation could be the use of recycled cotton, from pre- and/or post-consumer waste. Pre-consumer, also called post-industrial waste, includes the leftovers generated by spinning and fabric by-products that are discarded at the time of cutting and manufacturing garments and other by-products of the textile industry. On the other hand, post-consumer waste includes clothing in general, upholstery, towels, and household items, whose cotton fibres can be recycled and regenerated.

The largest volume of regenerated cotton is produced through pre-consumer waste since post-consumer waste is usually more difficult to process and classify due to the dyeing suffered by the fabrics and the fibre mixtures that make up each fabric. Both alternatives, even though they are undoubtedly more environmentally friendly than traditional cotton, still present certain aspects against them, such as their high recycling costs and their low production efficiency, as well as an obstacle: a consumer, so accustomed to the fast fashion model, that they will hardly be willing to pay more for a garment because it is made with resources from the circular economy.

In real data, according to Ecoembes (The Circular Lab, 2021) and Circular Economía, it is estimated that a total of 92 million tons of textile waste is generated in the world each year. And it has been calculated that, if the current rate of consumption of clothing and textile products remains unchanged, by the year 2030 this number could reach 134 million tons per year.

As is well known, this is the sad result of the phenomenon known as fast fashion—consumption of clothing at very low prices, which impels those who follow it to compulsively buy and discard clothing, reducing the useful life of the garments, generally by changes in fashion or lack of quality of products. This is a trend that is evident in the most advanced economies. If we take US data as an example, where this phenomenon is strongest, on average an American tends to throw away 37 kilos of clothing per year, in addition to the 12% of clothes that are never worn in their closets (The Circular Lab, 2021). In general, it is estimated that we only wear 40% of the clothes that are in our closets, which are becoming more and more crowded (Carrera i Gallissà, 2019).

The fast fashion strategy also tends to trivialize the real value of textile products and the agents involved in the long value chain of this industrial sector. One of the consequences of this problem is the increase in the generation of waste. The high levels of contamination in the fashion industry are due to two factors: firstly, due to the overproduction of clothing to which it contributes, and secondly, the mismanagement of the garments. According to figures presented by the awareness campaign Break the Pattern, promoted by Electrolux in the year 2022, a garment is used on average ten times and is discarded much sooner than it should be, being thrown into landfills, incinerated, or worse, ending up in nature reserves. Globally, approximately 73% of all textile waste ends up in landfills or is incinerated. According to data from this campaign, each year 21 million tons of clothing end up in landfills, with the Atacama Desert being the second largest landfill in the world, receiving approximately 39,000 tons of clothing per year.

To try to reuse all this waste, there are different mechanical or chemical methodologies for waste management in this sector, but which, according to the Technological Innovation Center of the Polytechnic University of Catalonia, still need more technological advances to improve their efficiency. However, research in this sector continues to advance, and an innovative process developed by Australian researchers from the Queensland University of Technology is already known. Together with the founders of Start Up BlockTexx and QUT, they have developed what is known as "tissue separation technology"—which stands out because it is designed to treat hybrid tissues. This process works with mixed fabrics, turning cotton into cellulose and polyester into flakes for industrial uses such as injection moulding. The hope with this technology is that it will help "close the loop on the cotton chain" by diverting textiles from landfills whilst simultaneously replacing virgin material.

To increase the magnitude of the problem, according to Carrera i Gallissà (2019), regulatory changes are also approaching that will affect the management of textile waste, such as the ban by Africa on the importation of second-hand clothing from Europe, or the European regulations which on 01 January 2030 will prohibit throwing into landfills any waste that can be reused and/or recycled. This will save a large amount of textile waste that was previously exported, burned and/or landfilled.

In order to try to reduce this unspeakable amount of waste, it is necessary to encourage research and proposals for the regeneration of the cotton used by the textile industry, in addition to replacing part of the space used for cotton cultivation in favour of food production.

Proposals for a Better Life The objective of this work is to establish improvements for a better life. Based on the aspects raised in the previous sections, an attempt is made to comply with Sustainable Development Goals 6, 7, 8, 9, 12 and 15, establishing a series of proposals that cover different areas of agribusiness (Table 2).

Table 2 Proposals for a better life

Proposal	Objective	Material in reuse	New products obtained	Advantages	Disadvantages	SDGs
Use of biomass residues in agribusiness	Make use of them as agro-industrial raw material	Organic waste/biomass	Raw material for animal feed and fertilizers	Promotes care for the environment and reduces pollution of the planet	Advanced technology is required	8, 12, 15
Generation of biofuel and electricity from biogas for agricultural use	Supply the growing demand for energy efficiency	Biodegradable waste	Biogas: Digestate, electric power, biomethane	Removes tons of organic substrate from the environment. Mitigates climate change	A guaranteed system is needed to ensure its viability	8, 7, 9, 12, 15
Reuse of treated wastewater for irrigation	Sustainably supply water to the population and crops	Water	Valuable products, irrigation water rich in agricultural nutrients	Sustainable and more stable access to water	Potential presence of pathogens and soil contaminants	8, 6, 12, 15
Decrease in cotton production and use of regenerated cotton by the textile industry	Reduce cotton production in favour of food production	Waste from the textile industry	New fibres, cellulose and polyester flakes for industrial uses	Transformation of cultivation areas destined for non-food products to produce food. Reduction in the amount of textile waste incinerated or taken to landfills is of great environmental importance	High cost of cotton regeneration technology and need for consumer awareness regarding the cost of clothing	8, 9, 12, 15

Source: Own elaboration

Thus, we propose the use of biomass waste to promote care for the environment by reducing pollution, for which advanced technology is necessary to reuse organic waste and generate biomass.

Another of our proposals is the generation of biofuels and electricity through the use of biodegradable waste. With this, not only are tons of organic substrate removed, but climate change is mitigated.

We also consider the reuse of water as a key element for improving life in agribusiness. Despite the presence of possible contaminants, efficiency in the use of water can be achieved by also providing new nutrients for soil fertilization.

Lastly, we also consider the cotton industry, which includes the agricultural and textile sectors as promoters of sustainable growth. This aspect seems to us to stand out because it implies a change of mentality of the consumer and great investments in the textile producers.

With these proposals, some ideas are collected to get closer to fulfilling the SDGs. With the intention of complying with Objective 8 of providing decent work with economic growth in the agro-industrial sector, an attempt is made to accommodate the fulfilment of another five objectives, namely: 6: clean water and sanitation; 7: affordable and clean energies; 9: industry, innovation and infrastructure; 12: responsible production and consumption and 15: the life of terrestrial ecosystems.

6 Discussion and Conclusions

The CE is establishing itself with force in global thinking; the impact and care of the environment through sustainable development are becoming more and more relevant, boosting more efficiency in the use of productive resources, and encouraging economic growth and promoting decent work.

For this reason, in 2015, the United Nations established some Objectives as a guide to follow in different parts of the world to achieve this efficiency and care for the environment. Although it is true that many times, the fulfilment and adoption of certain measures imply a change in the social, governmental and business conception, it is also true that certain transformations seem necessary to achieve long-term sustainability.

However, not all regions take these premises into account, either because their level of development makes it difficult to adopt some measures—because of the high cost involved and their vital needs—or because the change to be made is so profound that they look forward to seeing how they develop in their environment to take advantage of the technology synergies adopted and benefit from lower costs as these innovations expand.

The truth is that if we intend to continue with economic and social development in the long term, it is essential that progress be made in this area. Specifically, this work focuses on agribusiness as a promoter of change, with the primary purpose of achieving stable economic growth over time.

The contribution that this chapter offers is the compilation of a series of proposals aimed at providing a solution to the problem of how to improve sustainable development in the agricultural sector; more specifically aspects related to the reuse of resources such as water, cotton or organic and biodegradable waste, which allow substantial improvements in environmental life.

As a first proposal, the use of biomass residues from agro-industrial activity was presented. Its particularities are:

- Organic waste rich in nutrients can be reused through the preparation of composting, as a form of fertilizer for new crops.
- Dry matter residues from the transformation of some agricultural products can be used as alternative feed in animal nutrition, with acceptable results, in periods when the price of their ideal feed is very high.
- Biomass can be used as a source of green energy and is an alternative to the use of fossil fuels.
- The use of biomass promotes the removal of a large amount of organic waste from the environment, avoiding the proliferation of pests and diseases from the accumulation of fermented organic matter.
- By generating economic value, it encourages the removal of unwanted green/dry material from the field, thus favouring the clearing of forests and thus avoiding fires in times of high temperatures.

The second proposal was the generation of biofuel and electricity from biogas for agricultural use. Its most outstanding characteristics are:

- The production of biogas by anaerobic decomposition to treat biodegradable waste can replace the use of gases of fossil origin in their usual uses, since they are interchangeable, especially in those applications with high-energy consumption or difficult to electrify, such as heavy transport or intensive industry in the use of thermal energy.
- Biogas can also be used to produce electricity or to generate heat in ovens, stoves, dryers, boilers or any other gas combustion system used in rural areas.
- The sources of biomass to give rise to biogas are very versatile and can be from various types of organic substrate residues such as WWTP, MSW, amongst others. This is favourable to the environment since it eliminates sources of production of greenhouse gases and other polluting elements.

The third proposal mentions the reuse of treated wastewater for irrigation. This point is especially sensitive in territories that suffer from a lack of water for human consumption and food production. The most important points are:

- The use of this type of water in agriculture can provide, in addition to water as such, nutrients and micronutrients necessary for the proper development of the crop.
- What is sought is that water can be used and reused over and over again, just as it occurs in its natural cycle, thus materializing the pure concept of CE.

- However, the use of treated wastewater in irrigation must be strictly controlled, either because of the irrigation technique to be used or because of its characteristics, which must comply with rigorous quality control. The main factors that determine its adequacy are its salinity and the presence of heavy metals or pathogens that are harmful to health, plants or the soil.
- The use of this type of water requires extensive training on the part of the agents involved and very clear and strict legislation in this regard.

And finally, the fourth proposal invites a reduction in the production of cotton as an agricultural crop in favour of agriculture that is destined for human consumption. In addition, the use of regenerated cotton is encouraged to reduce the amount of garbage caused by the discarding of millions of tons of clothing around the world and by waste from the textile industry. In this sense, it should be noted:

- Allocating more land to growing food to the detriment of growing cotton may cause an economic crisis in certain regions in the short term, but in the long run it could be offset by sustainable and environmentally friendly food production.
- The largest volume of regenerated cotton is produced through pre-consumer waste, since fabrics that have undergone some types of dyeing are more difficult and expensive to regenerate.
- There are large cotton waste dumps in areas of special environmental value throughout the world. According to Ecoembes, it is estimated that around 92 million tons of textile waste are discarded worldwide every year.
- The production of clothing at very low prices (fast fashion) favours the consumption of new garments and the compulsive discarding of clothing that could have an extended useful life.
- There is a great need for investment in the research and development of new technologies that are more competitive for the use of discarded cotton in the textile industry.

All these proposals are designed to reduce the accumulation of waste and rubbish in the area closest to its use, thus eliminating unnecessary expenses with the displacement and wear and tear that the transport of heavy material can cause in existing transport systems.

This work aims to bring these suggestions to build a better life for governments, companies and consumers to face the sustainable development of the planet. For future lines of research, we recommend studying other aspects related to agribusiness and involving the fulfilment of other SDGs.

References

Achinas, S., Achinas, V., & Euverink, G. J. W. (2017). A technological overview of biogas production from biowaste. *Engineering, 3*(3), 299–307. https://doi.org/10.1016/J.ENG.2017.03.002

Aiginger, K., & Rodrik, D. (2020). Rebirth of industrial policy and an agenda for the twenty-first century. *Journal of Industry, Competition and Trade, 20*(2), 189–207. https://doi.org/10.1007/s10842-019-00322-3

Barros, M. V., Salvador, R., de Francisco, A. C., & Piekarski, C. M. (2020). Mapping of research lines on circular economy practices in agriculture: From waste to energy. *Renewable and Sustainable Energy Reviews, 131*, 109958. https://doi.org/10.1016/j.rser.2020.109958

Brand, U. (2016). "Transformation" as a new critical orthodoxy: The strategic use of the term "transformation" does not prevent multiple crises. *GAIA-Ecological Perspectives for Science and Society, 25*(1), 23–27. https://doi.org/10.14512/gaia.25.1.7

Carrera i Gallissà, E. (2019). *Los retos del reciclaje textil.* https://blog.cit.upc.edu/la-problematica-de-los-residuos-textiles/

Caselles-Osorio, A., et al. (2018). Tomato (Lycopersicum esculentum) production in sub surface flow constructed wetlands for domestic wastewater treatment in rural a Colombian community. *Ingeniería, investigación y tecnología, 19*(4), 1–10. https://doi.org/10.22201/fi.25940732e.2018.19n4.038

Coderoni, S., & Perito, M. A. (2020). Sustainable consumption in the circular economy. An analysis of consumers' purchase intentions for waste-to-value food. *Journal of Cleaner Production, 252*, 119870. https://doi.org/10.1016/j.jclepro.2019.119870

De Guimarães, J. C. F., et al. (2020). Governance and quality of life in smart cities: Towards sustainable development goals. *Journal of Cleaner Production, 253*, 119926. https://doi.org/10.1016/j.jclepro.2019.119926

Di Baldassarre, G., et al. (2019). Sociohydrology: Scientific challenges in addressing the sustainable development goals. *Water Resources Research, 55*(8), 6327–6355. https://doi.org/10.1029/2018WR023901

Eisenmenger, N., et al. (2020). The sustainable development goals prioritize economic growth over sustainable resource use: A critical reflection on the SDGs from a socio-ecological perspective. *Sustainability Science, 15*, 1101–1110. https://doi.org/10.1007/s11625-020-00813-x

Europeia, C. (2015). *COMUNICAÇÃO DA COMISSÃO AO PARLAMENTO EUROPEU, AO CONSELHO, AO COMITÉ ECONÓMICO E SOCIAL EUROPEU E AO COMITÉ DAS REGIÕES*. s.n..

Fallahi, A., et al. (2021). Interactions of microalgae-bacteria consortia for nutrient removal from wastewater: A review. *Chemosphere, 272*, 129878. https://doi.org/10.1016/j.chemosphere.2021.129878

FAO. (2011). *Manual de Biogás.* Proyecto CHI/00/G32 ed. Santiago de Chile: s.n.

FAOSTAD. (2020). *Perspectivas por sectores principales - Producción de Cultivos.* https://www.fao.org/3/y3557s/y3557s08.htm

Fernandes, T., et al. (2015). Use of dried waste of cassava starch extraction for feeding lactating cows. *Anais da Academia Brasileira de Ciências, Issue, 87*, 1101–1111. https://doi.org/10.1590/0001-3765201520140220

Geissdoerfer, M., Savaget, P., Bocken, N. M., & Hultink, E. J. (2017). The circular economy–a new sustainability paradigm? *Journal of Cleaner Production, 143*, 757–768. https://doi.org/10.1016/j.jclepro.2016.12.048

Goralski, M. A., & Tan, T. K. (2020). Artificial intelligence and sustainable development. *The International Journal of Management Education, 18*(1), 100330. https://doi.org/10.1016/j.ijme.2019.100330

Gouvêa, L. P., et al. (2023). Phenotypic plasticity in Sargassum forests may not counteract projected biomass losses along a broad latitudinal gradient. *Ecosystems, 26*(1), 2. https://doi.org/10.1007/s10021-022-00738-9

Hickel, J. (2019). The contradiction of the sustainable development goals: Growth versus ecology on a finite planet. *Sustainable Development, 27*(5), 873–884. https://doi.org/10.1002/sd.1947

Hua, F., et al. (2022). The biodiversity and ecosystem service contributions and trade-offs of forest restoration approaches. *Science, 376*(6595), 839–844. https://doi.org/10.1126/science.abl4649

Hysa, E., Kruja, A., Rehman, N. U., & Laurenti, R. (2020). Circular economy innovation and environmental sustainability impact on economic growth: An integrated model for sustainable development. *Sustainability, 12*(12), 4831. https://doi.org/10.3390/su12124831

Kinnunen, P., Karhu, M., Yli-Rantala, E., & Kivikytö-Reponen, P. (2022). A review of circular economy strategies for mine tailings. *Cleaner Engineering and Technology, Volumen, 8*, 100499. https://doi.org/10.1016/j.clet.2022.100499

Kirchherr, J., Reike, D., & Hekkert, M. (2017). Conceptualizing the circular economy: An analysis of 114 definitions. *Resources, Conservation and Recycling, 127*, 221–232. https://doi.org/10.1016/j.resconrec.2017.09.005

Lazarova, V., & Bahri, A. (2008). *Water reuse practices for agriculture. Available at: Water reuse: An international survey of current practice, issues and needs* (pp. 199–227). IWA Publishing.

Lehtonen, M. (2004). The environmental–social interface of sustainable development: Capabilities, social capital, institutions. *Ecological Economics, 49*(2), 199–214. https://doi.org/10.1016/S0921-8009(04)00076-X

Lewandowski, M. (2016). Designing the business models for circular economy—Towards the conceptual framework. *Sustainability, 8*(1), 43. https://doi.org/10.3390/su8010043

Marino, A., & Pariso, P. (2020). Comparing European countries' performances in the transition towards the circular economy. *Science of the Total Environment, 729*, 138142. https://doi.org/10.1016/j.scitotenv.2020.138142

Melgarejo-Moreno, J., López-Ortiz, M. I., & Fernández-Aracil, P. (2019). Water distribution management in south-East Spain: A guaranteed system in a context of scarce resources. *Science of the Total Environment, 648*, 1384–1393. https://doi.org/10.1016/j.scitotenv.2018.08.263

Ministerio para la Transición Ecológica y el Reto Demográfico. (2020). *Plan Nacional Integrado de Energía y Clima (PNIEC) 2021–2030,* s.l. https://www.miteco.gob.es/es/prensa/pniec.aspx.

Ministerio para la Transición Ecológica y el Reto Demográfico. (2022). *Hoja de Ruta del Hidrógeno*. MITERD.

OAS. (2023). *OAS (Organization of American States).* https://www.oas.org/en/about/who_we_are.asp

OECD. (2020). *OCDE-FAO Perspectivas Agrícolas 2020–2029.* https://www.oecd-ilibrary.org/sites/3a8abf59-es/index.html?itemId=/content/component/3a8abf59-es

Ofori, S., Puškáčová, A., Růžičková, I., & Wanner, J. (2021). Treated wastewater reuse for irrigation: Pros and cons. *Science of the Total Environment, 760*, 144026. https://doi.org/10.1016/j.scitotenv.2020.144026

Pichler, M., Schaffartzik, A., Haberl, H., & Görg, C. (2017). Drivers of society-nature relations in the Anthropocene and their implications for sustainability transformations. *Current Opinion in Environmental Sustainability, 26*, 32–36. https://doi.org/10.1016/j.cosust.2017.01.017

Ramírez, V. M., Peñuela, L. M., & Pérez, M. D. R. (2017). Los residuos orgánicos como alternativa para la alimentación en porcinos. *REVISTA DE CIENCIAS AGRÍCOLAS, 34*(2), 107–124. https://doi.org/10.22267/rcia.173402.76

Sachs, J. D., Woo, W. T., Yoshino, N., & Taghizadeh-Hesary, F. (2019). Importance of green finance for achieving sustainable development goals and energy security. In *Handbook of green finance* (pp. 3–12). Springer. https://doi.org/10.1007/978-981-13-0227-5_13

Salvia, A. L., Leal Filho, W., Brandli, L. L., & Griebeler, J. (2019). Assessing research trends related to sustainable development goals: Local and global issues. *Journal of Cleaner Production, 208*, 841–849. https://doi.org/10.1016/j.jclepro.2018.09.242

Santika, W. G., et al. (2019). From goals to joules: A quantitative approach of interlinkages between energy and the sustainable development goals. *Energy Research & Social Science, 50*, 201–214. https://doi.org/10.1016/j.erss.2018.11.016

Scarlat, N., Dallemand, J. F., & Fahl, F. (2018). Biogas: Developments and perspectives in Europe. *Renewable Energy, 129*, 457–472. https://doi.org/10.1016/j.renene.2018.03.006

Schot, J. & Steinmueller, W. E., 2018. Three frames for innovation policy: R &D, systems of innovation and transformative change. Research Policy, 47(9), pp. 1554–1567. https://doi.org/10.1016/j.respol.2018.08.011.

Shi, L., Han, L., Yang, F., & Gao, L. (2019). The evolution of sustainable development theory: Types, goals, and research prospects. *Sustainability, 11*(24), 7158. https://doi.org/10.3390/su11247158

Sousa, A. D., & Uceda-Maza, F. X. (2017). Más allá de los desafíos del decenio de la educación para el desarrollo sostenible: una reflexión necesaria. *HOLOS, 5*, 136–150. https://doi.org/10.15628/holos.2017.6107

The Circular Lab. (2021). *Montañas de residuos textiles en el mundo: Allanar la montaña del textil.* https://www.thecircularlab.com/allanar-la-montana-del-textil/#:~:text=Se%20estima%20que%20el%20mundo,y%20cualquier%20otro%20material%20textil

Tsoukalis, L. (2022). *The European community and its Mediterranean enlargement.* Taylor & Francis. https://doi.org/10.4324/9781003335337

UNEP, W. (2021). *UNEP food waste index report 2021.* United Nations Environment Programme.

United Nations. (2015). *United Nations.* Available at: https://www.un.org/sustainabledevelopment/es/objetivos-de-desarrollo-sostenible/

Van der Waal, J. W., & Thijssens, T. (2020). Corporate involvement in sustainable development goals: Exploring the territory. *Journal of Cleaner Production, 252*, 119625. https://doi.org/10.1016/j.jclepro.2019.119625

Vaz Junior, S. (2020). Aproveitamento de resíduos agroindustriais: uma abordagem sustentável.

WHO (2013). World health report 2013: research for universal health coverage. ISBN 9789240690837.

World Bank. (2020). Agricultural land. https://datos.bancomundial.org/indicator/AG.LND.AGRI.K2?end=2020&locations=1W&start=1961&view=chart

World Bank Group Water Global, s.f. (n.d.). *World Bank Group Water Global.* https://www.bancomundial.org/es/topic/water-in-agriculture#:~:text=Actualmente%2C%20la%20agricultura%20representa%20(en,la%20evapotranspiraci%C3%B3n%20de%20los%20cultivos)

The Social Dimension of Agribusiness and How It Has Been Measured

José Massuça, Ana Marta-Costa, and Maria Raquel Lucas

Abstract The social aspect of sustainability is commonly acknowledged as the least researched among the three fundamental pillars. Specifically in the context of agribusiness, the social themes cover a wide range of topics spanning from basic needs to cultural heritage. For these reasons, we decided to focus our attention on how researchers have been capturing the social aspects of agriculture using sustainability assessment indicators over the past two decades.

Based on a previously extensive literature review and on a comprehensive bibliometric study about agricultural social indicators, between 1999 and 2022, this work aims to classify those indicators according to their thematic scope, grouping them in distinct clusters, and subsequently sort the most frequently referenced indicators in each of the defined clusters. Given the data patterns found during our investigation, we also looked for changes in the distribution of social indicators per thematic cluster in two different periods. This research brought attention to the unique characteristics and developments within each identified cluster, revealing a clear increase in concerns related to personal and societal welfare. These concerns particularly revolve around satisfaction with work and active involvement in civic activities (i.e., civic participation). Furthermore, by classifying each indicator as either traditional or emerging, we have been able to confirm this trend toward more principles-driven

J. Massuça (✉) A. Marta-Costa
CETRAD – Centre for Transdisciplinary Development Studies, University of Trás-os-Montes e Alto Douro, Vila Real, Portugal
e-mail: josemassuca@utad.pt, e-mail: amarta@utad.pt

M. R. Lucas
Department of Management and CEFAGE- Center for Advanced Studies in Management and Economics, Universidade de Évora, Évora, Portugal
e-mail: mrlucas@uevora.pt

C. Rego et al. (eds.), *Entrepreneurship, Technological Change and Circular Economy for a Green Transition*, Studies on Entrepreneurship, Structural Change and Industrial Dynamics, https://doi.org/10.1007/978-3-031-48079-9_9

indicators. The identification of the top five indicators used in each cluster further highlights the multidimensional character of the assessments of social sustainability.

Both traditional and emerging social themes are important in assessing the impact of agribusiness demands on workers, communities, and the environment to inform the development of strategies for promoting sustainable and socially responsible practices in the agroindustry. Thus, social indicators covering both components should be dynamically monitored over time, ideally supporting and promoting stakeholder engagement in the development of sustainability evaluation tools. Given the specificities of each societal context where agribusiness takes place, creating assessment models for generic applications may fail to grasp the complexity of social variables. For this reason, we aim to present in the future a set of guidelines and key indicators to implement and integrate into each contextualized model, identifying specificities of communities, workers, and consumers in a particular agribusiness context.

Keywords Social sustainability · Sustainability assessment · Social indicators · Agribusiness

1 Introduction

Researchers, agricultural scientists, economists, and policy makers are increasingly interested in discussing Sustainable Agribusiness. Understanding that agriculture and related industries are considered sustainable when they are financially sound, socially acceptable, and ecologically beneficial is at the heart of this topic (Elkington, 2018). Nevertheless, the research on agricultural sustainability frequently ignores the social component of sustainability (Rasmussen et al., 2017).

Literature provides sustainability assessments of specific agricultural sectors using pre-formatted assessment models that generally incorporate a set of indicators (Chopin et al., 2021; Desiderio et al., 2022; Reddy et al., 2014; Triviño-Tarradas et al., 2020), although it is not always clear why certain indicators were chosen or how they were defined. When these sustainability assessment models are analyzed, it becomes recognizable that not only are the social dimension of sustainability indicators significantly outweighed by indicators for the economic and environmental dimensions (De Luca et al., 2018) but also that certain indicators within this set lack scientific validity, while others are often limited in their ability to capture realities beyond the scope of individual companies and lack the necessary standardization to allow for comparisons between assessment results (Boyer et al., 2016).

There are several possible causes for these flaws. Above all, agribusiness ecosystems and related agricultural systems are part of larger socio-ecological and socio-economical contexts, which complexity makes it more difficult to cover the social dimension in a thorough, evidence-based, and quantifiable way (Meul et al., 2008).

The three basic pillars of sustainability, formerly defined by Elkington (1994), should be considered while selecting a standard set of indicators for measuring sustainability. Yet, the specificities of each societal context in which the agribusiness develops make it hard to create indicators that can grasp the complexity of social variables (Havardi-Burger et al., 2021).

The proposed chapter will concentrate on how the social sustainability aspect of agriculture has been addressed over the past two decades. To do this, it will present information on the most relevant indicators applied to the various forms and models of social sustainability assessment, between 1999 and 2022. Resorting to previous research of Massuça et al. (2023) on a previously extensive literature review and on a detailed bibliometric analysis of social indicators theme for agriculture, this work aims to classify those indicators according to their thematic scope and group them into distinct clusters. In addition to this classification in clusters, the recognized indicators were further categorized as either Traditional or Emerging, aiming to identify their potential emphasis on contemporary concerns that necessitate monitoring and evaluation. This classification helps ensure that newer sustainability assessment models adequately address these evolving issues, which may not be adequately covered in older frameworks.

This work will contribute to produce potential guidelines on the most appropriate social indicators to apply in future sustainability assessment models in the agribusiness context. This will be of certain benefit to institutional experts, developers, and researchers in the fields of Agribusiness entrepreneurship, Social and Territorial cohesion, and Agribusiness Sustainability. Ultimately, the objective of the proposed chapter is to provide support for the socially oriented sustainable development of the agricultural sector.

2 Background

There are various perspectives on what social sustainability entails. According to Olakitan Atanda (2019), some authors define social sustainability as long-term connections between nature and society that contribute to the viability of society, while others believe that social sustainability rests on the key notions of equality and democracy (the core of society). Colantonio (2009) acknowledged earlier work by Polèse and Stren (2000) that promoted that social sustainability develops through balancing the progress of a civil society, where the development should produce a rich environment. In Colantonio's (2009) study it is also mentioned that authors like Biart (2002) have long advocated that social sustainability incorporates the alleged and pressing societal necessities for the sustainability of societies in the long run. Hence, attaining social sustainability is essential to society's long-term survival. This acknowledgment led to the definition of social sustainability as *"a positive condition within communities and a process within communities that can achieve that condition"* (Mckenzie, 2004, p. 23), defining it as a progression rather than a target to be attained in the future.

In the context of agriculture, Bacon et al. (2012) point out that previous authors (Goodman & Watts, 1997; Zimmerer & Bassett, 2003; and Wells, 2011) show that this activity has an intricate connection with society and ecology since the beginning of time. However, the continuous growth of the globalized food system, together with the constant development of agroindustry practices, have contributed to gradually dimming the relationship between "man and natural environment" or between "culture and nature," mainly due to the expanding gap concerning physical environments and knowledge distance. These expanding gaps are one of the reasons why it is important to understand the social dimension of agriculture, agriculture systems, and agribusinesses.

It is also acknowledged how crucial it is to consider powerful institutions and pertinent parties, such as farmers, farmer cooperatives and associations, local policymakers, and organized civic groups that play active roles in local agriculture. They should be included because it will help determine the factors that have the greatest influence on the context of agribusiness, affecting how agri-scoial policies are implemented in the future, reshaping markets for more sustainable goods and - most importantly - promoting the sharing of knowledge on a topic that is personally important to them (Rasmussen et al., 2017).

There are several models for assessing sustainability that include the social dimension, focused on the measurement of indicators that must comply with certain characteristics. Each indicator employed in sustainability evaluation models should be manageable by researchers, economically feasible, socially structured, analytically sound, and measurable (Popovic et al., 2017). It should also be adequate for multiple spatial levels and sensitive to changes in the environment and management. De Olde et al. (2017) recurs to Reed et al. (2006) to add two more broad requirements for creating indicators: (i) indicators should be able to measure the achievement of sustainable development goals objectively and precisely and (ii) indicators should be user friendly so that they can be used by local users. These indicators, as referred by several authors, must also be based on data that is readily available, have some degree of uniformity, and have scientific validity in order to allow for comparisons across different time periods and regions (Bedeau et al., 2021; Binder et al., 2010; Bonisoli et al., 2018; Corvo et al., 2021; Popovic et al., 2017).

Similar to other fields of study, the rationale behind the selection of sustainability assessment indicators in agriculture is not always available (Meul et al., 2008), despite the extensive discussions found in the literature on this topic (Havardi-Burger et al., 2021). Kühnen and Hahn (2017) emphasize the necessity for a rigorous and dependable process to choose from the many possible social indicators. With this objective in mind, Binder et al. (2010) propose three principles for indicator selection: vision or goal alignment, representation of the system, and availability of data.

Ul Haq and Boz (2017) emphasized the significance of accurately describing the environment in which the assessment models will be implemented and of choosing indicators that are pertinent to the biophysical and socio-economic condition of the study area regarding this specific situation. Moreover, Littig and Griessler (2005) reflected on social sustainability absence from the sociological theory and proposed that, for better comprehension, it is appropriate to construct social sustainability indicators in the light of these theories. These authors (Littig & Griessler, 2005) additionally supported the idea that these social sustainability indicators must be organized into three core categories: (1) satisfaction of basic needs and quality of life, (2) social justice, and (3) social coherence. Subsequent studies, such as that conducted by Colantonio (2009), present arguments for categorizing social indicators according to two fundamental principles: (1) the basic need, which concentrates on the material aspects of life and society (health, housing, and food) and (2) equity, which focuses on social disparities and equitable access to services such as education and justice.

3 Methodology

Massuça et al. (2023) selected 158 articles published in the period between 1999 and 2022, using the "Preferred Reporting Items for Systematic Reviews and Meta-Analyses" (PRISMA) 2020 protocol (Page et al., 2021). To identify scholarly papers that address the discussion, listing, or empirical testing of social sustainability indicators, a comprehensive screening procedure was implemented. These articles were subjected to both quantitative and qualitative analysis utilizing automated technologies (Rx64 4.1.2 and RStudio Version 2 software and the Bibliometrix 3.1 application and NVIVO Version Release 1.4.1 (815) software and its analysis functionalities), software along with their respective analysis functionalities.

From this second screening stage, the entire set of identified social sustainability indicators identified in these papers were compiled into a list and prompted as the basis of our research.

With all the identified social sustainability indicators listed, eight distinct clusters were proposed, drawing on prior research of Sachs (1999), Colantonio (2007), and Desiderio et al. (2022). However, this clustering process proved to be a complex task due to two main factors: (i) the uneven scope of the different studies and (ii) the multidimensional nature of some of the indicators.

For this reason, namely regarding the recurrence and equivalence of the topics that the indicators identify and also the possible areas of overlap and complementarity of these same indicators, careful consideration was given to previous frameworks presented by various authors to have a guideline and supported decision on choosing the most useful set of indicators to work with (Vanclay, 2002; Pope et al., 2005; Ahmadvand et al., 2011; Boström, 2012; Ripoll-Bosch et al., 2012; De Olde et al., 2017; Rasmussen et al., 2017; Bonisoli et al., 2018; Bachev et al., 2019; Mollayosefi et al., 2019; Janker & Mann, 2020; Maisarahmokti & Hamzah Sharaai, 2020).

In this chapter, using the database of articles gathered by Massuça et al. (2023), a list of all the social sustainability indicators identified in the reviewed literature was created, followed by case-by-case harmonization of the description of the indicators, so to make sure that simple differences in spelling did not mistakenly count as different indicators when they actually referred to the same. Then the indicators were grouped and analyzed in each cluster (Table 1) and afterward they were also categorized using Colantonio's (2007) distinction of Traditional and Emerging themes (Table 2), thus capturing the underlying dynamics of the sustainability assessment models covered in the analysis.

The cluster disaggregation employed is supported by the literature as exposed in Table 1, and a similar structure was proposed by Rasouli and Kumarasuriyar (2016)

Table 1 Employed cluster codification and scope

Cluster	Designation	Scope	Sources
1	Community relations, participation, and engagement	Indicators that identify and quantify the relational interactions at the community level among the various stakeholders involved in the agribusiness to be evaluated. It includes the indicators that somehow question and evaluate the community participation and civic involvement of stakeholders in the production (and decision) process.	Cuthill (2010) Binder et al. (2010) Reddy et al. (2014) Doğu and Aras (2019) Mills et al. (2021)
2	Demographic dynamics	Indicators from a general demographic context (e.g., average age of stakeholders, average size of families involved in production processes, and size of agribusiness-related households) or more specific to population dynamics within the geography of study (e.g., migration data, demographic potential of the region where agribusiness is located, information on succession in agribusiness, among others).	Colantonio (2007) Jane Dillon et al. (2016) Anastasova-Chopeva (2019)
3	Education and training	Indicators that aggregate information on education, access to training, and extension services to the stakeholders in agribusiness. Contrary to the objectivity of the information related to education, the evaluation of capacity building and training programs incorporates innovative views on new techniques, methods, and technologies related to agribusiness, such as evaluation of access to information, levels of innovation achieved, investment in R&D, access to computer and info literacy, Internet and information platforms availability.	Spangenberg (2004) Åhman (2013) Allahyari et al. (2016) Rasmussen et al. (2017) Mills et al. (2021)
4	Employment and labor conditions	Traditionally validated indicators such as the level of employment or the mean age of the worker, as well as more recent indicators related to gender parity in agribusiness, freedom of association, or the measurement of labor mobility levels within the sector.	Anastasova-Chopeva (2019) Prasara-A and Gheewala (2021)

Table 1 (continued)

Cluster	Designation	Scope	Sources
5	Equity and social justice	Entails the largest number of indicators worked on and used in literature (along with Cluster 4), covering aspects such as social security issues and concerns related to transparency, inclusion, fair competition, non-discrimination, gender equality, and recognition of "Women Empowerment."	Mckenzie (2004) Vallance et al. (2011) Rasmussen et al. (2017) Röös et al. (2019) Landert et al. (2020)
6	Health and safety	Basilar theme in any model for development assessment, Health and Safety are transversal to all social evaluation conditions. Its scope spans from traditional topics covering the availability of Health services, to new assessments of ease of access to these services; from hard topics such as life expectancy to more emergent assessments of safety and health-promoting activities.	Romero et al. (2020) Singh and Nayak (2020) St Flour and Bokhoree (2021)
7	Living conditions and wellbeing	Range of indicators have evolved with the reinforcement of new welfare concerns, such as the existence of community amenities (e.g., cultural centers, libraries, museums, among others), or the existence of functional forms of waste treatment and pollution monitoring. Most of these indicators are grounded in a geographical context.	Weingaertner and Moberg (2014) Doğu and Aras (2019) Romero et al. (2020) Dos Santos and Guarnieri (2021)
8	Productivity and food security	Mirroring the evolution of societal and business interests, these cover metrics such as quantities, productivity ratios, the quality of products, the financial survival of the producer, and the farmers' sense of non-privation, the evaluation of food availability and the assessment of food quality.	Van Cauwenbergh et al. (2007) Singh and Nayak (2020) Mouratiadou et al. (2021)

Source: Authors' elaboration, adapted from Massuça et al. (2023)

and Olakitan Atanda (2019) or even by the main groups in the Sola model of Pieper et al. (2019).

The following step involved an empirical methodology for selecting the most frequent indicators from the resulting Excel list and convoluted in the creation of "PivotTables." Validation was then performed using lookup and reference formulas ("Sort") and statistical formulas ("Countif") on the data and identify the most frequent variable, based on objective criteria.

Several methodologies were considered to identify the most frequent indicators, namely word clouds, other data visualizations, and natural language processing techniques. Given the size of the list, the available resources, and the objectives of the analysis, we decided to start by making a frequency count in Excel followed by pivot charts to get a first list of the most referenced indicators in each of the eight clusters identified.

Table 2 Distinction between traditional and emerging social sustainability key themes and their indicators characteristics

	Traditional	Emerging
Key themes description	. Basic needs, including housing and environmental health	. Demographic change (aging, migration, and mobility)
	. Education and skills	. Social mixing and cohesion
	. Employment	. Identity, sense of place, and culture
	. Equity	. Empowerment, participation, and access
	. Human rights and gender	. Health and safety
	. Poverty	. Social capital
	. Social justice	. Wellbeing, happiness, and quality of life
Indicators's characteristics	. Static	. Intergenerational and incorporating uncertainty
	. Predominantly quantitative	. Hybrid (quali/quanti)
	. Descriptive	. Strategic
	. Mono-dimensional	. Multidimensional
	. Target oriented	. Principles and objective driven
	. Top-down selection	. Deliberative and reiterative selection
	. Product	. Process

Source: Adapted from Andrea Colantonio (2011)

4 Results

The dataset used in our analysis, gathered from Massuça et al. (2023), comprised very few publications up until 2008, followed by an increasing trend in a number of studies in the years that followed. Both the increase in scientific output and the diversification of the publishing platforms reflect the increasing recognition of the multidisciplinary nature of sustainability science. The theme of "social indicators" is noticeably the latest and most dynamic tendency among those relating to the social sustainability dimension in agriculture, according to an examination of the authors' keyword evolution (Massuça et al., 2023).

Following the authors, the Rio + 20 Summit, which was held in June of 2012, may have stimulated efforts to encourage social engagement toward the development and implementation of sustainability initiatives, which, in turn, stimulated scientific research into social sustainability, social sustainability indicators, and social sustainability assessment frameworks. This may account for the identifiable overall increase in thematic publications since the referenced summit year.

In evaluating the themes and indicators of the selected literature, Massuça et al. (2023) note the lack of standardization of definitions, domains, and indicators for assessing social sustainability. However, the authors also point out that it is no longer reasonable to argue that the social dimension of sustainability is being ignored by investigators.

The analysis also identifies an expanding pool of social indicators covering an increasingly wide range of social issues. By separately interpreting and categorizing the social sustainability indicators mentioned in the selected papers into eight thematic clusters (Massuça et al., 2023), we were able to show trends in the scientific debate.

Using the Massuça et al. (2023) database of publications, we were able to create a preliminary database of 1249 distinct social indicators that were most commonly researched in the context of agribusiness by year of publication. The thematic clustering exercise applied to these indicators resulted in the matrix depicted in Table 3, summarizing the number of indicators collected from each article (both numerically and graphically). Gathered 193 indicators in Cluster#1, 73 in Cluster#2, 165 in Cluster#3, 280 in Cluster#4, 242 in Cluster#5, 63 in Cluster#6, 156 in Cluster#7, and 77 in Cluster#8. These numbers represent the sum of the indicators employed by the different authors in their research, and although there may be some similarities between the indicators, their definitions vary, just as the context of each research work varies.

Ninety-one different authors are listed, bringing research that covers different agribusiness contexts and ecosystems, using different methodologies used for the assessment of social sustainability, as well as distinct frameworks guiding their evaluation and, consequently, the selection of the indicators. Nevertheless, all identified indicators clearly regard social matters of sustainability.

Two additional aspects are noteworthy: (i) 2021 was the most prolific year in terms of number of publications (20 articles selected, covering 258 social sustainability indicators) and (ii) few authors have more than one work on the list (Zhen in 2003 and 2009, and ul Haq in 2017 and 2020).

Table 3 shows also articles that focus on indicators belonging to only one or two clusters while, at the opposite extreme, there are articles that explore indicators in seven or eight clusters simultaneously, in accordance with the breadth of the research topic.

Despite these differences, Clusters 4 and 5 ("Employment and labour conditions" and "Equity and social justice," respectively) hold the indicators that are most frequently mentioned throughout the selected literature. Two possible reasons for this have been suggested by Massuça et al. (2023): (i) Cluster 4 is where the most "traditional" indicators are found since they help define a grounded view of the problem and (ii) Cluster 5 embraces the most purposeful and straightforward socio-economic indicators that are key to a more immediate analysis of social sustainability of any production ecosystem (simply put, profitability, financial fairness and trade transparency translate into sustainability for the stakeholders), and even more interesting in this cluster are the new emergent sub-themes of non-discrimination, gender equality and recognition of "Women Empowerment."

In contrast, Clusters 2 and 6 ("Demographic Dynamics" and "Health and Safety," respectively) have the least number of indicators. The specificity and technical expertise involved in both themes may, to some degree, be an explanation, but

Table 3 Number of social indicators per reference and cluster, ordered by Author and date

	Cluster 1	Cluster 2	Cluster 3	Cluster 4	Cluster 5	Cluster 6	Cluster 7	Cluster 8	Total
Yli-Viikari (1999)	1				1				**2**
Zhen and Routray (2003)	3		3	1	7		1	3	**18**
Lehtonen et al. (2005)				2					**2**
Munda (2005)				1	1		4		**6**
Adams and Ghaly (2006)	2		1	1	1		2		**7**
Hua-jiao et al. (2007)		1		1	5			1	**8**
Van Cauwenbergh et al. (2007)	3		2	1	3	1	4	6	**20**
Alkan Olsson et al. (2009)	1	5	2	1			4		**13**
Zhen et al. (2009)		3	3		5	2	2		**15**
Ahmadvand et al. (2011)	2				2		2		**6**
Manos et al. (2011)		1	4	7		1			**13**
Vilei (2011)	2	1	3	1	5	1	2	4	**19**
Bacon et al. (2012)					3	1			**4**
Atanga et al. (2013)					4				**4**
Kadir and Jamaludin (2013)	2	1		1	5	1	3		**13**
Liu and Zhang (2013)					2			1	**3**
Ochieng et al. (2013)				4	2	1	4		**11**
Vasileiadis et al. (2013)	2		2	4	2				**10**
Jamshidi et al. (2014)	1		3	1			1		**6**
Manara and Zabaniotou (2014)	1		1	2	2		1	1	**8**
Reddy et al. (2014)	16	1	3	2	12	2	11		**47**
Yegbemey et al. (2014)	2						2	1	**5**
Bélanger et al. (2015)	6	4	5	3			1		**19**
Chand et al. (2015)				4	1				**5**
Venkatesh (2015)			1		1		4		**6**
Allahyari et al. (2016)	4	3	8	5	4		2	3	**29**

Dabkiene (2016)	1	5	2	10	1		2		**21**
Gaviglio et al. (2016)	2		3	2	1				**8**
Haileslassie et al. (2016)			1		3		1	1	**6**
Jane Dillon et al. (2016)		1	1	1	1		1		**5**
Rasouli and Kumarasuriyar (2016)	7		2	1	6	1	3	1	**21**
Schader et al. (2016)	1		1	5	6	1	1	1	**16**
Kühnen and Hahn (2017)	10	1	1	9	5	1	3		**30**
Rasmussen et al. (2017)	8		11	5	14	3	6		**47**
ul Haq and Boz (2017)	6	4	4	9	4	1	5	5	**38**
Ajmal et al. (2018)	3		3	2	6	2			**16**
De Luca et al. (2018)	1				2	1	1		**5**
Dietz et al. (2018)	1			3	1				**5**
Escribano et al. (2018)	4	3	3	9	1	1	2	3	**26**
Hossain et al. (2018)	1		1	4	1	1	1		**9**
Lijó et al. (2018)				1			3		**4**
Ludden et al. (2018)					3				**3**
Mutyasira et al. (2018)	1			1	1		1		**4**
Pelletier (2018)	3		1	6	3	1	2		**16**
Sokil et al. (2018)			1	6	3			2	**12**
Anastasova-Chopeva (2019)	4	6	2	18	3				**33**
Bachev et al. (2019)	2	3	3	6	2				**16**
Martínez et al. (2019)	3		4		1	1	3		**12**
Doğu and Aras (2019)	11		2				8		**21**
Martucci et al. (2019)	6	1	2	11	6	2	2		**30**
Mastronardi et al. (2019)				1	3		1	3	**8**
Mili and Martínez-Vega (2019)		1	1	1			1		**4**

(continued)

Table 3 (continued)

	Cluster 1	Cluster 2	Cluster 3	Cluster 4	Cluster 5	Cluster 6	Cluster 7	Cluster 8	Total
Mollayosefi et al. (2019)	2	1	3	1		1	1		**9**
Munyaneza et al. (2019)	2	1	2	5	3		2		**15**
Pieper et al. (2019)	4		1	2	4	2	3		**16**
Röös et al. (2019)	5		3	13	9	4	4	3	**41**
Ataei et al. (2020)	3	1	1	1	1	1		1	**9**
Boimah et al. (2020)	2	5	2	9	1		1	3	**23**
Guliyeva and Lis (2020)	3		1	4	2			2	**12**
Heredia-R et al. (2020)	2		3	4	2	2			**13**
Hilde et al. (2020)	1		1	6	6		1	1	**16**
Krishna et al. (2020)								1	**1**
Landert et al. (2020)	1		1	6	5		1	1	**15**
Machado et al. (2020)			2			1	5		**8**
Paul et al. (2020)			4	1	1				**6**
Romero et al. (2020)	1	1	2		2	1	8	1	**16**
Dos Santos et al. (2020)	1					1	2	1	**5**
Singh and Nayak (2020)		1	2	1	1	7		6	**18**
Stylianou et al. (2020)		4	3	4	1		3	1	**16**
Triviño-Tarradas et al. (2020)		1	1	2					**4**
ul Haq and Boz (2020)	2	2	1		2				**7**
Beşen and Olhan (2021)	2	2	1		1		1		**7**
Chopin et al. (2021)	4	1		1	4	1	1	2	**14**
Corvo et al. (2021)	4		1	1	3				**9**
Gosnell et al. (2021)	1		1		2	1			**5**
Hammond et al. (2021)	2		4		3			3	**12**
Havardi-Burger et al. (2021)	1		1	6	1	1			**10**
Herrera Sabillón et al. (2021)	1			3					**4**

Iocola et al. (2021)	5		3	3	1				**12**
Kowaleski et al. (2021)	2		7	1					**10**
Mills et al. (2021)	9	1	11	1	2				**24**
Mouratiadou et al. (2021)	1	3		2	4	3	4	7	**24**
Pereira et al. (2021)	1		1	11	5	1		1	**20**
Prasara-A and Gheewala (2021)	1	1	1	15	3	1	2		**24**
Roesch et al. (2021)				1			1		**2**
Sandhu (2021)	1				3	1	1	3	**9**
Dos Santos and Guarnieri (2021)	2		2	7	4	1	11		27
St Flour and Bokhoree (2021)	2	1	3		7	6	6		25
Torkayesh et al. (2021)		1	4	4	5				14
Yakar Pritchard and Çalıyurt (2021)	1			1			1		3
Zhao et al. (2021)				2	1				3
Desiderio et al. (2022)	1		2	4	6	1		2	16
Devkota et al. (2022)				2	1			2	5
Volkov et al. (2022)		1		3	1				5
Grand Total	**193**	**73**	**165**	**280**	**242**	**63**	**156**	**77**	**1249**

Source: Authors' elaboration

further investigation is required on this matter. Obtaining reliable data for these indicators represents a challenge, and the required methodological harmonization with administrative data producers (to allow comparisons and cross analysis) adds to the challenge.

Overall, and regardless of the different breadths in terms of a number of clusters covered and indicators used for each cluster, the scientific production on the social sustainability theme has been steadily increasing in the last years. There has also been an increase in the number and variety of indicators used to investigate this topic.

Also, as described by Massuça, et al. (2023), the number of thematic clusters observed in each publication follows a symmetric unimodal distribution, with the most frequent number of clusters being 4 to 5 (34 articles). Only 12 articles focus on indicators belonging to only one or two clusters while, at the opposite end, there are 17 articles that explore indicators in seven or eight clusters simultaneously.

Further to this analysis, we have also looked for changes in the distribution of indicators per cluster in two different periods: "up to 2012" and "since 2013," so to explore possible differences in social sustainability assessment beliefs prior and post Rio +20 Summit held in 2012. This is relevant because Rio + 20 Summit has been identified by Massuça et al. (2023) to have been a catalyst for an overall increase in social sustainability research publications. Figure 1 illustrates the changing patterns within each thematic cluster over the two periods, emphasizing the increasing focus on the wellbeing of individuals and society. Specifically, there has been a rise in concerns related to work-related satisfaction, civic participation, and health and safety in recent years.

By categorizing each indicator as either "Traditional" or "Emerging," we discovered growing concerns and also found that even established themes like "Equity and Social Justice" incorporate a considerable number of emerging indicators. Notably, the indicators related to Demographic Dynamics, Community Relations/Participation/Engagement, and Health and Safety represent current issues that require monitoring and evaluation but were not necessarily addressed in previous sustainability assessment models. Productivity and Food Security, however, are firmly grounded in widely accepted conventional indicators. This analysis demonstrates the existence of a growing collection of social indicators that address a variety of social issues. Colantonio (2007) observed that traditional social sustainability themes like employment, social justice, or poverty reduction are gradually giving way to more ephemeral themes like civic participation, access, quality of life (satisfaction), and wellbeing. This gradual shift is a result of efforts on the part of researchers, planners, and practitioners to "address and integrate social aspects of sustainability" (Boström, 2012, p. 3) (Fig. 2).

Further to the previous analysis, the following phase consisted of identifying the five most mentioned indicators in each of the eight clusters. The objective was to identify the most suitable ground for further research. The resulting list is observable in Table 4, which shows the five most frequently referenced Social Indicators in each cluster. The balance between Traditional and Emerging topics is one of the main findings, revealing the complementarity of both dimensions.

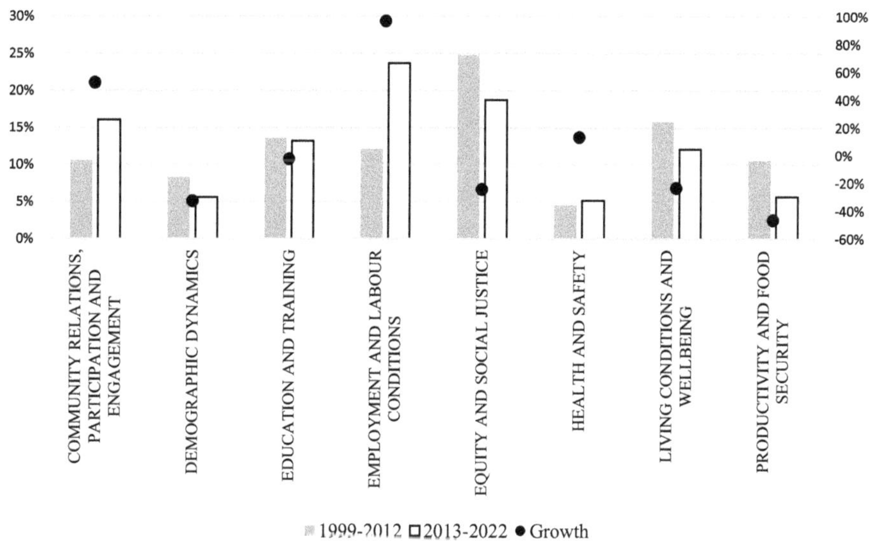

Fig. 1 Distribution of social indicators per Cluster and Period, and respective growth (secondary axis)

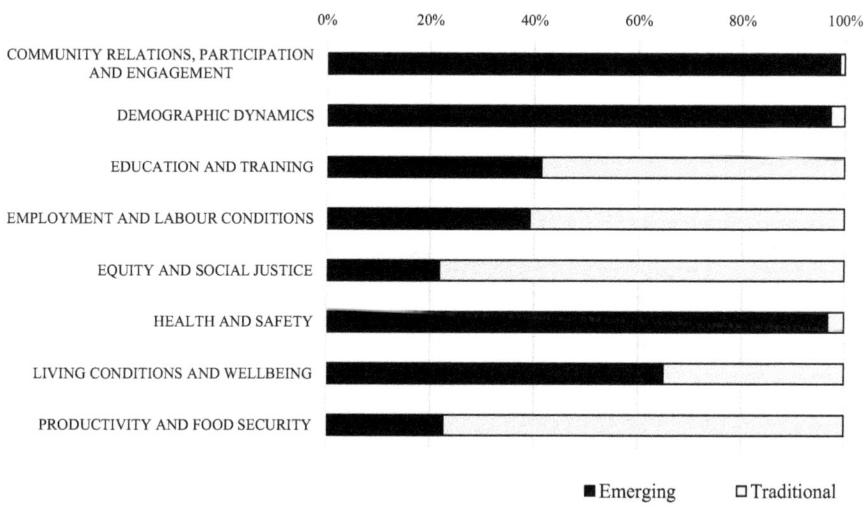

Fig. 2 Percentual distribution of emerging and traditional social indicators per Cluster

This shows that the gradual shift (previously identified) from traditional to emerging indicators might have achieved an equilibrium at this stage and for that reason it is important and relevant to keep monitoring the dynamics of these two timeframe perspectives in future analysis and characterization of the achieved results, and during the design stage of new social sustainability assessment models. Moreover, Cluster 4 (Employment and Labour Conditions) remains more

Table 4 Top five most referenced Social Indicators in each Cluster

Clusters	Social Indicators	Traditional	Emerging
1 – Community relations, participation, and engagement	• Community engagement	X	
	• "Indigenous" knowledge		X
	• Participation in decision-making	X	
	• Social involvement and inclusion		X
	• Transparency		X
2 – Demographic dynamics	• Farmers age	X	
	• Number of family workers	X	
	• Continuity of farming activity		X
	• Migration		X
	• Population density	X	
3 – Education and training	• Adult education level		X
	• Agricultural training	X	
	• Access to technical support and rural extension	X	
	• Level of computer use		X
	• Technology development		X
4 – Employment and labor conditions	• Agricultural employment rate	X	
	• Equal opportunities		X
	• Local employment	X	
	• Fair salary	X	
	• Working hours	X	
5 – Equity and social justice	• Equitable income	X	
	• Fair competition	X	
	• Gender equality		X
	• Non-discrimination		X
	• Social security	X	
6 – Health and safety	• Existence of Healthcare and Safety	X	
	• Access to Healthcare and Safety		X
	• Life expectancy	X	
	• Health index		X
	• Safety and health training/promoting activities		X
7 – Living conditions and wellbeing	• Quality of life (satisfaction).		X
	• Access to water, electricity, and hygienic services	X	
	• Housing quality		X
	• Satisfaction with amenities availability (cultural, sport, housing, transportation, etc.)		X
	• Risk perception		X
8 – Productivity and food security	• Access to food	X	
	• Auto-consumption rate		X
	• Quality of food improvements		X
	• Agricultural productivity • Farm expansion	X X	

Source: Authors' elaboration

traditional, in contrast with Cluster 7 (Living Conditions and Wellbeing) which is clearly more emergent.

Overall, both traditional and emerging social indicators are important in assessing the impact of agribusiness on workers, communities, and the environment. Tracking these indicators over time, it will be of great help to identify trends and areas of improvement in agribusiness operations and will support the development of strategies to promote and support sustainable and socially responsible practices in the agribusiness ecosystems.

5 Discussion and Remarks

Agribusiness is a complex and multifaceted activity that involves the production, processing, and distribution of agricultural products. Social sustainability indicators are important measures that provide insight into the social aspects of agribusiness, including its impact on communities, workers, and consumers.

Despite the multiple definitions of social sustainability found in the literature review, there is relevant common ground in the discussions, namely the fact that employment, food security, demographics, and social justice (i.e., basic needs and wellbeing) are consistently considered as starting point in the traditional definitions of social sustainability. Subsequent divergences show that some authors also address wider development goals that involve scales beyond the initial basic needs and wellbeing (Vallance et al., 2011), meaning that those cover the relationship between human communities and their environment, and then they consider an even wider scale where preservation of culture and tradition together with community relations and engagement represent the principal assessment goal. Future investigation in this field must cover the successive widening of the scope of analysis, considering interdependencies and possible precedency links between factors.

Overall, both emerging and traditional social indicators play a critical role in measuring the social impact of agribusinesses. Our analysis shows how both these dimensions are present in all the eight thematic clusters defined previously by Massuça et al. (2023). While traditional social indicators have been in use for a long time, emerging social indicators are becoming increasingly important as consumers and stakeholders demand more transparency and accountability for the sector.

Traditional social indicators in agribusiness include (i) employment (number of jobs in agribusinesses), (ii) income (both for the farmer and for workers in the agricultural sector), and (iii) food security (the basic access to affordable and nutritious food), ultimately aiming at poverty reduction in rural areas where agribusinesses operate. Complementarily, Emerging social indicators comprise wider themes such as (i) social responsibility (translating the level of commitment of agribusinesses toward their impact on society, which includes ethical business practices, transparency in operations, and sustainable practices), (ii) human capital development (the level of investment made by agribusinesses toward the development of their employees, which includes training, education, and skill development), (iii) community development (agribusinesses commitment toward the development of the communities in which they operate, which includes supporting local businesses, creating jobs, and providing social amenities), and finally (iv) stakeholder engagement (representing agribusinesses involvement with their stakeholder network, which includes farmers, consumers, employees, and communities, in decision-making processes).

Notably, both traditional and emerging social themes are important in assessing the impact of agribusiness demands on workers, communities, and the environment. Thus, indicators covering both dimensions should be tracked over time, allowing the early identification of trends and areas of improvement in agribusiness operations, so to inform the development of strategies for promoting sustainable and socially responsible practices in the industry. Moreover, as highlighted by Chopin et al. (2021), besides this dynamic monitoring, both the relevance and reliability of sustainability assessment tools depend largely on the active involvement of stakeholders in the design of indicators.

The current work seeks to set the foundation for further empirical research toward defining an evidence-based set of core indicators deemed pertinent for the evaluation of social sustainability. Next, we propose applying a participatory, decision-making, multi-stage methodology to determine the most significant social sustainability indicators for each identified core cluster.

6 Conclusions

Together traditional and emerging social themes are important in assessing the impact of agribusiness demands on workers, communities, and the environment. Thus, indicators covering both dimensions should be tracked over time, allowing the early identification of trends and areas of improvement in agribusiness operations, so to inform the development of strategies for promoting sustainable and socially responsible practices in the industry. Moreover, as highlighted by Chopin et al. (2021), in addition to this suggestion of constant monitoring, the effectiveness and credibility of sustainability assessment instruments are largely dependent on stakeholder engagement in selecting and weighing the indicators to be included in the model.

For that, we intend to use a participatory approach to engage with agribusiness stakeholders. Previous evidence from consultation-based assessments (such as RISE) and other assessments based on a more active engagement of stakeholders in the selection of indicators, can be used as a reference. We are planning a mixed methods approach consisting of a structured survey to identify stakeholders, followed by workshops with those stakeholders in which the Delphi Method and the Hierarchical Analysis Process (AHP) will be applied to prioritize and rank social sustainability indicators within each cluster.

This ranking process and weighting of the indicators selected it is of most importance regarding their future use in the definition of a new model for social sustainability assessment in the context of agribusiness.

This weighing, if correctly carried out, will allow the elaboration of an assessment model that truly reflects the interests and knowledge of the different stakeholders in the sector. To achieve this weighted "ranking" of the selected indicators it will be essential to involve and consult in a participatory manner the widest possible spectrum of stakeholders in the agribusiness to be evaluated.

Engaging with farmers, processors, consumers, and policymakers will likely provide value-added information on the social sustainability challenges faced by the agroindustry and help refine future agribusiness sustainability assessment models. All players in the fields of agribusiness entrepreneurship, the social and territorial cohesion governance, and surely the implementation of agribusiness sustainability policies will benefit from this work.

By updating and improving certain indicators (both traditional and emerging) and identifying which are more suitable for application in specific agribusiness models, it will be possible to a more rigorous and pertinent sustainability assessment. This will provide knowledge to the producer and support the decision-making process of the Entrepreneur toward the global impact on business, particularly at the level of implementing new technologies, adopting functional policies of Circular Economy, or even investing in a Green Transition.

Acknowledgments This work was supported in the frame of the R&D project "Vine & Wine PT - Driving Sustainable Growth Through Smart Innovation," with the financial support of the PRR—Recovery and Resilience Plan and by the European NextGeneration EU Funds, within the scope of the "Agendas Mobilizadoras para a Reindustrialização," and by FCT—Fundação para a Ciência e Tecnologia under the projects UIDB/04011/2020; UIDB/04007/2020.

References

Adams, M. A., & Ghaly, A. E. (2006). An integral framework for sustainability assessment in agro-industries: Application to the Costa Rican coffee industry. *International Journal of Sustainable Development and World Ecology, 13*(2), 83–102. https://doi.org/10.1080/13504500609469664

Ahmadvand, M., Karami, E., & Iman, M. T. (2011). Modeling the determinants of the social impacts of agricultural development projects. *Environmental Impact Assessment Review, 31*(1), 8–16. https://doi.org/10.1016/j.eiar.2010.06.004

Åhman, H. (2013). Social sustainability – society at the intersection of development and maintenance. *Local Environment, 18*(10), 1153–1166. https://doi.org/10.1080/13549839.2013.788480

Ajmal, M. M., Khan, M., Hussain, M., & Helo, P. (2018). Conceptualizing and incorporating social sustainability in the business world. *International Journal of Sustainable Development and World Ecology, 25*(4), 327–339. https://doi.org/10.1080/13504509.2017.1408714

Alkan Olsson, J., Bockstaller, C., Stapleton, L. M., Ewert, F., Knapen, R., Therond, O., Geniaux, G., Bellon, S., Correira, T. P., Turpin, N., & Bezlepkina, I. (2009). A goal oriented indicator framework to support integrated assessment of new policies for Agri-environmental systems. *Environmental Science and Policy, 12*(5), 562–572. https://doi.org/10.1016/j.envsci.2009.01.012

Allahyari, M. S., Daghighi Masouleh, Z., & Koundinya, V. (2016). Implementing Minkowski fuzzy screening, entropy, and aggregation methods for selecting agricultural sustainability indicators. *Agroecology and Sustainable Food Systems, 40*(3), 277–294. https://doi.org/10.1080/21683565.2015.1133467

Anastasova-Chopeva, M. (2019). Evaluation of social sustainability of bulgarian agriculture. *Bulgarian Journal of Agricultural Science, 25*(6), 1092–1098.

Ataei, P., Sadighi, H., Chizari, M., & Abbasi, E. (2020). In-depth content analysis of conservation agriculture training programs in Iran based on sustainability dimensions. *Environment, Development and Sustainability, 22*(8), 7215–7237. https://doi.org/10.1007/s10668-019-00484-4

Atanga, N. L., Treydte, A. C., & Birner, R. (2013). Assessing the sustainability of different small-scale livestock production systems in the Afar region. *Ethiopia. Land, 2*(4), 726–755. https://doi.org/10.3390/land2040726

Bachev, H., Ivanov, B., & Toteva, D. (2019). Assessment of sustainability of agro-ecosystems in Bulgaria. *Bulgarian Journal of Agricultural Science, 25*(4), 607–624.

Bacon, C. M., Getz, C., Kraus, S., Montenegro, M., & Holland, K. (2012). The social dimensions of sustainability and change in diversified farming systems. *Ecology and Society, 17*(4). https://doi.org/10.5751/ES-05226-170441

Bedeau, J. V., Rezaei, M., Pera, M., & Morrison, J. (2021). Innovation and sustainability of Agrifood system in the Mediterranean area. Mediterranean Journal of Economics, Agriculture and Environment, September.

Bélanger, V., Vanasse, A., Parent, D., Allard, G., & Pellerin, D. (2015). DELTA: An integrated indicator-based self-assessment tool for the evaluation of dairy farms sustainability in Quebec. *Canada. Agroecology and Sustainable Food Systems, 39*(9), 1022–1046. https://doi.org/10.1080/21683565.2015.1069775

Beşen, T., & Olhan, E. (2021). Assessment of agricultural sustainability in Sarikum lake basin, Sinop Province, Turkey. *Tarim Bilimleri Dergisi, 27*(1), 42–49. https://doi.org/10.15832/ankutbd.547463

Biart, M. (2002). Social sustainability as part of the social agenda of the European community, in Ritt, T. (Ed.): Soziale Nachhaltigkeit: Von der Umweltpolitik zur Nachhaltigkeit? Arbeiterkammer Wien, Informationen zur Umweltpolitik 149, Wien, pp.5–10.

Binder, C. R., Feola, G., & Steinberger, J. K. (2010). Considering the normative, systemic and procedural dimensions in indicator-based sustainability assessments in agriculture. *Environmental Impact Assessment Review, 30*(2), 71–81. https://doi.org/10.1016/j.eiar.2009.06.002

Boimah, M., Mensah-Bonsu, A., Osei-Asare, Y., & Sarpong, D. B. (2020). Measuring sustainability of conservation and conventional practices in maize production in Ghana. *Journal of Agriculture and Environment for International Development, 114*(2), 21–40. https://doi.org/10.12895/jaeid.20202.1222

Bonisoli, L., Galdeano-Gómez, E., & Piedra-Muñoz, L. (2018). Deconstructing criteria and assessment tools to build Agri-sustainability indicators and support farmers' decision-making process. *Journal of Cleaner Production, 182*, 1080–1094. https://doi.org/10.1016/j.jclepro.2018.02.055

Boström, M. (2012). A missing pillar? Challenges in theorizing and practicing social sustainability: Introduction to the special issue. *Sustainability: Science, Practice, and Policy, 8*(1), 3–14. https://doi.org/10.1080/15487733.2012.11908080

Boyer, R. H. W., Peterson, N. D., Arora, P., & Caldwell, K. (2016). Five approaches to social sustainability and an integratedway forward. *Sustainability (Switzerland), 8*(9), 878. https://doi.org/10.3390/su8090878

Chand, P., Sirohi, S., & Sirohi, S. K. (2015). Development and application of an integrated sustainability index for small-holder dairy farms in Rajasthan, India. *Ecological Indicators, 56*, 23–30. https://doi.org/10.1016/j.ecolind.2015.03.020

Chopin, P., Mubaya, C. P., Descheemaeker, K., Öborn, I., & Bergkvist, G. (2021). Avenues for improving farming sustainability assessment with upgraded tools, sustainability framing and indicators. A review. *Agronomy for Sustainable Development, 41*(2), 1–20. https://doi.org/10.1007/s13593-021-00674-3

Colantonio, A. (2007). Social sustainability: An exploratory analysis of its definition, assessment methods metrics and tools. *EIBURS working paper series, July 2007*, 1–37. http://www.eprints.lse.ac.uk/35947/

Colantonio, A. (2009). Social sustainability: A review and critique of traditional versus emerging themes and assessment methods. Sue-mot conference 2009: Second international conference on whole life urban sustainability and its assessment, 865–885. http://eprints.lse.ac.uk/35867/

Colantonio, A. (2011). Social sustainability: Exploring the linkages between research, Policy and Practice. 35–57. https://doi.org/10.1007/978-3-642-19202-9_5.

Corvo, L.; Pastore, L.; Antonelli, A.; Petruzzella, D. (2021). Social impact and sustainability in short food supply chains: An experimental assessment tool. New Medit, September.

Cuthill, M. (2010). Strengthening the "social" in sustainable development: Developing a conceptual framework for social sustainability in a rapid urban growth region in Australia. *Sustainable Development, 18*(6), 362–373. https://doi.org/10.1002/sd.397

Dabkiene, V. (2016). The scope of farms sustainability tools based on Fadn data. *Scientific Papers-Series Management Economic Engineering in Agriculture and Rural Development, 16*(1), 121–128.

de Luca, A. I., Falcone, G., Stillitano, T., Iofrida, N., Strano, A., & Gulisano, G. (2018). Evaluation of sustainable innovations in olive growing systems: A life cycle sustainability assessment case study in southern Italy. *Journal of Cleaner Production, 171*, 1187–1202. https://doi.org/10.1016/j.jclepro.2017.10.119

de Olde, E. M., Bokkers, E. A. M., & de Boer, I. J. M. (2017). The choice of the sustainability assessment tool matters: Differences in thematic scope and assessment results. *Ecological Economics, 136*, 77–85. https://doi.org/10.1016/j.ecolecon.2017.02.015

Desiderio, E., García-Herrero, L., Hall, D., Segrè, A., & Vittuari, M. (2022). Social sustainability tools and indicators for the food supply chain: A systematic literature review. *Sustainable Production and Consumption, 30*, 527–540. https://doi.org/10.1016/j.spc.2021.12.015

Devkota, K. P., Beebout, S. E. J., & Sudhir-Yadav, & Bunquin, M. A. (2022). Setting sustainability targets for irrigated rice production and application of the sustainable Rice platform performance indicators. *Environmental Impact Assessment Review, 92*(October 2021), 106697. https://doi.org/10.1016/j.eiar.2021.106697

Dietz, T., Auffenberg, J., Estrella Chong, A., Grabs, J., & Kilian, B. (2018). The voluntary coffee standard index (VOCSI). Developing a composite index to assess and compare the strength of mainstream voluntary sustainability standards in the global coffee industry. *Ecological Economics, 150*(April), 72–87. https://doi.org/10.1016/j.ecolecon.2018.03.026

Doğu, F. U., & Aras, L. (2019). Measuring social sustainability with the developed MCSA model: Güzelyurt case. *Sustainability (Switzerland), 11*(9), 7–11. https://doi.org/10.3390/su11092503

dos Santos, L. P., Schmidt, C. M., & Mithöfer, D. (2020). Impact of collective action membership on the economic, social and environmental performance of fruit and vegetable farmers in Toledo, Brazil. *Journal of Co-Operative Organization and Management, 8*(1), 100107. https://doi.org/10.1016/j.jcom.2020.100107

dos Santos, R. R., & Guarnieri, P. (2021). Social gains for artisanal agroindustrial producers induced by cooperation and collaboration in Agri-food supply chain. *Social Responsibility Journal, 17*(8), 1131–1149. https://doi.org/10.1108/SRJ-09-2019-0323

Elkington, J. (1994). Towards the sustainable corporation: Win-win-win business strategies for sustainable development. *Corporate Environmental Responsibility, 36*, 109–119.

Elkington, J. (2018). 25 years ago I coined the phrase triple bottom line. Here's why It's time to rethink it. Harvard Business Review, 1–5. https://hbr.org/2018/06/25-years-ago-i-coined-the-phrase-triple-bottom-line-heres-why-im-giving-up-on-it

Escribano, M., Díaz-Caro, C., & Mesias, F. J. (2018). A participative approach to develop sustainability indicators for dehesa agroforestry farms. *Science of the Total Environment, 640–641*, 89–97. https://doi.org/10.1016/j.scitotenv.2018.05.297

Gaviglio, A., Bertocchi, M., Marescotti, M. E., Demartini, E., & Pirani, A. (2016). The social pillar of sustainability: A quantitative approach at the farm level. *Agricultural and Food Economics, 4*(1), 1–19. https://doi.org/10.1186/s40100-016-0059-4

Goodman, D., & Watts, M. (1997). *Globalising food: agrarian questions and global restructuring*. Routledge.

Gosnell, H., Emard, K., & Hyde, E. (2021). Taking stock of social sustainability and the u.S. beef industry. *Sustainability (Switzerland), 13*(21), 11860. https://doi.org/10.3390/su132111860

Guliyeva, A. E., & Lis, M. (2020). Sustainability management of organic food organizations: A case study of Azerbaijan. *Sustainability (Switzerland), 12*(12), 1–20. https://doi.org/10.3390/su12125057

Haileslassie, A., Craufurd, P., Thiagarajah, R., Kumar, S., Whitbread, A., Rathor, A., Blummel, M., Ericsson, P., & Kakumanu, K. R. (2016). Empirical evaluation of sustainability of divergent farms in the dryland farming systems of India. *Ecological Indicators, 60*, 710–723. https://doi.org/10.1016/j.ecolind.2015.08.014

Hammond, J., van Wijk, M., Teufel, N., Mekonnen, K., & Thorne, P. (2021). Assessing smallholder sustainable intensification in the Ethiopian highlands. *Agricultural Systems, 194*(April), 103266. https://doi.org/10.1016/j.agsy.2021.103266

Havardi-Burger, N., Mempel, H., & Bitsch, V. (2021). Framework for sustainability assessment of the value chain of flowering potted plants for the German market. *Journal of Cleaner Production, 329*, 129684. https://doi.org/10.1016/j.jclepro.2021.129684

Heredia-R, M., Torres, B., Cayambe, J., Ramos, N., Luna, M., & Diaz-Ambrona, C. G. H. (2020). Sustainability assessment of smallholder agroforestry indigenous farming in the amazon: A case study of ecuadorian kichwas. *Agronomy, 10*(12), 1–25. https://doi.org/10.3390/agronomy10121973

Herrera Sabillón, B., Gerster-Bentaya, M., & Knierim, A. (2021, August). Measuring farmers' Well-being: Influence of farm-level factors on satisfaction with work and quality of life. Journal of Agricultural Economics, 73, 1–20. https://doi.org/10.1111/1477-9552.12457.

Hilde, W., Thielen, M., & Messeley, L. (2020). *Review of current metrics, indicators and tools for monitoring the socio-economic performance of FAB solutions. FABulous Farmers*.

Hossain, M. U., Poon, C. S., Dong, Y. H., Lo, I. M. C., & Cheng, J. C. P. (2018). Development of social sustainability assessment method and a comparative case study on assessing recycled construction materials. *International Journal of Life Cycle Assessment, 23*(8), 1654–1674. https://doi.org/10.1007/s11367-017-1373-0

Hua-jiao, Q. I. U., Wan-bin, Z. H. U., Hai-bin, W., & Xu, C. (2007). Analysis and Design of Agricultural Sustainability Indicators System. *Agricultural Sciences in China, 6*(4), 475–486.

Iocola, I., Colombo, L., Guccione, G. D., De Vita, P., Palumbo, M., Ritunnano, V., Sciacca, F., Virzì, N., & Canali, S. (2021). A multi-criteria qualitative tool for the sustainability assessment of organic durum wheat-based farming systems designed through a participative process. *Italian Journal of Agronomy, 16*(1), 1–15. https://doi.org/10.4081/ija.2021.1785

Jamshidi, O., Asadi, A., & Motiee, N. (2014). Assessment of sustainability of greenhouse culture and identifying factors affecting in alborz province. *Middle – East Journal of Scientific Research, 19*(5), 625–630. https://doi.org/10.5829/idosi.mejsr.2014.19.5.13622

Jane Dillon, E., Hennessy, T., Buckley, C., Donnellan, T., Hanrahan, K., Moran, B., & Ryan, M. (2016). Measuring progress in agricultural sustainability to support policy-making. *International Journal of Agricultural Sustainability, 14*(1), 31–44. https://doi.org/10.1080/14735903.2015.1012413

Janker, J., & Mann, S. (2020). Understanding the social dimension of sustainability in agriculture: A critical review of sustainability assessment tools. *Environment, Development and Sustainability, 22*(3), 1671–1691. https://doi.org/10.1007/s10668-018-0282-0

Kadir, S. A., & Jamaludin, M. (2013). Universal design as a significant component for sustainable life and social development. *Procedia – Social and Behavioral Sciences, 85*(June 2015), 179–190. https://doi.org/10.1016/j.sbspro.2013.08.349

Kowaleski, T., Brizolla, M. M. B., Felipin, R., Hedlund, P. R., Zaluski, F. C., & Baggio, D. K. (2021). Indicadores de avaliação para melhoria do desempenho sustentável em organizações rurais: um estudo de caso. *Holos, 37*(1), 1–17. https://doi.org/10.15628/holos.2021.10228

Krishna, V. R., Paramesh, V., Arunachalam, V., Das, B., Elansary, H. O., Parab, A., Reddy, D. D., Shashidhar, K. S., El-Ansary, D. O., Mahmoud, E. A., & El-Sheikh, M. A. (2020). Assessment of sustainability and priorities for development of indian west coast region: An application of sustainable livelihood security indicators. *Sustainability (Switzerland), 12*(20), 1–19. https://doi.org/10.3390/su12208716

Kühnen, M., & Hahn, R. (2017). Indicators in social life cycle assessment: A review of frameworks, theories, and empirical experience. *Journal of Industrial Ecology, 21*(6), 1547–1565. https://doi.org/10.1111/jiec.12663

Landert, J., Pfeifer, C., Carolus, J., Schwarz, G., Albanito, F., Muller, A., Smith, P., Sanders, J., Schader, C., Vanni, F., Prazan, J., Baumgart, L., Blockeel, J., Weisshaidinger, R., Bartel-Kratochvil, R., Hollaus, A., Mayer, A., Hrabalová, A., Helin, J., et al. (2020). Assessing agro-ecological practices using a combination of three sustainability assessment tools. *Landbauforschung, 70*(2), 129–144. https://doi.org/10.3220/LBF1612794225000

Lehtonen, H., Aakkula, J., & Rikkonen, P. (2005). Alternative agricultural policy scenarios, sector modelling and indicators: A sustainability assessment. *Journal of Sustainable Agriculture, 26*(4), 63–93. https://doi.org/10.1300/J064v26n04_06

Lijó, L., Frison, N., Fatone, F., González-García, S., Feijoo, G., & Moreira, M. T. (2018). Environmental and sustainability evaluation of livestock waste management practices in Cyprus. *Science of the Total Environment, 634*, 127–140. https://doi.org/10.1016/j.scitotenv.2018.03.299

Littig, B., & Griessler, E. (2005). Social sustainability: A catchword between political pragmatism and social theory Beate Littig * and Erich Grießler. *Sustainable Development, 8*(1/2), 65–79.

Liu, F., & Zhang, H. (2013). Novel methods to assess environmental, economic, and social sustainability of main agricultural regions in China. *Agronomy for Sustainable Development, 33*(3), 621–633. https://doi.org/10.1007/s13593-012-0131-8

Ludden, M., Welsh, R., Weissman, E., Hilchey, D., Gillespie, G., & Guptill, A. (2018). The progressive agriculture index: Assessing the advancement of Agri-food systems. *Journal of Agriculture, Food Systems, and Community Development, 8*(3), 1–27. https://doi.org/10.5304/jafscd.2018.083.003

Machado, N. G., Dos Santos, G. T., Biudes, M. S., Da Silva, J. L., Bacarji, A. G., Da Costa, M. E. L., & De Souza Bilio, R. (2020). Sustainable development index of municipalities in Mato Grosso. *Revista Brasileira de Gestao e Desenvolvimento Regional, 16*(1), 222–234.

Maisarahmokti, M., & Hamzah Sharaai, A. (2020). Systematic literature review of social sustainability approach in palm oil industry. *International Journal of Advanced Science and Technology, 29*(9 Special Issue), 1770–1777.

Manara, P., & Zabaniotou, A. (2014). Indicator-based economic, environmental, and social sustainability assessment of a small gasification bioenergy system fuelled with food processing residues from the Mediterranean agro-industrial sector. *Sustainable Energy Technologies and Assessments, 8*, 159–171. https://doi.org/10.1016/j.seta.2014.08.007

Manos, B., Bournaris, T., & Chatzinikolaou, P. (2011). Impact assessment of CAP policies on social sustainability in rural areas: An application in northern Greece. *Operational Research, 11*(1), 77–92. https://doi.org/10.1007/s12351-010-0078-y

Martínez, F. C. B., Cruz, P. Z., & Taco, R. P. (2019). Social sustainability of the agricultural system of hard yellow corn in the Pativilca Valley – Lima. *IDESIA (Chile), 37*, 107–114.

Martucci, O., Arcese, G., Montauti, C., & Acampora, A. (2019). Social aspects in the wine sector: Comparison between social life cycle assessment and VIVA sustainable wine project indicators. *Resources, 8*(2), 69. https://doi.org/10.3390/resources8020069

Massuça, J., Marta-Costa, A., & Lucas, M. R. (2023). Social dimension of sustainability: Assessment in the agribusiness context. *New Medit, 22*(2), 63–80. https://doi.org/10.30682/nm2302e

Mastronardi, L., Marino, D., Giaccio, V., Giannelli, A., Palmieri, M., & Mazzocchi, G. (2019). Analyzing alternative food networks sustainability in Italy: A proposal for an assessment framework. *Agricultural and Food Economics, 7*(1), 1–19. https://doi.org/10.1186/s40100-019-0142-8

Mckenzie, S. (2004). *Working paper series no 27 social sustainability : Towards some definitions University of South Australia* (Vol. 27, pp. 1–31). University of South Australia.

Meul, M., Van Passel, S., Nevens, F., Dessein, J., Rogge, E., Mulier, A., & Van Hauwermeiren, A. (2008). MOTIFS: A monitoring tool for integrated farm sustainability. *Agronomy for Sustainable Development, 28*(2), 321–332. https://doi.org/10.1051/agro:2008001

Mili, S., & Martínez-Vega, J. (2019). Accounting for regional heterogeneity of agricultural sustainability in Spain. *Sustainability (Switzerland), 11*(2), 299. https://doi.org/10.3390/su11020299

Mills, J., Chiswell, H., Gaskell, P., Courtney, P., Brockett, B., Cusworth, G., & Lobley, M. (2021). Developing farm-level social indicators for Agri-environment schemes: A focus on the agents of change. *Sustainability (Switzerland), 13*(14), 1–22. https://doi.org/10.3390/su13147820

Mollayosefi, M. M., Hayati, B., Pishbahar, E., & Nematian, J. (2019). Selecting weighting methodologies for evaluating agricultural sustainability in Iran. *Proceedings of the Institution of Civil Engineers: Engineering Sustainability, 172*(4), 207–217. https://doi.org/10.1680/jensu.18.00004

Mouratiadou, I., Latka, C., van der Hilst, F., Müller, C., Berges, R., Bodirsky, B. L., Ewert, F., Faye, B., Heckelei, T., Hoffmann, M., Lehtonen, H., Lorite, I. J., Nendel, C., Palosuo, T., Rodríguez, A., Rötter, R. P., Ruiz-Ramos, M., Stella, T., Webber, H., & Wicke, B. (2021). Quantifying sustainable intensification of agriculture: The contribution of metrics and modelling. *Ecological Indicators, 129*(July 2020), 107870. https://doi.org/10.1016/j.ecolind.2021.107870

Munda, G. (2005). "Measuring sustainability": A multi-criterion framework. *Environment, Development and Sustainability, 7*(1), 117–134. https://doi.org/10.1007/s10668-003-4713-0

Munyaneza, C., Kurwijila, L. R., Mdoe, N. S. Y., Baltenweck, I., & Twine, E. E. (2019). Identification of appropriate indicators for assessing sustainability of small-holder milk production systems in Tanzania. *Sustainable Production and Consumption, 19*, 141–160. https://doi.org/10.1016/j.spc.2019.03.009

Mutyasira, V., Hoag, D., Pendell, D., Manning, D. T., & Berhe, M. (2018). Assessing the relative sustainability of smallholder farming systems in Ethiopian highlands. *Agricultural Systems, 167*(November 2017), 83–91. https://doi.org/10.1016/j.agsy.2018.08.006

Ochieng, B. O., Hughey, K. F. D., & Bigsby, H. (2013). Rainforest Alliance certification of Kenyan tea farms: A contribution to sustainability or tokenism? *Journal of Cleaner Production, 39*, 285–293. https://doi.org/10.1016/j.jclepro.2012.07.048

Olakitan Atanda, J. (2019). Developing a Social Sustainability Assessment Framework. *Sustainable Cities and Society*. https://doi.org/10.1016/j.scs.2018.09.023

Page, M. J., McKenzie, J. E., Bossuyt, P. M., Boutron, I., Hoffmann, T. C., Mulrow, C. D., Shamseer, L., Tetzlaff, J. M., Akl, E. A., Brennan, S. E., Chou, R., Glanville, J., Grimshaw, J. M., Hróbjartsson, A., Lalu, M. M., Li, T., Loder, E. W., Mayo-Wilson, E., McDonald, S., et al. (2021). The PRISMA 2020 statement: An updated guideline for reporting systematic reviews. *Journal of Clinical Epidemiology, 134*, 178–189. https://doi.org/10.1016/j.jclinepi.2021.03.001

Paul, S., Das, T. K., Pharung, R., Ray, S., Mridha, N., Kalita, N., Ralte, V., Borthakur, S., Burman, R. R., Tripathi, A. K., & Singh, A. K. (2020). Development of an indicator based composite measure to assess livelihood sustainability of shifting cultivation dependent ethnic minorities in the disadvantageous northeastern region of India. *Ecological Indicators, 110*(November 2019), 105934. https://doi.org/10.1016/j.ecolind.2019.105934

Pelletier, N. (2018). Social sustainability assessment of Canadian egg production facilities: Methods, analysis, and recommendations. *Sustainability (Switzerland), 10*(5), 1–17. https://doi.org/10.3390/su10051601

Pereira, S. A., Kimpara, J. M., & Valenti, W. C. (2021). Sustainability of the seaweed Hypnea pseudomusciformis farming in the tropical southwestern Atlantic. *Ecological Indicators, 121*, 107101. https://doi.org/10.1016/j.ecolind.2020.107101
Pieper, R., Karvonen, S., & Vaarama, M. (2019). The SOLA model: A theory-based approach to social quality and social sustainability. *Social Indicators Research, 146*(3), 553–580. https://doi.org/10.1007/s11205-019-02127-7
Polèse, M., & Stren, R. E. (2000). *The social sustainability of cities: Diversity and the management of change*. University of Toronto Press.
Pope, J., Morrison-Saunders, A., & Annandale, D. (2005). Applying sustainability assessment models. *Impact Assessment and Project Appraisal, 23*(4), 293–302. https://doi.org/10.3152/147154605781765436
Popovic, T., Kraslawski, A., Barbosa-Póvoa, A., & Carvalho, A. (2017). Quantitative indicators for social sustainability assessment of society and product responsibility aspects in supply chains. *Journal of International Studies, 10*(4), 9–36. https://doi.org/10.14254/2071-8330.2017/10-4/1
Prasara-A, J., & Gheewala, S. H. (2021). An assessment of social sustainability of sugarcane and cassava cultivation in Thailand. *Sustainable Production and Consumption, 27*, 372–382. https://doi.org/10.1016/j.spc.2020.11.009
Rasmussen, L. V., Bierbaum, R., Oldekop, J. A., & Agrawal, A. (2017). Bridging the practitioner-researcher divide: Indicators to track environmental, economic, and sociocultural sustainability of agricultural commodity production. *Global Environmental Change, 42*(January 2017), 33–46. https://doi.org/10.1016/j.gloenvcha.2016.12.001
Rasouli, A. H., & Kumarasuriyar, D. A. (2016). The social Dimention of sustainability: Towards some definitions and analysis. *Journal of Social Science for Policy Implications, 4*(2), 23–34. https://doi.org/10.15640/jsspi.v4n2a3
Reed, M. S., Fraser, E. D. G., & Dougill, A. J. (2006). An adaptive learning process for developing and applying sustainability indicators with local communities. *Ecological Economics, 59*, 406–418.
Reddy, K. R., Sadasivam, B. Y., & Adams, J. A. (2014). Social sustainability evaluation matrix (SSEM) to quantify social aspects of sustainable remediation. ICSI 2014: Creating infrastructure for a sustainable world – proceedings of the 2014 international conference on sustainable infrastructure, 831–841. https://doi.org/10.1061/9780784478745.078.
Ripoll-Bosch, R., Díez-Unquera, B., Ruiz, R., Villalba, D., Molina, E., Joy, M., Olaizola, A., & Bernués, A. (2012). An integrated sustainability assessment of mediterranean sheep farms with different degrees of intensification. *Agricultural Systems, 105*(1), 46–56. https://doi.org/10.1016/j.agsy.2011.10.003
Roesch, A., Nyfeler-Brunner, A., & Gaillard, G. (2021). Sustainability assessment of farms using SALCAsustain methodology. *Sustainable Production and Consumption, 27*, 1392–1405. https://doi.org/10.1016/j.spc.2021.02.022
Romero, E. M., Canto Saenz, M., Jiménez Dávalos, J. E., Blas Sevillano, R., Vargas Winstanley, S. E., & Julca Otiniano, A. (2020). Sostenibilidad social de un sistema productivo familiar con granadilla (Passiflora Ligularis Juss.) en la provincia de Oxapampa, Pasco-Perú. *Agricultura Sociedad y Desarrollo, 17*(2), 217–232. https://doi.org/10.22231/asyd.v17i2.1343
Röös, E., Fischer, K., Tidåker, P., & Nordström Källström, H. (2019). How well is farmers' social situation captured by sustainability assessment tools? A Swedish case study. *International Journal of Sustainable Development and World Ecology, 26*(3), 268–281. https://doi.org/10.1080/13504509.2018.1560371
Sachs. (1999). *Social sustainability and whole development: Exploring the dimensions of sustainable development*. Zed Books.
Sandhu, H. (2021). Bottom-up transformation of agriculture and food systems. *Sustainability (Switzerland), 13*(4), 1–13. https://doi.org/10.3390/su13042171
Schader, C., Baumgart, L., Landert, J., Muller, A., Ssebunya, B., Blockeel, J., Weisshaidinger, R., Petrasek, R., Mészáros, D., Padel, S., Gerrard, C., Smith, L., Lindenthal, T., Niggli, U., & Stolze, M. (2016). Using the sustainability monitoring and assessment routine (SMART) for the

systematic analysis of trade-offs and synergies between sustainability dimensions and themes at farm level. *Sustainability (Switzerland), 8*(3), 274. https://doi.org/10.3390/su8030274

Singh, S., & Nayak, S. (2020). Development of sustainable livelihood security index for different agro-climatic zones of Uttar Pradesh. *India. Journal of Rural Development, 39*(1), 110–129. https://doi.org/10.25175/jrd/2020/v39/i1/125991

Sokil, O., Zhuk, V., & Vasa, L. (2018). Integral assessment of the sustainable development of agriculture in Ukraine. *Economic Annals-XXI, 170*(3–4), 15–21. https://doi.org/10.21003/ea.V170-03

Spangenberg, J. H. (2004). SUSTAINABILITY AND GROWTH : CRITERIA. *INDICATORS, POLICIES., 86*(January), 74–86.

St Flour, P. O., & Bokhoree, C. (2021). Sustainability assessment methodologies: Implications and challenges for SIDS. *Ecologies, 2*(3), 285–304. https://doi.org/10.3390/ecologies2030016

Stylianou, A., Sdrali, D., & Apostolopoulos, C. D. (2020). Integrated sustainability assessment of divergent mediterranean farming systems: Cyprus as a case study. *Sustainability (Switzerland), 12*(15), 6105. https://doi.org/10.3390/su12156105

Torkayesh, A. E., Ecer, F., Pamucar, D., & Karamaşa, Ç. (2021). Comparative assessment of social sustainability performance: Integrated data-driven weighting system and CoCoSo model. *Sustainable Cities and Society, 71*(April), 102975. https://doi.org/10.1016/j.scs.2021.102975

Triviño-Tarradas, P., Carranza-Cañadas, P., Mesas-Carrascosa, F. J., & Gonzalez-Sanchez, E. J. (2020). Evaluation of agricultural sustainability on a mixed vineyard and olive-grove farm in southern Spain through the INSPIA model. *Sustainability (Switzerland), 12*(3), 1090. https://doi.org/10.3390/su12031090

ul Haq, S., & Boz, I. (2017). Developing sustainable agriculture indicators framework for tea farms in Rize province Turkey. International conference on food and agricultural economics, April, 9.

ul Haq, S., & Boz, I. (2020). Measuring environmental, economic, and social sustainability index of tea farms in Rize Province, Turkey. *Environment, Development and Sustainability, 22*(3), 2545–2567. https://doi.org/10.1007/s10668-019-00310-x

Vallance, S., Perkins, H. C., & Dixon, J. E. (2011). What is social sustainability? *A clarification of concepts. Geoforum, 42*(3), 342–348. https://doi.org/10.1016/j.geoforum.2011.01.002

Van Cauwenbergh, N., Biala, K., Bielders, C., Brouckaert, V., Franchois, L., Garcia Cidad, V., Hermy, M., Mathijs, E., Muys, B., Reijnders, J., Sauvenier, X., Valckx, J., Vanclooster, M., Van der Veken, B., Wauters, E., & Peeters, A. (2007). SAFE-A hierarchical framework for assessing the sustainability of agricultural systems. *Agriculture, Ecosystems and Environment, 120*(2–4), 229–242. https://doi.org/10.1016/j.agee.2006.09.006

Vanclay, F. (2002). Conceptualising social impacts. *Environmental Impact Assessment Review, 22*(3), 183–211. https://doi.org/10.1016/S0195-9255(01)00105-6

Vasileiadis, V. P., Moonen, A. C., Sattin, M., Otto, S., Pons, X., Kudsk, P., Veres, A., Dorner, Z., van der Weide, R., Marraccini, E., Pelzer, E., Angevin, F., & Kiss, J. (2013). Sustainability of European maize-based cropping systems: Economic, environmental and social assessment of current and proposed innovative IPM-based systems. *European Journal of Agronomy, 48*, 1–11. https://doi.org/10.1016/j.eja.2013.02.001

Venkatesh, G. (2015). Sustainable development as a single measure: Case study of some developing Asian countries | Zrównoważony rozwój jako kompleksowe narzędzie pomiarowe: Przykład wybranych Azjatyckich krajów rozwijających. *Problemy Ekorozwoju, 10*(2), 31–42.

Vilei, S. (2011). Local perceptions of sustainability of farming systems on Leyte, Philippines – divergences and congruencies between different stakeholders. *International Journal of Sustainable Development and World Ecology, 18*(4), 291–303. https://doi.org/10.1080/13504509.2011.555112

Volkov, A., Morkunas, M., Balezentis, T., & Streimikiene, D. (2022). Are agricultural sustainability and resilience complementary notions? Evidence from the north European agriculture. *Land Use Policy, 112*(October 2021), 105791. https://doi.org/10.1016/j.landusepol.2021.105791

Weingaertner, C., & Moberg, Å. (2014). Exploring social sustainability: Learning from perspectives on urban development and companies and products. *Sustainable Development, 22*(2), 122–133. https://doi.org/10.1002/sd.536

Wells, S. (2011). *Pandora's seed: the unforeseen cost of civilization.* Random House.

Yakar Pritchard, G., & Çalıyurt, K. T. (2021). *Sustainability reporting in cooperatives. Risks, 9*(6), 1–15. https://doi.org/10.3390/risks9060117

Yegbemey, R. N., Yabi, J. A., Dossa, C. S. G., & Bauer, S. (2014). Novel participatory indicators of sustainability reveal weaknesses of maize cropping in Benin. *Agronomy for Sustainable Development, 34*(4), 909–920. https://doi.org/10.1007/s13593-014-0214-9

Yli-Viikari, A. (1999). Indicators for sustainable agriculture – A theoretical framework for classifying and assessing indicators. *Agricultural and Food Science in Finland, 8*(3), 265–283. https://doi.org/10.23986/afsci.5628

Zhao, W., Wang, S., Zhang, Y., Xu, M., Qie, H., Lu, Y., Lu, H., & An, R. (2021). A un sdgs-based sustainability evaluation framework for globally important agricultural heritage systems (Giahs): A case study on the kuaijishan ancient chinese torreya community. *Sustainability (Switzerland), 13*(17), 9957. https://doi.org/10.3390/su13179957

Zhen, L., Cao, S., Wei, Y., Dilly, O., Liu, X., Li, F., Koenig, H., Tscherning, K., & Helming, K. (2009). Comparison of sustainability issues in two sensitive areas of China. *Environmental Science and Policy, 12*(8), 1153–1167. https://doi.org/10.1016/j.envsci.2009.03.002

Zhen, L., & Routray, J. K. (2003). Operational indicators for measuring agricultural sustainability in developing countries. *Environmental Management, 32*(1), 34–46. https://doi.org/10.1007/s00267-003-2881-1

Zimmerer, K., & Bassett, T. (2003). *Political ecology: an integrative approach to geography and environmentdevelopment studies.* Guilford.

Urban Green Infrastructure: Circular Socioeconomy with Nature Based Solutions

Julián Briz, Isabel de Felipe, and Teresa Briz

Abstract In a broader sense, we can consider the circular socioeconomy, which considers "material and immaterial" products and the corresponding services. In this context, to traditional material products, we could incorporate human health and well-being, which can also be recovered or recycled.

As a general introduction, we describe urban strategies with environment and pollution dimensions, recycling, and insulated actions. Business and Nature Based Solutions (NBS) have to be coordinated with urban health.

Research on social perception of green urban infrastructure was developed with a national online survey of the main actors: academics, entrepreneurs, functionaries, and others. The more significant opinions are analyzed. Case studies on energy and urban agriculture in Spain are complemented by some projects in Madrid.

Special attention is given to the project gas recycling harm waste into urban benefits. The main objective are to identify and promote activities for social welfare throughout nature based solution, where circular economy throughout gas recycling is a useful instrument.

Finally, the results include health impact, evaluation, and errors in socioeconomic circular actions.

Keywords Urban · Green infrastructure · NBS · Circular socioeconomy · Environment · Recycle · Business

---

J. Briz (✉)
PRONATUR, Universidad Politécnica Madrid, Madrid, Spain
e-mail: julian.briz@upm.es

I. de Felipe
itdUPM, Madrid, Spain
e-mail: isabel.defelipe@upm.es

T. Briz
ETSIAAB, Universidad Politécnica Madrid, Madrid, Spain
e-mail: teresa.briz@upm.es

C. Rego et al. (eds.), *Entrepreneurship, Technological Change and Circular Economy for a Green Transition*, Studies on Entrepreneurship, Structural Change and Industrial Dynamics, https://doi.org/10.1007/978-3-031-48079-9_10

1 Introduction

The current city model is not sustainable, and it requires making decisions for the urban community to achieve an adequate quality of life. Useful instruments to carry out this task include green infrastructures, which must be promoted by both the public and private sectors and endorsed by citizens. This requires training and objective information on its effects, cost, and benefits, as well as information on the degree of implementation.

The challenge of this chapter is to expand the traditional approach of the circular concept focused on economic aspects, incorporating areas such as health, natural resources, and social relations in what we call circular socioeconomics.

To do this, in addition to recycling products, it is necessary "recycle" health, provided the deterioration of human relationships or loss of natural resources. This chapter is based on existing works and projects developed by the authors, through interviews with the actors to know the reality and perception that society has of the impact of nature on our lives.

The concept of recycling is frequently used. Besides residual organic products, recovery of rainwater and gray water, heat energy, and polluting gases. We include the scenario of intangibles, health, community, and personal relationships as well as heritage of natural resources.

Traditional development is based on a linear model: produce, transform, buying, selling, and throwing away residuals. However, besides economic inefficiency, there is a negative environmental impact.

An innovative system is the circular economy, which coordinates, shares, recycles, and reuses products as many times as possible, which gives greater sustainability to the system. The benefits at the local level are several, from the reduction of the consumption of natural resources to the lower environmental impact, creating new business models. In the international framework, we have to mention the 2030 Agenda, with the SDG, especially numbers 3 (good health and well-being), 7 (clean energy), 8 (economic growth), 9 (innovation), 11 (sustainable cities), and 12 (responsible production and consumption).

Human well-being is the result of a healthy environment (air quality, temperature, humidity) and the ability to interact socially, which facilitates recovery and prevents the deterioration of mental and physical health, and the consequent saving of medical and hospital costs.

The interaction of the different circular models requires coordination to achieve a stable and sustainable balance.

In similar terms, circular bioeconomy relies on healthy, biodiversity, and resilient ecosystems, with sustainable well-being services and a sustainable environment, structured in the Circular Bioeconomy Alliance.

The term circular has been used in various scenarios with the idea of productivity, sustainability, economy, and sociology with nature in ecosystems durable. The basic idea was to coordinate economic growth, employment, socio environmental conditions in a healthy environment. To optimize the use of natural resources for a longer life cycle.

2 General Overview of the Situation and Possible Strategies

In this section, some of the most significant dimensions of the role of nature in urban areas are described, as well as the importance of each variable.

2.1 *Urban Planning Strategies*

Urban strategies must be "glocal," with a global vision but with local actions, combining the Sustainable Development Goals (SDGs) and the 2030 Agenda with a friendly environment. This leads to a profound social transformation. The multifunctional productive urban landscape must be valued, which implies an urban food and ornamental agriculture, where in addition to food, also air quality, human relations, biodiversity or climate change, and water management are improved. Technicians must recycle their knowledge in the face of new environmental challenges and simultaneously know how to transmit to citizens in popular language the existing problems, the capacity to solve them, the available resources and their cost, and the possible benefits obtained. The utopia of the future city must reconcile migratory demographic trends caring about the environment, consolidating a civic awareness that channels political and socioeconomic decisions.

Urban life has evolved through cycles caused by exogenous (wars, pandemics, natural disasters, climate) and endogenous (site-specific) factors. The ability to react is based on realistic and utopian situations.

We are in the stage of eco-urban utopia that has been brewing in recent years, with variants that are not always complementary. The new consciousness is expressed through language, where the terms green, naturation, or naturalization, overlap with the language of the city.

Facing this situation, the panorama leads us to consider whether the eco-urban is strong enough to make the utopia of living with nature in a sustainable way a reality.

In the face of the challenges posed, a resilient society must be able to adapt, through the appropriate changes. Innovative spirit involves discovering new horizons (energy sources such as hydrogen or the most recent clean fusion) or recovering existing ones, but which have been marginalized. This is the case of Nature Based Solutions (NBS), as a valid instrument, with millennial experience, driven by biomimicry. For example, the circular economy, a symbol of eco-urbanism, is found in the foliar laboratory of the chlorophyll function, with optimal performance and total recycling of all its components. Another NBS scenario is biodiversity, as a weapon against pandemics of certain species and promotion of pollination.

In recent centuries, urban life has turned its back on nature, creating jungles of crystal and cement. Citizens have lived a series of stages, based on utopias that have been diluted over time. Agriculture encourages self-sufficiency, with the utopia of becoming independent from the outside, but they have forgotten the exponential growth of the city, creating unsustainable economic, environmental, and human ghettos. The socialists rely on the myth of the factory and its social gear as an integrating element. The functionalists, based on the engine, encouraged the transport and connection of the city with the outside. The technologies that allow us to expand the existing limits, have led us to an unviable nucleus that requires utopian ecocity, where the citizen can surround himself with nature in his daily life, in a viable and sustainable way.

2.2 *Landscape*

Landscape has been a traditional dimension of urban nature. The various styles of gardening have left their mark on the urban space. In the Western world, from prehistory, antiquity in Greece and Egypt, the Roman Empire, the Middle Ages, the Italian Renaissance, French rationalism, with oriental and American influences, contemporary art has been reached (Páez de la Cadena, 1998). Utopia has been shaped through art.

It is necessary to adopt measures to improve the situation, reducing emissions and increasing immissions in the process of urban growth and temperature increase. Nature is offered as a solution through corridors, walls and green roofs, water sheets, and others.

Green infrastructure will continue to increase especially in areas where available space is limited and the cost of land does not allow the construction of parks, stormwater ponds, and other green and blue facilities.

Municipalities have to claim that maintenance of the green infrastructures has to be performed by the property owner in coordination with the neighbors' organizations, firms, and individuals living in the district. As an example, we have a positive experience in Tokyo, with interaction of all the stakeholders living in a community district.

2.3 *Environment: Climate and Pollution*

For a theoretical discussion, we should look at the real urban screen. For example, the temperature in Spanish cities has risen twice as high as the world average in the last 50 years (Royé et al., 2020).

Global warming means a rise in average temperatures, although there are areas with cooling. In cities, the role of gray water is essential due to the soil being waterproofed with cement and asphalt. The pavements decrease evaporation and water filtration, causing runoff. They trap solar radiation and hinder air currents favoring the increase of heat and pollution caused by vehicles, heating, and air conditioning. That is why global warming is joined by the urban heat island.

There is a contrast between the center and the periphery in the cities, caused by the greater demographic concentration and heat sources. This causes an increase in urban morbidity, especially in chronic diseases, in children and the elderly.

Air pollution decreases life expectancy by 2 years, due to the impact of chemicals from agriculture, industry, and traffic. PM2.5 fine particles are especially harmful, which advises a tightening of current legislation, with maximum tolerated limits. In the UE it is 25 micrograms per m^3 but the WHO recommends 10. Green infrastructures serve as a filter for polluting gases and particles. A tree can absorb 1 kg of CO_2 per year, a medium-sized fir tree sets 20 kg of annual tile and transpires 350 liters of water daily, contributing to environmental improvement.

Food supply is another challenge in metropolitan areas. However, the dilemma is producing and/or ecological neighborhoods (Hernandez Aja et al., 2009).

The concept of eco-neighborhood means interaction of the city with the environmment and the rights and duties of citizens and their ecological responsibility in a model for sustainable urban rehabilitation.

The challenge is how to reverse the current urbanization process, with negative effects on the ecosystem, hindering its autonomous regeneration. The air quality, the soil quality, and the management of rainwater and gray water are destroyed, with waste discharges, waterproofing of soils, and difficulties for their recycling.

We need a circular model that incorporates the social fabric into decision-making, defining urban criteria. The objectives must be concrete and in addition to the environment, food production, waste and energy recycling material for use in the construction of buildings must be incorporated.

The eco-neighborhood aims to optimize the management of water, energy, and products, improving the environment and social relations through NBS. All this must be complemented by other aspects such as the transport system and the availability of basic services in a 15-min environment.

For this, it is necessary to achieve coherence between urban strategies with social responsibility, biodiversity and naturation, mobility and metabolism, connecting to the existing reality.

Comparing eco-neighborhood and producer neighborhoods models both promoted in urban planning, the former emphasized nature and the environment, involving the whole society. The producing neighborhoods have a productivism approach, with food generation and employment, essentially affecting marginal groups. The rehabilitation of the sustainable urban model requires the modalities that may arise, as they are not opposed but complementary, with active citizen participation that replaces the usual heteronomy.

2.4 Recycle, Insulation, and Waste

Waste recycling is another line of action. Half of urban waste is organic matter from food, gardens, and pruning. Properly located composting plants can be recycled and use the compost as fertilizers for soils and agricultural activities.

The key to the success of the eco-neighborhoods is global citizen participation, both in the design and in the development of the project and its monitoring and evaluation, all within the environmental education program.

Another function of urban metabolism is the recycling of contaminated gases (CO_2, NO_2), waste products, and energy. Purpose-built buildings are being constructed in several cities, with tax benefits for their developers (Fig. 1).

Events such as COVID, climate change, high levels of stress and pollution, and wars, such as the one in Ukraine, with shortages of basic materials such as cereals,

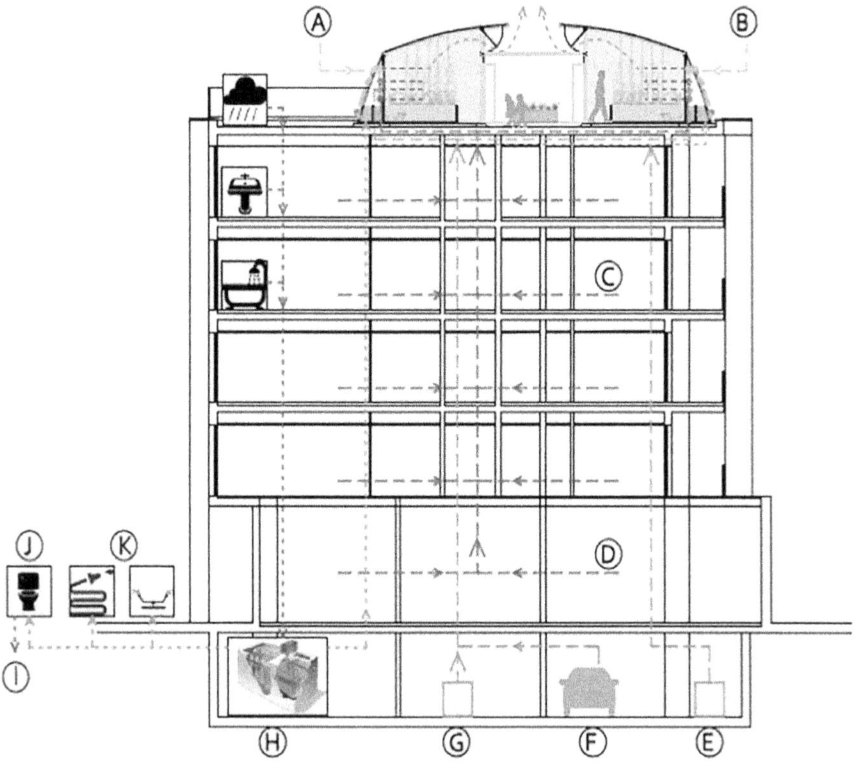

Fig. 1 (**a**). External air with CO_2 polluting particles. (**b**). External air with dust, fumes, and other gases. (**c**). Office and residential air with CO_2 and heat, from human respiration. (**d**). Air of commercial premises, shops, and other activities. CO_2 and heat. (**e**). Air purifiers of dust, microbes, dust, and pathogenic germs. (**f**). Car smoke car park air. (**g**). Heating boiler with combustion fumes. (**h**). Gray water recycling treatment. (**i**). Urban sanitation. (**j**). Black water tank for sewerage. (**k**). Irrigation water. Source: Authors' elaboration

are fostering a consciousness that demands preciously nature based solutions. This is how urban agriculture is silently penetrating the city.

In the case of organic products, the recovery of organic waste is carried out through composting and its subsequent use as organic fertilizer in organic farming.

Insulation as energy saving is another dimension to take into consideration and green infrastructures can be used in this area, especially in walls and green roofs.

More complex and sophisticated is the question of recovery and improvement of human health and well-being, which can be achieved through NBS.

2.5 Business Strategies and NBS

NBS are an innovative source for companies willing to solve society's problems. New construction techniques and materials coupled with agricultural and gardening practices, as well as management of available resources, are valuable tools. As an example, the new urban agriculture in addition to providing food improves the environment and requires a new profession. Local crops, with added value, as biological control, or without fertilization or agrochemicals, are a challenge for new professionals. This creates new businesses and employment opportunities.

The social pressure on climate and pollution to change the current unsustainable model of life is leading all actors involved in the *rurban* (rural and urban) world to rethink their plans and strategies to come. Scientists and academics redirect their lines of research toward issues of environmental sustainability, carbon footprint, and pollution. National and local administrations seek solutions through regulations and fiscal policies, consistent with the above-mentioned objectives.

Companies, especially large corporations, focus their strategies on carbon neutrality, energy saving, and issues sensitive to public opinion, as a sign of their corporate social responsibility. To do this, they maintain their production and work lines in their operational places and try to compensate for their deficiencies with the development of NBS in other places, planting trees, developing green spaces, and promoting existing ones with a more ecological management.

With this practice, at the global level, the balance of zero impacts on the carbon footprint and balance in climate change is met. However, a series of questions arise about the net efficiency of these strategies, without producing discrimination and geographical imbalances. Instead of innovating and introducing new techniques that reduce in situ and globally the harmful effects of the carbon footprint, pollution and energy waste, and natural resources, they move the problem to another place, where these problems are reproduced.

As an example, the proliferation of electric cars in the city reduces air and noise pollution in that environment but transfers certain problems to the place of production of that electrical energy in case the appropriate technology is not applied. It would be more coherent to plan the city, with green spaces, where citizens were at an adequate distance from the workplace and access to basic services so that they could do so as pedestrians. Teleworking can be a very useful tool in many of these cases.

The problems derived from the production process itself represent a scenario to consider. Publications such as “No to nature (based solutions) dispossession” (https://www.biofuelwatch.org.uk/2021/nbs-cso-statement) draw attention to this field, where large electricity companies (ENEL in Italy), food companies (Nestlé), technology companies (Microsoft), propose neutrality in the carbon balance.

In Madrid, an entrepreneurial area with a social focus, we may comment on the evolution of urban agriculture, which initially had only ornamental objectives and social cohesion of the neighborhood, but the new approach has an economic dimension, through the offer of products to the urban market. There are the so-called “Barrios Productores” (Producing Neighborhoods) as well as plots in the Metropolitan Forest that border the city, with a productive vision.

Producing neighborhoods are located in polygons constructed in the 1960s and 1970s. Its action focuses on the neighborhoods of Manoteras and Entrevías, an area of 678 hectares with a population of 1.2 million inhabitants. Its approach contemplates an expansion of traditional urban gardens focused on social improvement and community life with an economic vision of the market.

In recent times the Madrid metropolis has been promoting new NBS, in its different dimensions through urban agriculture, both food and ornamental, taking advantage of underused or marginal spaces on the ground and in height. Its multifunctional character is manifested in fruit and vegetable food production and floriculture, social cohesion in leisure and recreation spaces, fight against pollution, environmental improvement (temperature and humidity), biodiversity, and pollination.

In the development of large urban projects, there is also a three-dimensional collaboration: private, public, and academic, as is the case of Madrid Rio and Madrid Nuevo Norte.

Madrid Rio is a blue-green project along 10 km of Manzanares river constructed in 1974. The Park is covering a former road traffic.

Madrid Nuevo Norte is one of the largest urban projects in Europe. It represents an investment of more than 25,000 million euros, 90% is a private investment. It acts on two million square meters, existing a large central park, with vertical green infrastructures. It is expected the creation of 10,500 homes and 350,000 jobs, being the business center of Madrid.

2.6 Urban Health

In 2050, more than 75% of the world’s population will live in cities and we must establish indicators for urban development that promote social equity in a healthy environment for housing, transport, and food. Air pollution is estimated to cause 1.3 million deaths annually, due to levels of CO_2, NO_2, ozone, and PM10/2.5 particulate matter (UN. Rio Conference, 2012).

Greening improves health, environment, and socioeconomic conditions with a holistic approach that facilitates decision-making for urban planners. The reduction of air pollution is a fact and with it, the diseases caused (diabetes, asthma, blood circulation, heart attacks, and cardiovascular problems). There is also a correlation between newborn weight and immune systems (Kumar et al., 2019).

Medicinal plants, herbs, and spices, are part of the tradition and should be incorporated into natural areas, similar to aromatic ones, which provide comfort and are a natural barrier to mosquitoes and other insects.

Naturation mitigates urban heating, saves energy, and improves the thermal and humid comfort of the environment. Several experiences show up to a thermal reduction of 4.9 degrees in warm periods, with energy savings of 40 kW per day in winter and 24.7 k W in summer (Wang et al., 2019).

On the other hand, a human being needs 300–600 liters of oxygen per hour, that is, 22 trees are needed, depending on the size, species, and type of leaf (Burgos, OK Diario 24/ 91 2021). Similar calculations have been made for grass surface. This indicates that oxygen is a public good that citizens obtain free of charge from nature and is basic to our survival.

Pollution, stress, violence and isolation, are part of the urbanite plague, affecting their physical and mental health. The current city model is not viable. Therefore, it is necessary to propose affordable strategies, such as naturation (incorporation of nature into our environment), with an impact on health and socioeconomics.

Due to the heterogeneity of urban scenarios, we have focused the methodology of this chapter on the combination of fact and perception dimensions, with bibliography publications, empirical analysis, and interviews of stakeholders.

To sum up, this chapter has described the variety of roles that Nature Based Solutions may perform in the cities. Thus, it has to be developed a series of planned strategies which include landscape, impact on climate and pollution, and the need to recycle waste. Business and health are also special areas to pay attention to in our society.

3 Methodology

Following, the methodology is presented. It is a combination of theory and practice. The knowledge of urban actors' opinions is a relevant factor; thus, it analyzed their social perception in a research project. As an experimental design, we also include an ongoing project on gas recycling. Agrifitness is a promising area with recreation and human activities. Finally, there is a description of some urban green infrastructure in Madrid city, where we may identify some practical examples.

3.1 Social Perceptions and Gas Recycling

Urban nature is complex due to its multifunctionality (Briz et al., 2019), and needs to be analyzed from both a quantitative and a qualitative perspective. In both cases, the availability of financial and human resources and the opportunity to employ a multidisciplinary approach and collaboration is a determining factor, for which we have carried out several surveys of groups of experts and people involved in the operation of the project (Briz, 2022).

In social perception, the following research offers a reflection on the urban nature recycling process in Spain, with a comprehensive vision that considers the difficulties of statistical and documentary sources in a sector often neglected by society. The administrative fragmentation of the regulations and the variety of institutions and professions involved, have led us to supplement the available documentation with qualitative information, by including the perceptions of some of the main actors in the sector. An online survey was carried out, focused on different profiles (academics, businesspeople, civil servants, neighborhoods), seeking their opinion on the role of urban green infrastructure. Their opinions provide a qualitative value that can be useful for more in-depth and extensive studies in the future, which would serve as guidance for the actions of leaders, investors, and politicians in the field of urban green infrastructure.

We describe some of the most significant comments in the mentioned social research work:

- Level of importance of the BENEFITS of green infrastructure for cities and buildings in the near future. The level of importance given to all the aspects measured is high, with ratings above 3 on a scale of 1–4. But the improvement of the microclimate and the increase in the quality of life stand out in a special way.
- Degree of INTEREST in green infrastructures. Landscape and garden designers together with landscape architects and designers are the groups that are perceived as the most interested in green infrastructures. However, it is the construction companies and real estate developers who are perceived as the least interested.
- EFFECTS PRODUCED by urban naturation. Academics, experts, designers, and promoters consider that the main effects produced by urban nature are landscapes, filtering pollution, biodiversity, and improving the work environment.
- Measures to PROMOTE urban naturation. Academics, officials, experts, and designers consider that the main measures to promote urban nature would be inclusion in urban plans, followed by citizen awareness, administrative regulations, and subsidies.
- Degree of ATTRACTION that arouses. It is an attractive market. However, the fact that it generates stable and quality employment is not so clear, but it is an open market, flexible to the incorporation of new ideas and possibilities.

The project gas recycling of CO_2, N_2, and dust, contributes to improve air quality and mitigate climate change. It was developed by Foro Agrario, Pronatur, itdUPM, and other institutions.

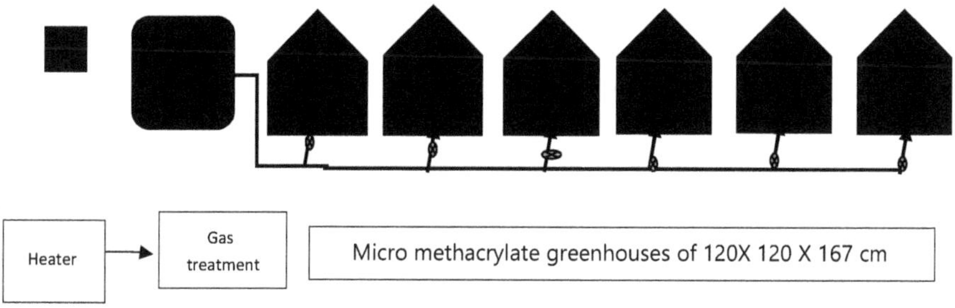

Fig. 2 Experimental prototype scheme. Source: Authors' elaboration

The chronological program includes several steps. The first one is to build experimental prototypes (figure below) that are testing future applications for heating gas in green roof houses.

The second step is to observe the growth and health of ornamental and food plants. The third objective is the evaluation of reducing harmful products from the environment.

Finally, we are on the way to observe the interaction of heating gases with soil microbiota, especially on symbiotic nitrogen (rhizobium) (Fig. 2).

3.2 *Urban Food Production and Green Actions*

In recent times, all metropolises have been promoting new NBS, in its different dimensions through urban agriculture both food and ornamental, taking advantage of underused or marginal spaces both on the ground and height (Briz et al., 2017). Its multifunctional character is shown in horticultural food production and floriculture, social cohesion in leisure and recreation spaces, fight against pollution, environmental improvement (temperature and humidity), biodiversity, and pollination.

There are several scenarios of action according to the orography, demographic concentration, type of planning and construction of buildings, which we can identify in: Producing neighborhoods, metropolitan forests, new blue-green projects, axis of naturation in the central core, and attitude of stakeholders.

The level of importance of the benefits offered by green infrastructure is very high. Particularly noteworthy is an improvement in the microclimate and an increase in quality of life. The greening of walls with self-climbing plants is what green infrastructure experts have the greatest knowledge of and experience in.

The main obstacles to installing green roofs, green walls, and green interior walls are the barriers to construction, the low level of knowledge of potential clients, the difficulty in maintenance, and the high costs compared to the advantages and benefits.

The main measures to promote urban nature would be its inclusion in urban plans, followed by citizen awareness, administrative regulations, and subsidies.

The evolution of urban nature is very positive: what has increased the most are urban gardens, green roofs, and walls. However, what has increased the least are trees in squares and streets and trees in interior spaces.

The above-mentioned activities and remarks must be complemented by the appropriate studies that provide new information, as well as the correction of errors that may be detected in greening as a dynamic process of adaptation to existing needs. The subsequent contributions will identify the characteristics of urban green infrastructure in Spain, evaluating the impact of the successive measures applied. The interdisciplinary collaboration of professions and the multifunctionality of urban greening (Briz et al., 2019) will provide a global vision, highlighting its importance in the sustainability of urban well-being.

The idea of having greener cities in Europe continues to lead various European countries to prepare reports, sponsored by both public and private initiatives (https://greenmarketreport.eu/en/). It is in this context, we have established a platform for collaboration between institutions, universities, the private sector, and the social sector in Spain to analyze the field of green infrastructure.

3.3 Agrifitness: Energy Consumption and Urban Agriculture Recreation Tasks

Direct or indirect methods can be used to evaluate physical activity and energy expenditure (López Fontan, 2003). The usual direct methods can be the recording of activities, calorimetry, accelerometers among others. By indirect methods, we consider metabolic measurements, heart rate, surveys, anthropometry, etc. (Park Sin, 2012). The use of one or the other depends on the objectives indicated and the human and financial resources available, having as a common denominator the identification of the muscles used and the energy consumption.

Some authors like Martinsen (Martinsen, 1987) have found that low-intensity exercises such as gardening have beneficial effects on mental health (depression, anxiety) and physical, and it is advisable to practice them regularly for a relatively long period.

Given the growing interest in the activities carried out in urban agriculture, it is considered appropriate to objectively evaluate some relevant parameters in physical exercise, such as energy consumption. To this end, a pilot study has been carried out at the Universidad Politecnica Madrid (Martinez Ruiz & Aparicio, 2017). Five tasks were analyzed (digging, raking, watering, planting, and carrying wheelbarrows), in the organic garden of the ETSIAAB, located in the University Campus of Madrid. Energy expenditure was measured in activities performed by six people (three men and three women).

The measurement was performed using a heart rate monitor with a chest band and a wrist heart rate sensor bracelet. The heart rate monitor shows 25.8% more energy expenditure with 497 Kcal / h than the wristband with 395 Kcal / h. The activities carried out can be considered of medium type, in relation to others with lower energy expenditure, such as walking (300 kcal/h) and others of a higher consumption such as tennis (750 Kcal/h).

The pilot study shows the applications of evaluation techniques of physical exercise to urban agriculture activities, based on existing experience in sports practices.

The challenge for later stages includes the expansion of the experience, with a wider and more diverse sample, in different locations, through agreements with institutions and the Network of Urban Gardens. The studies should incorporate more vulnerable people such as retired people, the elderly, and those who are in active medical recovery or with specific problems (drug addicts, social maladjustment). The period of application would be variable, as well as the intensity of the work, from lower to higher degree. All this would allow us to know the improvement experienced.

The physiological knowledge of the muscle tissues that are used in the tasks and the load they support in each of them is another scenario of analysis when programming the activities to be developed.

It is also of interest to know the necessary activities to be carried out in the agronomic practices, duration, and time of application. Each task can be classified into different scales, according to time and intensity. For example, transport by forklift will have as factors the weight supported, travel, and duration of the application.

We can also identify a positive dimension of urban agriculture in a contemplative way, which does not involve physical exercise. We refer to the mere observation of the landscape. In hospital residences, there is evidence of the best recovery of patients who have a vision of green areas and evolution of nature. As an example, Japan promotes the use of green areas both in hospitals and in offices with different degrees of involvement. In the first case, patients who simply contemplate the garden or do some maintenance activities are compared with those with sedentary lives watching only TV or reading. The recovery of the latter is slower. In Tokyo, there are companies that include in the employment contract, a few hours of work in the maintenance of the community garden, in coordination with experts from the municipality and groups of volunteers from the neighborhood.

3.4 *Case Study: Green Urban Infrastructure in Madrid*

As a practical analysis of the former questions, we describe examples applied in urban green infrastructures in Madrid.

We can identify several lines of action following the NBS and circular economy in the Madrid metropolitan area.

Recycling of products is carried out in 12 districts, with the objective of reusing and recycling half of the domestic and commercial wastes.

For adaptation to climate change and reduction of CO_2 emissions, in 2017 the Quality and Climate Change Plan was established with initiatives such as "Madrid + natural," integrating into various international associations in the network of cities for climate change. It has developed more than a hundred green roofs and walls throughout the city.

In the innovative aspect, the Polytechnic University of Madrid, through itdUPM, PRONATUR, and Foro Agrario, is promoting national and international projects on air decontamination, CO_2 sequestration, and rainwater management.

Madrid is a city with green areas that have a great artistic value and heritage (Zárate, 2003), and which have become popular, as seen in the new expansions in the upcoming urban plans.

We provide a brief description of the green infrastructure in Madrid, which demonstrates the dynamics in recent decades:

- The Spain 2050 Strategy (May sixth, 2021) on the need for green spaces to improve health and sustainability.
- Law on Climate Change and Ecological Transition (Law 7/2021) with Articles (17 and 24) on the importance of green infrastructure and connectivity.
- The EU initiative in 2020, through the EIT Climate KIC aimed at climate mitigation, created the Deep Demonstration platform that seeks Nature Based Solutions within the LIFE-PACT Project (LIFE, 2020).
- Simultaneously, in 2020, the Madrid City Council launched the Green Infrastructure and Biodiversity Plan with 180 specific actions, including the ARCE Network to connect green areas through corridors, where vertical agriculture must be an essential element.
- Madrid 360 is a sustainability plan, reducing pollution emissions, especially NOx, and increasing gas immission factors, dust particles, and heat, through the Metropolitan Forest as a 600 ha. green belt, and the planting of one-hundred thousand trees of native species, and other green infrastructure such as orchards, gardens, and recreation areas.
- In 2021, the roadmap toward climate neutrality in 2030 of the city of Madrid was presented, expecting to reduce emissions by 65% in 2030 compared to 1990 and reach climate neutrality in 2050.
- The "Plan A for air quality and climate change" is another of the instruments that include actions such as the "Madrid + natural" program, focusing on three levels: buildings, neighborhood, and city. At the first level, special attention is paid to green roofs and facades in municipal buildings, and a census is kept of these and of the achievements. (Comunidad de Madrid, 2014). At the neighborhood level, action is taken through environmental corridors that connect existing or future natural spaces, creating the appropriate green spaces through roofs and facades, which until now have been marginalized. On a greater scale, the greening of the river Manzanares is worth highlighting, as part of the blue-green project "Madrid Rio."

With this comprehensive vision, we can include the subsequent high-impact projects such as the "Madrid Metropolitan Forest" (75 km green ring and 100,000 new trees, available at https,//estrategiaurbana.madrid.es). Another series of actions address specific issues such as food supply in "Madrid Alimenta" which is based on existing infrastructure, such as the Network of Community Urban Gardens, considering the guidelines set out in the Milan Pact and the Network of Cities for Agroecology. The "Madrid City of Care" plan boosts community cohesion and environmental health with projects to improve schoolyards (MICO).

Significant aspects in the application of green infrastructures in Madrid have been the transfer from the research centers to both public and private sectors. The university and research centers, specifically itdUPM, UCM, and CESIC, have developed research programs with an ongoing application.

Training and information, essential components for entrepreneurship, are carried out through courses, conferences, and awards. There is a Massive Open Online Course (MOOC) on green cities. Also developed annual academic and business awards about green infrastructure, and seminars on urban agriculture in schools and municipal facilities.

Institutions such as the mentioned PRONATUR and Foro Agrario encourage mixed groups of researchers and entrepreneurs to facilitate activities supported by the City Council, Community of Madrid, and international organizations such as the EU.

Cultural centers, such as Matadero, are serving as a platform for international projects, with hybrid participation of researchers, officials, and entrepreneurs, as is the case of the Adaptive Cities through the NBS project. In parallel, the social connection is made through the Red de Huertos Urbanos (Network of Urban Gardens), and school programs.

4 Results

In the results, we focus the attention on several scenarios in which NBS may impact urban life: health, green infrastructure, socioeconomy in ecocities, and errors in the circular economy.

By 2050, millions of people will live in unhealthy and poor suburbs and marginal areas affected by climate change, pandemics, and labor imbalances due to new technologies (Gill & Handley, 2017). This forces us to take appropriate measures to try to correct the coming and current drama, seeking Nature Based Solutions within a socioeconomic action.

The challenge for the social actors of naturation (researchers, officials, managers, entrepreneurs, disseminators) is to design strategies that show the positive effects on health and socioeconomic aspects of urban naturation, creating an awareness favorable to its implementation (Bai, 2018).

- Health nature interaction must be promoted following acceptable objectives in different scenarios.
- Equity and social justice. It is appreciated that the presence of nature improves sanitary conditions. Slums are those that show a greater deficit in green areas and worse healthy conditions.

- Climate change and environment: The existence of green belts and areas, facilitates recreational activities in climatologically tempered and healthier areas.
- The governance of urban planning must account for the reduction of costs derived from naturation, encouraging public and private investments.
- Finally, naturation has a correlation with the socioeconomic status of the neighborhood, hence an increase in it is considered progressive.

4.1 How to Minimize Health Impact in Urban Areas?

As it has been previously mentioned, agricultural activities require work and exercises that are included in a discipline called agrofitness. Improving health by reducing air and noise pollution implies reducing the emission of polluting sources (traffic, heating boilers, industries) and increasing immissions, through greening. For this, some guidelines must be followed, looking for Nature Based Solutions and instead of having "a garden in the city, the best solution is having the city in a garden" (Wang & Bakker, 2014).

Some issues dealing with this statement assess the effects of greening on their environment, and they may have a wide range of results, depending on the resources, time, and specific characteristics of each urban area:

- Ecosystem services may influence air quality, climate, energy saving, and landscape. Climate change includes temperature, radiation, wind, humidity, and rain, among other elements.
- Ambient temperature influences human well-being, while surface temperature influences heat exchange (Whitworth & Handley, 2001), which is to be taken into account according to the objective sought.
- The adaptation of naturation to the required conditions requires a selection of suitable plants and fauna creating a microclimate, as well as a filter to air and noise pollution, offering food, aesthetics, and landscaping (Von Stulpnogel & Harber, 1990). In addition to health, nature environment has socioeconomic and energy saving impacts. Thus, green covers purify the air, promote biodiversity, save energy, and retain water (www.sempergreen.com). The vegetation and substrate absorb 50% of sunlight and reflect 30%, which means an energy saving of 33% in summer over air conditioning and reduced urban temperature. The combination of solar panels and plants increases the efficiency of the panels by up to 15%. Useful life of the roof is extended to 60 years, which allows to recover the investment in a period of 8–21 years.
- Mental and affective health interacts with recreational spaces, where the increase in biodiversity is reflected in the proliferation of wild native plants with greater pollination and the presence of butterflies, insects, and birds. By maintaining humidity, it acts as a fire barrier, while aesthetics and landscaping create a healing environment, with a positive impact on hospitals and residences. The presence

of recreational green spaces fosters social cohesion, facilitating common meetings and tasks (Burgos, OK Diario 24/ 9/ 2021).

- However, naturation has some negative aspects that must be considered. We refer to allergies and gases that give off and have biogenic volatile organic components. The location on the streets can act as barriers to drafts for their renewal. Therefore, the type and size of the trees matter, being advisable for the parterres and the green facades. Green covers also generate cooler, denser air that, as it slides down, displaces the most polluted hot air.

4.2 Evaluation of Green Urban Infrastructure

The complexity of urban greening makes it difficult to understand for the urbanite who needs to be able to appreciate and value the resources allocated to improve his or her environment in a sustainable way. Technical, socioeconomic, and environmental evaluations should be proposed using consistent criteria, as well as expert opinions summarizing the current situation and highlighting strengths and weaknesses, which are presented below.

Urban Green Infrastructure (UGI) is multifunctional in nature, allowing us to adapt it to the specific place and time, or to existing needs, which explains why this sector has had to evolve throughout history. The agroclimatic conditions shape the UGI model. Therefore, in rainy, dry or extremely wet climates, water management is essential to reduce flooding, or maintain the water supply and humidity. If there is air or noise pollution, the substrate and type of plants must be adapted to it. Water management is also necessary for the production of food, ornamental plants, or to facilitate recreational spaces with landscaping.

For this reason, UGIs have been identified as one-dimensional elements that have subsequently been expanded to include urban agriculture and bioclimatic architecture, among others. Today, it is seen as a helpful instrument to achieve the objectives of urban sustainability and resilience, fostering well-being and improving citizens' lives. A more complete vision is provided by authors such as Frantzeskaki and McPhearson (2022) who consider UGIs to be a system of interconnected ecosystems through technological and ecological integration and the construction of infrastructures that provide socioeconomic and environmental benefits.

In order to get a sense of the UGI of a city and its functions, such as photosynthesis and other services provided, a Vegetation Index is recommended, which would make it possible to raise awareness among citizens and city officials when prioritizing the policies to be followed.

The informative green guide for Spain considers four elements of UGI: green areas (parks, gardens, facades, and roofs), ecological corridors (connections between green areas), priority conservation areas (protection of flora and fauna), and buffer zones (transition between rural and urban areas). A model approach is ecocities.

4.3 Ecocities: A Prototype for Circular Urban Socioeconomic

The behavioral model of the urban ecosystem is like living organisms: they breathe (use exogenous or endogenous energy), grow (modify over time in their structure, size, and shape), reproduce (development and planning, level of training and information), move (land, air, water transport), feed (water, air, physical food) and generate waste (organic, wastewater, contaminated materials).

For their development, they need a series of resources, both material and human. Naturation provides, through plants and substrates, nutrients, chlorophyll function, recreation areas, and food. The city's services should include socioeconomic and cultural activities that are enhanced by green infrastructure.

As a result, an ecocity is a self-sustainable human settlement and the potential recycle of products, health, and natural resources, with natural ecosystems that provide healthy conditions to its inhabitants, which have a positive balance in the production–consumption relationship of products and resources, without surplus waste or negative externalities in their environment. Its inhabitants maintain adequate social relations with equity and justice.

Ecocities Builders among other organizations, have been working in this field, and have developed standards, the "International Ecocity Framework Standard Initiative." For its implementation, detailed information is needed on the sociodemographic and urban environmental conditions, based on a bidirectional integral management system, from top to bottom and vice versa, consulting all the actors of the city, from neighborhoods, businessmen, officials, politicians, academics among others, which will provide us with the bases to develop technical-socioeconomic tools that make cities alive, green, healthy, and sustainable. Ecocities are the prototype of socioeconomic and bio-circular activities.

4.4 Errors in Circular Urban Socioeconomic

There are errors that may occur in the recycled process to integrate nature into our environment.

In the supplier group, residential owners may experience disagreements in the availability of spaces to be greened for economic or social reasons. Urban communities face many problems and decisions, making this a social barrier.

In the industrial and service group, errors can be technical, and design related. The selection of suitable plants and substrates, their maintenance, or the project itself affects its viability. Therefore, professionalism is necessary in the sector since the mistakes that are made may discourage future investments and achievements.

Professional associations of architects, agronomists, and other professions must provide clear and acceptable UGI guidelines and quality controls.

Among users, errors are derived from a lack of awareness of the advantages and disadvantages of greening, a lack of maintenance or the lack of technical and financial resources.

In the government sector, errors occur when there are false contracts awarded, and in the control and monitoring of economic resources, not to mention the misalignment between objectives and the resources facilitated, corruption in the process of contracting and monitoring projects.

As a summary, the results are very heterogeneous, as we may expect from the variety of problems in urban areas and the possibilities offered by nature. Therefore, it is possible to mention some dimensions which have interaction between them. Better health has a positive correlation with efficient green infrastructures and socioeconomic situations.

5 Conclusions

Doctor Oliver Sacks, a prestigious British neurologist and writer states that only two types of non-pharmaceutical "therapy" have an essential relevance for patients with chronic neurological diseases: music and gardens. As he states "in many cases, gardens and nature are more powerful than any medicine. The experience is showing the amazement at the calming and organizing power, at the restorative and healing power that the gardens exerted on the brains of his patients, many of them with serious neurological disorders" (Sacks, 2007).

The multitude and variety of urban problems require finding multiple solutions, through multidisciplinary teams that in a coordinated way provide viable and sustainable responses, both locally and globally. Municipal corporations and local governments establish regulations adapted to their geographical and urban conditions, depending on their specific problems and available resources. On the world stage, there are actions such as the UN2030 Agenda for sustainable development which defined 17 targets with 189 goals in economics, sociology, and environment.

Broad scenarios are poverty, well-being, education, growth, gender, in specific areas such as water, energy, food, and climate change.

The Sustainable Development Goals (SDG) are interconnected, and cities and rural areas (*rurban*) are key elements for achieving them. Nature Based Solutions (NBS) are instruments for facing some of the *rurban* problems, considering their lower cost and sustainability.

A special challenge is the coordination of NBS with other actions as a circular socioeconomy in metropolitan areas.

Green Infrastructures (GI) have high correlation with high socioeconomic status. Therefore, increasing green areas improves health and environmental conditions in marginal communities, with lower heat island and water runoff.

In results, we have final conclusions in expert analysis, about the priorities in green urban infrastructure development. However, in relation to the project on gas recycling, there are only partial conclusions, as the project is going on. However, similar studies in other countries emphasize the positive effect of circular Nature Based Solutions in urban environments.

Nature adaptation to climate change includes the introduction of drought-resistant plants and native flora and fauna. It is also important the economies of scale, thus, green area dimensions of 1 hectare, are the minimum for having an adequate microclimate change.

We consider the role of green infrastructure (GI) in urban areas to have a significant impact on economic, social, and environmental scenarios.

As there is evidence of the positive effects of GI in city environment, we need solid arguments to convince the urban citizens about it, with objective data of the impact on our environment for a moderate climate change. The question is to clearly describe the type of action and their locations.

In technical and social scenarios, pollution and climate problems of large cities are experiencing a revolution in its urban planning, with the framework of the NBS, with innovative technical aspects, business and municipal entrepreneurship, and citizen participation, through the formation and information of an ecosystem consciousness.

Considering the climate change during last decades, we may look for new strategies in urban design based on biophysical features comparing urban and rural areas. Thus, we analyze urban biophysical evolution as a consequence of the substitution of nature by buildings. Green areas facilitate rainwater infiltration and cooling effects with a direct impact on the urban environment. Heat waves and storm flooding have economic and social impacts on urban society, with millions of euros in catastrophic losses and human lives.

As economic actions, urban planners need evidence to develop strategies for climate mitigation, health, and recreation. However, there is a lack of long-term studies on urban climates and their impacts. It is important to look for cross-disciplinary projects between regions, with research focused on local and single disciplines.

In the coming future, city stakeholders should incentive universities and research centers for long-term research in coordination with policy makers, businessmen, and social leaders for a sustainable horizon.

References

Bai, X. (2018). Six research priorities for cities and climate change. *Nature, 555*, 23–25.

Briz, J. (2022). *Green urban infrastructure Spain*. Ed. Agricola.

Briz, J., Kohler, M., & De Felipe, I. (2017). *Vertical urban agriculture*. Ed. Agrícola.

Briz, J., Kohler, M., & De Felipe, I. (2019). *Multifunctional urban green infrastructure*. Ed. Agricola.

Burgos, F. (2021). OK daily 24/ 9/2021.

Comunidad de Madrid. (2014). *Agricultura urbana en la ciudad verde, aspectos legales, educativos y de ocio*. http://www.foroagrario.com

Frantzeskaki, N., & McPhearson, T. (2022). Mainstream nature-based solutions for urban climate resilience. *Bioscience, 72*(2), 113–115. https://doi.org/10.1093/biosci/biab105

Gill, S. E., & Handley, A. R. (2017). *Adapting cities for climate change: The role of green infrastructures*. WHO.

Hernandez Aja, A., Velazquez, I., & Verdaguer, C. (2009). "Ecobarrios" for better cities. *City and Territory, Territorial Studies, XLI*, 161–162.

Kumar, P., Druckman, A., Gallagher, J., Gatersleben, B., Allison, S., Eisenman, T. S., Hoang, U., Hama, S., Tiwari, A., Sharma, A., Abhijith, K. V., Adlakha, D., McNabola, A., Astell-Burt, T., Feng, X., Skeldon, A. C., Lusignan, S., & Morawska, L. (2019). The nexus between air pollution, green infrastructure and human health. *Environment International, 133*(Part A), 105181.

López Fontan, I. (2003). Obesity, energy metabolism and measurement of physical activity. *Spanish Journal of Obesity, 1*, 29–36.

Martinez Ruiz, F. J., & Aparicio, J. A. (2017). *Measurement of energy expenditure in young people in horticulture*. Sports Department, Faculty of Sports Physical Activity Sciences INEF..

Martinsen, E. W. (1987). The role of aerobic exercise in the treatment of depression. *Stress Medicine, 3*, 93–100.

Páez de la Cadena, F. (1998). *History of styles in gardening*. Akal Ed.

Park Sin, A. (2012). Metabolic cost of horticulture. Activities in older adults.

Royé, D., Íñiguez, C., & Tobias, A. (2020). Comparison of temperature–mortality associations using observed weather station and reanalysis data in 52 Spanish cities. *Environmental Research, 183*(2020), 109237. https://doi.org/10.1016/j.envres.2020.109237. ISSN 0013 9351. https://www.sciencedirect.com/science/article/pii/S0013935120301298

Sacks, O. (2007). Musicophilia: Tales of music and the brain.

UN. Rio Conference on Sustainable Development. (2012, May). Health indicators in sustainable cities.

Von Stulpnogel, A., & Harber, M. (Eds.). (1990). *The importance of vegetation for urban climate urban ecology*. SPB Publishers.

Wang, Y., and Bakker, F. (2014). Effects of ecosystem services provided by urban green infrastructures.

Wang, Y., Zhuobiao, N., Shaoqing, C., & Beicheng, X. (2019). *Microclimate regulation and energy saving potential in urban green infrastructures cleaner production*. Elsevier.

Whitworth, V., & Handley, T. F. (2001). City form and natural process. *Landscape and Urban Planning, 57*(2), 91–103.

Zárate Martín, M. A. (2003). "Efectos de la globalización urbana de Madrid" *Boletín de la Real Sociedad Geográfica*, ISSN 0210-8577, N°. 139–140, 2003–2004, 386–388.

Alentejo Wines and Olive Oil Sustainability Programs: Background of Absorptive Capacity in the Development of Circular Economy Business Models

Ibrahim Prazeres, Vlademir Silva, Pedro Henriques, Isabel Joaquina Ramos, Conceição Rego, and Maria Raquel Lucas

Abstract The Alentejo sustainability programs for wine and olive oil seek different topics, such as the sectors' integration in a philosophy of social, environmental, and economic well-being at both local and regional level; the business cooperation throughout the identification and comparison of the producers' performance; the identification of improvements in production practices; and the statement of the certification scheme supported on sustainable production benchmark and the consequent recognition of the quality of Alentejo wines and olive oils for their sustainability performance. However, most companies find it difficult to integrate circular economy (CE) strategies and practices into their business models. To fill this gap, this chapter introduces an exploratory study investigating the backgrounds or ante-

I. Prazeres (✉) · M. R. Lucas
Department of Management and CEFAGE - Center for Advanced Studies in Management and Economics, Universidade de Évora, Évora, Portugal
e-mail: mrlucas@uevora.pt

V. Silva
CEFAGE- Center for Advanced Studies in Management and Economics, Universidade de Évora, Évora, Portugal

P. Henriques
Department of Economics and MED - Mediterranean Institute for Agriculture, Environment and Development, Universidade de Évora, Évora, Portugal
e-mail: pdamiao@uevora.pt

I. J. Ramos
Department of Landscape, Environment and Planning and CICS.NOVA.UÉvora - Interdisciplinary Centre of Social Sciences, Universidade de Évora, Évora, Portugal
e-mail: iar@uevora.pt

C. Rego
Department of Economics and CEFAGE- Center for Advanced Studies in Management and Economics, Universidade de Évora, Évora, Portugal
e-mail: mcpr@uevora.pt

C. Rego et al. (eds.), *Entrepreneurship, Technological Change and Circular Economy for a Green Transition*, Studies on Entrepreneurship, Structural Change and Industrial Dynamics, https://doi.org/10.1007/978-3-031-48079-9_11

cedents of absorptive capacity in successful CE business cases, using a qualitative multiple-case study, field observation, and focus groups.

Centering on the wine and olive oil sectors, 12 companies were interviewed, six for each program, using NVivo 12 for the analysis. The data were classified into first-order and second-order categories and then sorted into absorptive capacity dimensions. The results and their discussion highlight the relevant contribution of absorptive capacities to the CE in WASP and OASP firms.

Keywords Circular economy · Organization/management · Wine · Olive oil · Sustainability · Alentejo region

1 Introduction

The Alentejo programs for wine sustainability (WASP) and olive oil sustainability (OASP) started as two projects supported by European Union funds, through the Alentejo Regional Operational Programme. Both intend to support the three pillars of sustainability (environmental, social, and economic) and to promote their performance as a tool to face global challenges and barriers and promote Alentejo's brand in internal and external markets. Initiated at different times (WASP in 2017 and OASP in 2021) and, consequently, displaying different stages of development, both are voluntary initiatives organized by two institutions in charge of the respective sectors (CVRA—Alentejo Regional Wine Growing Commission and Olivum—Association of Olive Growers and Presses of Portugal) which target the producers (grapes and wine; olives and olive oil). Collective sustainability programs are common practice in some world wine regions and are gaining importance in markets, similarly to what occurred with Alentejo wines with sustainability certification, which have also been gaining a competitive position. In olive oil regions, the application of sustainability principles, in a broad sense, does not exist, with the OASP being a pioneer initiative in international and domestic markets. Considering their dependency on natural resources, the two programs assumed the protection and enhancement of these resources (solar energy, appropriate weather conditions, clean drinking water, healthy soils, and biodiversity, among others) through sustainable practices performed by a well-trained and highly skilled workforce.

The sustainability program member companies are induced to adopt both new green technologies and circular economy (CE) strategies within their business models. In a broader sense, these changes in technology mentality and business models, which seek to move to a more CE and to a better sustainability level in all dimensions, have been gaining momentum within the scientific, political, and managerial agenda. Particularly, the wine sustainability program, which is at a more advanced stage, has embraced CE as a driving force, putting into practice several strategies of eco-innovation through a regeneration production system, based on the use of renewable sources, respect for biodiversity, minimizing waste, protection of the environment and landscape, and generation of socio-economic benefits to the territory and the local community.

Despite OASP and WASP's focus on CE, enterprises face several difficulties to find transitional pathways to implement it. Previous studies, particularly those developed under the research project "Alentejo Circular", which aimed to promote CE in the Alentejo region (Magalhães et al., 2017), displayed simultaneously suitable technological and management practices, opportunities provided by the CE as a solution to the current need to reduce the environmental negative impacts, and solutions for the efficient use of resources as well as for the use and valorization of waste and other surplus.

The transition is not easy (Magalhães et al., 2017), especially for organizations that follow the economic linear model (Centobelli et al., 2020). Due to several difficulties of CE model adoption and inclusion in organizations' operations and processes, tackling of sustainability challenges slows down (Kirchherr et al., 2018). The resistance to change can be also explained by some controversy surrounding the CE concept (Murray et al., 2017). Additionally, there is a lack of research on the approaches for the successful implementation of CE business models (CEBM), which causes a gap between theory and practice (Geissdoerfer et al., 2017; Ghisellini et al., 2016), particularly in the field of agribusiness. To fill in this gap, this chapter introduces an exploratory study which investigates the backgrounds or antecedents of the absorptive capacity theory (ACAP) in successful CE business cases, using a qualitative multiple-case study, field observation, and focus group discussions.

Given the above, further exploration of the research work on these fields is required. Thus, following Marrucci et al. (2022 the research questions were:

RQ1. How do organizations acquire external knowledge on CE practices and opportunities?
RQ2. How do organizations assimilate external knowledge on CE practices and opportunities?
RQ3. What elements are fundamental to integrate newly acquired knowledge within the organizational processes and activities in order to boost the CE?
RQ4. What resources and processes are usually exploited by the organization to address newly transformed CE practices and opportunities?

This chapter attempts to provide answers to these questions. To achieve that, the chapter continues with a background section followed by the methodology used. Then, the results obtained are presented and the text ends with a brief section of conclusions.

2 Background

The current linear economy model, in which natural resources are taken and used for a certain period and discarded when they are no longer needed or wanted—take-make-use-dispose-pollute—is no longer functioning for the planet, the people, and, consequently, the firms. Instead, the CE emerged as an innovative model for the economic systems and business models, having the natural system as a reference, which produces-consumes-reintegrates, not generating waste and taking advantage of everything in each cycle (Ellen Macarthur Foundation, 2017).

CE is defined by the Ellen Macarthur Foundation as "an economic activity that builds and rebuilds overall system health" (Ellen Macarthur Foundation, 2017). It sustains the recognition of the economic efficiency in a broad and durable sense, for small and large business, individuals, and organizations at global and local contexts. However, CE is not only the reduction of the negative impacts of the linear economy, where natural resources are temporarily taken and used, being discarded in nature when not required. CE is an innovative model which represents a systemic change on producing-consuming-reintegrating, without generating or decreasing the waste level and trying to take advantage of everything in each cycle. It means a new way of (i) looking at the relationships between economic goods, wastes, markets, customers, and the use of resources; (ii) building long-term resilience; (iii) generating economic and business opportunities; and (iv) providing environmental and social benefits (Ellen Macarthur Foundation, 2017).

The CE is an ambitious challenge for the companies, a disruptive innovation, requiring a transition that should be applied not only to the material components but also to the natural, cultural, and social resources. The purpose is to redefine a new business model where nothing is considered waste or surplus and everything can become a resource, closing the loops, in synergy and symbiosis among the different sectors and with positive benefits for the whole society (Barrera, 2019). In this model, the consumers also play an important role since it proposes both a transformation of the production model and a paradigm shift in consumer behavior, which implies a change in people's mind-set toward a more responsible consumption behavior. Reduce, reuse, recycle, and recover are the basic principles that people must engage so as to achieve and accomplish the CE framework (Venâncio et al., 2022).

The literature review highlighted both how the wine and olive oil sectors have incorporated the premises of CE in their activities and the CE agribusiness models. In the first, the main concerns addressed were water, solid waste, energy efficiency, chemical use, land use, and ecosystems, in different steps of the chain such as viticulture or oliviculture, winemaking, or olive oil fabrication and distribution, respectively (Berardi & Dias, 2019, Sehnem et al. 2020; Chebbi et al., 2021). Similarly, the biomass valorization, resulting from pruning of live trees and plant removal to produce electricity, is also an important issue (Sales et al., 2021). Concerning the agribusiness models, despite the strong global goal to move to a CE, there is a lack of models that adopt strategies to improve the circularity of the production system, particularly because CE requires collaboration with other companies along the entire value chain and demands the implementation of suitable policies and tools (Hamam et al., 2021).

When analyzing the current wine and olive oil value chains in Portugal, which are two of the most important bioeconomic sectors in the country, it is disclosed that, despite some firms following a linear economy model, there are two important drivers inducing a CE. These are the Wine of Alentejo Sustainability Program (WASP) and the Alentejo Olive oil Sustainability Program (OASP), tools created and developed under two projects supported by European Union funds through the Alentejo Regional Operational Programme, assuming the CE philosophy and principles.

Especially in the WASP context, but also in the OASP, despite still being in development, much has been done to promote sustainable development and to align their activities with the optimal use of resources and the regeneration and restoration of the natural system, in agreement with the premises set by the CE. Their economic actors take natural and patrimonial resources, exploiting them in accordance with sustainability dimensions, leading to its conservation and preservation, as well as the social, cultural, and economic levels. They lead to the sustainable and competitive development of wine and olive oil sectors under a CE model that involves creating environmental quality, economic prosperity, and social equity for current and future generations.

Collective sustainability plans are common practice in some world wine regions, such as California, USA, and Chile, and are gaining importance in markets. The same occurred with the WASP pioneer initiative in Portugal, with the sustainability certification being a final goal, already confirmed by 12 certificated firms that are gaining sustainable competitive market positions.

3 Methodology

3.1 Research Approach and Theoretical Framework

The ACAP concept was defined by Cohen and Levinthal (1990) as "the ability of a firm to recognize the value of new, external information, assimilate and apply it, and apply it to commercial ends." The authors also claimed that ACAP, which is focused on the four mentioned key dimensions (acquisition, assimilation, transformation, and exploitation), is critical for the innovative capabilities and new technological developments of a firm with a view to improving competitiveness and market positioning.

Despite the critics that absorptive capacity concept means different things to different people (Kirchherr et al., 2017), it mainly refers to the capacity to assimilate value and use the knowledge transferred (Carayannis, 2012).

The capacity to recognize and understand the value of the external knowledge and the need to get it is largely dependent on the firm's human capital (Marrucci et al., 2022). To this contribute the results of investments in R&D; the contacts and partnerships with science and innovation institutions; or linkages with management actions, such as monitoring technological opportunities, market changes, consumer needs, suppliers, and competitor acquisitions (Scarpellini et al., 2020, 2020). Assimilation occurs when a firm has the capacity to integrate external knowledge through different procedures and routines which allow us to recognize, analyze, process, decode, and understand information obtained from external sources (Zahra & George, 2002). In this cognitive process, human capital is also a pertinent factor in relation to R&D and innovation (Scarpellini et al., 2017).

If a firm combines procedures involving the new assimilated knowledge and existing expertise, there is transformation (Zahra & George, 2002). In this context, universities and technology centers, and their respective knowledge transfer processes, play an important role in spreading knowledge and building links with the private sector and society (Scarpellini et al., 2012).

To meet the firm's need for the continuous regeneration of skills and resources exploitation requires the capacity, competencies, and efficiency of different organizational dimensions (human resources, operations, production, marketing, finance, among others), which must use new external knowledge to reach goals and gain competitive advantages. New knowledge on financial resource management appears to have a particularly positive effect in stimulating CE business initiatives (Aranda-Usón et al., 2019).

The improvement of ACAP is strongly related to the skills of workers and managers; thus, it is imperative to identify the company's needs and appeal to higher education institutions and research organizations to find the best solutions to those constraints, consequently enhancing the evolution of companies. For this to happen, companies need to be able to identify the knowledge-producing entities that can respond to the problems they encounter. Trust and familiarity, between companies and organizations that produce innovative knowledge, are crucial for improving the absorptive capacity. However, it is also known that there are some barriers to improving absorptive capacity. These difficulties are related, on the one hand, to the differences in cultural and expertise backgrounds between knowledge-producing organizations and companies and, on the other hand, to communication difficulties: on the companies' side it is sometimes difficult to identify and communicate the constraints, preventing improvement; on the part of knowledge institutions, in many cases, it is challenging to integrate the practical questions presented by companies in the context of the new knowledge produced (Kempton et al., 2021).

ACAP has already been used to study management strategies or environmental management systems (Todaro et al., 2020), process innovation (Jahan Khan et al., 2021), and environmental orientation in the consumption phase (Chan & Ma, 2020). Consequently, it was considered an important contribution to organizations faced with dynamic markets (Awan et al., 2021), such as those of wine and olive oil.

In successful CE businesses, some methodological evidence can be pointed out from the literature: (i) the scarcity of applied CE studies on business models, (ii) the limitation of a theoretical and conceptual approach, (iii) the prevalence of single-case studies (Marrucci et al., 2019) or quantitative large-scale research (Khan et al., 2020), and (iv) the scarce successful research exploring a multiple-case methodology (Daddi et al., 2019, Khan et al., 2020, Marrucci et al., 2022). Thus, following Stake (2005) and Marrucci et al. (2022), this exploratory research used a qualitative multiple-case methodology supported by an in-depth interview tool for data collection complemented with field observation and focus group discussion. The purpose was to capture business representativeness through the collection of vast comprehensive information on reasonably checked cases and apprehend the theory to investigate the backgrounds or antecedents of absorptive capacity in successful CE business cases. Additional methodological options were difficult to manage due to the length and complexity of developing a typical representative large-scale quantitative study; limited data about many cases or subjects; and the lack of firms' diversity using only a single-case study. Subsequently, the qualitative data collected was employed to support the ACAP theory, previously framed with sustainability topics (Pacheco et al., 2018; Walton et al., 2020) to explore the application and contribution of the CE in WASP and OASP firm's context.

3.2 *Data Collection and Analysis*

The primary and secondary data were collected following, respectively, an exploratory research approach with an in-depth interview tool plus field observation and focus group discussion, and a literature review which allowed the construction of the interview guidelines. The interviews were conducted during 2 months of research (January and February 2023) referentially with the top management of the firms. To have a more complete and holistic perspective of the integration of CE principles and process implementation within the firm, the CE project manager and/or the WASP and OASP managers were also interviewed. The purpose of this was to capture the breadth and totality of the organization's perspective on the acquisition, assimilation, transformation, and exploitation of the CE in the WASP and OASP context. At the end of the interview, the interviewers were encouraged to share added general comments about CE.

The interviews were supported by a semi-structured interview guide, organized into five topics, four of which related to the research questions and one of which contained the socio-demographic characteristics of the respondent. The structure of the interview guide was based on the literature on CE and ACAP (Scarpellini et al., 2019; (Camisón & Forés, 2010; Marrucci et al., 2022). The questions were the same for all participants, allowing for some level of consistent perception elicitation between them.

The selection criteria for the firm interviewees were (a) participation in WASP and OASP; (b) interest and availability to participate in the research; (c) being among the leading producers (recognized by other sectoral firms), (d) belonging to diverse specific geographical areas (Baixo Alentejo, Alto Alentejo, and Central

Table 1 Characteristics of the WASP and OASP organizations

Characteristics organization	Longevity (N° Years)	Employees (N°)	Area (ha)	Core business	Geographic Region	CE practices description
O1	60	374	2000	Agribusiness (olive oil, wine, cattle)	Alentejo Central	Water and energy consumption reduction; use of by-products (stalks, bagasse (wine); leaves and cuttings (olive oil) in animal feed)
O2	10	45	800	Agro-industrial crops (olive oil, wine, almonds)	Baixo Alentejo	Water and energy consumption reduction; use of tartaric acid salts in the wine stabilization process
O3	26	49	275	Agro-industrial crops (olive oil, wine)	Alentejo Central	Organic production, sustainable buildings (green roof), use of secondary raw materials
O4	27	97	700	Organic wine and olive oil	Alentejo Central	Water and energy consumption reduction; energy recovery of olive stone biomass, pruning
O5	15	39	500	Agro-industrial crops (olive oil, wine)	Alto Alentejo	Water consumption reduction Energy consumption reduction
O6	6	25	470	Agro-industrial crops (olive oil, wine, almonds)	Alentejo Central	Recovery of by-products, transformation of plant waste into soil

(continued)

Table 1 (continued)

Characteristics organization	Longevity (N° Years)	Employees (N°)	Area (ha)	Core business	Geographic Region	CE practices description
O7	300	65	680	Agro-industrial crops (olive oil, wine, almonds)	Baixo Alentejo	Water and energy consumption reduction; agronomic valuation of by-products (stalks, bagasse, pruning, leaves, WWTP sludge)
O8	18	128	2000	Olives/grapes integrated production system	Alto Alentejo	Conversion of olive by-products into thermal biomass and electrical energy generation Reduction in the use of water and chemicals
O9	25	25	600	Olives/grapes integrated production system	Baixo Alentejo	Improvement in the use of recycle materials; water and energy consumption reduction
O10	22	13	150	Tourism/ olives/grapes integrated production system	Alto Alentejo	Recover of organic waste (feed, energy, and fertilizer); reducing consumption of fossil fuels
O11	25	9	300	Olives/grapes integrated production system	Baixo Alentejo	Water and energy consumption reduction (irrigation and cleaning systems)
O12	25	250	8000	Agro-industrial crops (olive oil, wine, almonds)	Alto Alentejo	Water and energy consumption reduction; agronomic valuation of by-products (stalks, bagasse, pruning, leaves)

Source: Author's elaboration

Alentejo), and (d) answering with clarity and comprehensiveness. Table 1 sums up the organization's characteristics participated in the research identified from 1 to 12 (O1 to O12).

Overall, 12 organizations were interviewed face-to-face and audio-recorded with participants' consent. All of them held an academic degree and multidisciplinary expertise. Additionally, if necessary, the information collected was harmonized through telephone and WhatsApp. Basic information about the study context and motivation and the preservation of participants anonymity were provided before the interviews. Generally, most of them didn't answer exclusively to the questions, providing additional information on other issues such as qualified human resource scarcity, R&D partnerships, lack of sectoral public policies and strategies, logistics, and market access, among others. Apart from ACAP topics, all these subjects were also discussed during the 2 focus groups conducted until March of 2023 with 12 participants. Regarding the data analysis, the interviews were transcribed verbatim, and the transcripts were used to undertake content analysis using NVivo 12 software for the analysis and classification into different order-categories, which were sorted into absorptive capacity dimensions and development potential. Some of the additional comments and information collected under the focus groups were also integrated and discussed with the results.

The information obtained was structured and classified into the four ACAP dimensions, as shown in Fig. 1. They include first- and second-order categories. The first brings together relevant information extrapolated from the interviews. The second was obtained by dividing the central themes and relating the first category with each other through a combination of inductive and deductive thinking. This step, based on the ACAP theory, was essential to convert the data into the theoretical framework background and identify concepts related to the theoretical and practical adoption of each ACAP dimension. In the end, the second-order categories were aggregated into the four ACAP dimensions. Each background was linked to acquisition, assimilation, transformation, and exploitation (Fig. 1).

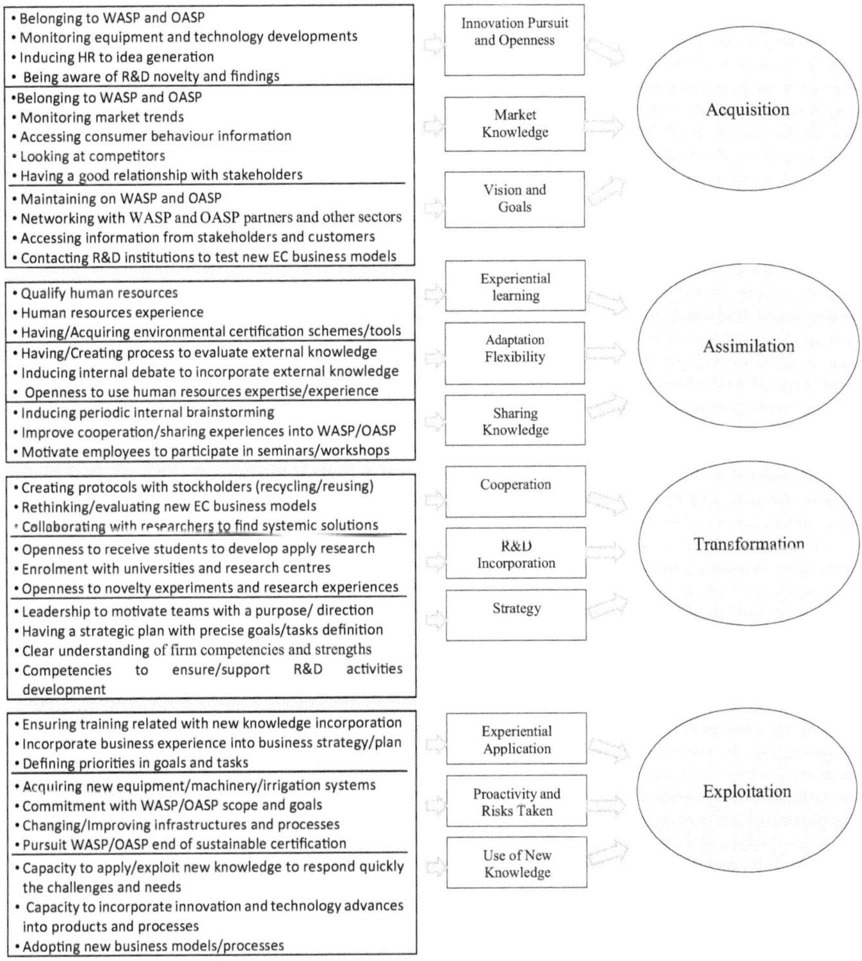

Fig. 1 Data structuration process. Source: Author's elaboration

4 Results

Figure 1 synthesizes the key dimensions of ACAP in CE gathered during the empirical work, which are detailed in the following results' subsections.

4.1 Acquisition

Regarding the acquisition, three antecedents were identified: market knowledge, openness and search for innovation, and the vision and objectives of the organization/entrepreneur. Most companies claim to be simultaneously aware of market trends and dynamics and competitors' performances in relation to consumer satisfaction. Even so, not everyone has a comprehensive view of the market, either because they sell wine and/or olive oil in bulk through previously established contracts or because they only sell it at the local market. However, the vast majority are up to date with the functioning of the national and international market, follow European directives in terms of climate neutrality, CE, and sustainability among other sectoral policy measures, and are aware of future evolution trends.

As for the acquisition of knowledge, all organizations mention the fact of belonging to the WASP and OASP and being committed to complying with their respective references, as the main sources of technically and scientifically supported information on sustainability in all its dimensions, including CE.

Regarding market information, the acquisition by companies can be aggregated into different groups. One group included those that continue to mention belonging to WASP and OASP (O1, O10, O11). Another group of the firms refers to the pressures from the market (especially external) and from consumers (O2, O3, O4, O5, O8) and the fact of being attentive to its evolution, dynamics, and trend prediction. The competition was only mentioned by another group of two companies (O8, O11) and following good practices from more experienced producers only by O10. Having visionary leadership and a highly qualified, up-to-date, and experienced multidisciplinary team (O2, O6, O7, O8, O12) was mentioned by another group. Additionally, protocols with teams and research centers and partnerships in studies on the carbon footprint of the product were also antecedents mentioned by other group of organizations (O1, O9, O8).

"We have almost real-time knowledge of the market (monitoring of prices of products and factors, distribution and communication processes, trends and needs of consumers and competitors) and we follow technological evolution (digitization of technology)" (O2).

"Observation of markets and consumers and the legislation produced in the meantime and to be complied with, as well as participation in projects, training and visits to specialty fairs are our main sources of information" (O3, O4).

"Information is obtained through visits to fairs and through the visibility of some of the competitors' CE practices" (O11).

"Contacts with partners, including distributors and suppliers, observation of market trends and competitors and having protocols with some research centers to study the carbon footprint of the product are our sources" (O8).

It should be noted that the companies under study are at an advanced level of technological use, following innovation and the latest developments, whether in irrigation systems, soil management, pest and disease management, energy efficiency, productivity optimization and cost minimization operations, or human resources management, among others. All can be considered benchmarks in the production of grapes and wine and olives and olive oil from vineyards and wineries and olive groves and modern mills, respectively, with quality and traceability monitored in all processes. In general, they combine tradition with new technologies and innovative agricultural techniques, including CE practices, for both environmental and economic reasons, such as the reduction of production costs and the need to find solutions imposed by identified problems with surpluses and residues, resultant from wine and olive oil production. Also in this case, the contribution of WASP and OASP networking, the monitoring of markets and consumers, the observation of competitors, and the positive relations with stakeholders were elements highlighted as positive. In addition, the companies recognized the importance of R&D in the process of acquiring knowledge and finding business opportunities. According to Litchfield and Gentry (2010), these skills not only facilitate the integration of knowledge but are essential to learning and organizational innovation. O1 and O6 are very oriented toward technological digitization and automation, although, like the others, they recognize the high investment associated with R&D and innovation as well as, in many situations, the slowness in obtaining results.

"The automation and monitoring of all processes are constant. They range from temperature control to sanitary management, degree of maturity at the reception of the raw material and subsequent temperature control, classification and storage conditions" (O6).

In the results obtained, the relevant role of human resources management in this whole process, either for its multidisciplinary, experience and qualification (O6, O7, O8, O12), or for its ability to generate innovative ideas and solutions and to communicate them to superiors, is evident. O2, O3, O5, and O7 affirm that they motivate and encourage their internal employees to practice all premises, in particular those related to sustainability.

"It is through our multidisciplinary, highly qualified, and up-to-date team, with profiles that complement each other, from agronomists to technicians with more than 25 years of experience and, above all, the ideas of younger people that we make progress. The difficult thing is often to manage and evaluate all the ideas communicated" (O3).

Not all companies have a formally defined business model or strategy, although all have identified goals and objectives. It operates throughout the entire chain, from the acquisition of materials and other products from suppliers, including raw material grapes and olives in some cases (O3, O4, O6), to the sale of wine and olive oil, to managing complaints, and in some cases (O3 and O4) to following up on the after-sales product. The O2, O6, O8, O9, O11, and O12 companies also provide services abroad for a stipulated payment at market price.

Co-creation and product development processes involving stakeholders and customers take place at O4, O2, and O12. O1 and O7 have been adjusting their business models and organizational strategies, becoming structurally more flexible and with greater agility in decision-making processes. O2 usually collaborates with partner companies in other agri-food and agri-industrial sectors, such as packaging producers, so as to develop materials and components that are more suited to CE, while ensuring quality and safety. It also establishes partnerships with machinery and equipment representatives to develop or adapt them to their needs.

4.2 Assimilation

Also in assimilation, the human resources management is a key factor in all antecedents, due to its qualification, expertise, experience, motivation to participate in events and internal debates, and cooperation, in addition to the adoption of environmental certification schemes and analysis tools and understanding of external information. Experiential learning, knowledge sharing, and adaptation flexibility were identified as antecedents of assimilation.

"The assimilation of knowledge is done through training, qualification, experience, motivation, and commitment of internal employees. Without them, despite the considerable investments in reducing water consumption, in a water treatment plant for reuse in irrigation, in the agronomic valorization of pruning residues, leaves, bagasse and sewage sludge and in the energetic valorization of biomass, it would not have been possible to achieve results or change/incorporate new washing and cleaning techniques, such as "Clean in Place", which includes reducers and a closed circuit and inertization of vats with argon, nitrogen or steam" (O4).

"The production director, although new to the company and with a long experience working in both sectors, has an open mind to innovation and a high pride and commitment to sustainability. He is an agronomist with deep knowledge of production systems and technologies, soil management, irrigation and pests and diseases (avoiding them) and the use of equipment, always motivated and motivating his teams to the importance of knowledge and its update" (O2).

"Training is something very important insofar as behavioral barriers are often the impediments to further development of CE" (Q9).

The employees' experience is also praised by O3 and O8, who additionally mention the importance of a fair management system, performance evaluation, and rewarding of human resources. All companies are unanimous in regard to the difficulty in recruiting qualified human resources due to scarcity in the labor market and emphasize that the success of many practices relies on their striving commitment to sustainability and a future healthier planet while learning from previous experiences. Companies O1, O9, and O8 carried out studies on the carbon footprint of the product (O1, O9, O8) and O12 carried out an LCA to assess the environmental performance. O3 is in the process of obtaining environmental certification.

In terms of human resources recruitment policy, not all companies have one. Even so, seeking to respond positively to the WASP and OASP in the criteria relating to socioeconomics and regional development, they seek to hire in the region (graduated or not graduated) although there is a shortage of qualified personnel committed to sustainability. Nonetheless, according to O4, it is possible to get good technicians, with an open mind to innovation, especially when borne by personal values such as sustainability.

Collaboration, knowledge sharing, and participation in colloquiums, seminars, congresses, and other similar events are other decisive elements for assimilation. For O6, knowledge sharing takes place through meetings that adopt an open and iterative form of communication where knowledge and experiences are shared and internal learning is induced. When necessary, ask the service or equipment provider for a clarification session and specific training for employees. Continuous training on emerging topics takes place regularly in the organization's life. CE began to be thought of to be integrated into the company in this way and due to legal obligations.

Participation in technical and scientific events, particularly on sustainability and CE, is evidenced by all organizations. O5 emphasizes that the reduction in the consumption of fossil fuels, with the substitution of diesel for biomass and the use of solar thermal systems for water heating, as well as the recovery of washing and cleaning water, resulted from employee training actions and participation in specialized events.

"Employee training and participation in events on CE principles resulted in the creation of specific skills in identified areas and in the enhancement of collaboration in knowledge networks resulting from these events" (O5).

Collaborative work processes within functional structures for exchanging experiences in periodic or sporadic meetings between internal employees to discuss specific or general matters, disseminate new knowledge, or generate ideas of organizational interest are current practices in several organizations (O2, O3, O8, O12).

"Employees are motivated to carry out weekly research on innovations and cutting-edge technologies, which they discuss as opportunities to be taken by the company in periodic weekly meetings, as well as being encouraged to participate in congresses, workshops, fairs and international events" (O6).

"We consider it is important not only to train our employees to balance quality, sustainability, and profitability, but also other socio-economic agents and public authorities on issues related to CE (in my opinion, most agave no skills) and also to promote concrete actions to mitigate threats and enhance opportunities. Hence the importance of including these agents in events organized by the company" (O6).

Adaptations are often made based on technologies designed or already tested in other sectors.

"In fact, we transferred or adapted some technologies and techniques successfully used in the vineyard and winery to the olive grove and mill" (O3, O4).

"Some of the CE solutions applied derived directly from the work developed in the WASP and OASP references and in other environmental certification processes that we have in progress" (O11).

4.3 Transformation

In the transformation dimension, the results allowed the identification of three antecedents: the incorporation of R&D, the strategy, and the cooperation. The importance of the relationships between higher education institutions and research and technological development centers, as creators and disseminators of knowledge, is of crucial importance to the positive evolution and performance of organizations. This cooperation is an essential mechanism to promote common and societal interests (Orazbayeva et al., 2019).

Almost all companies studied have links to R&D through partnerships with universities and research units, within the scope of specific CE projects or other developed actions and established collaboration protocols. Only O5 and O11, except for participating in WASP and OASP, had no other experiences of this nature. O5 considers that there are barriers of technological nature and reliable data on the real impacts of CE on companies, whether in terms of financial uncertainty due to the resulting revenues, legal issues, or its framework. Moreover, the need to invest in new procedures and assets, together with difficulties in accessing finance, are barriers that are difficult to overcome.

"*Many of the decisions taken resulted from participating in the Alentejo Circular project, with IQA and the University of Évora.*" (O4).

"*Some of the energy efficiency options we made resulted from experimentation with the university*" (O7).

Strategy, another antecedent of transformation, is an essential tool to ensure organization, innovation, stability, focus, and alignment of procedures and actions in any context. Even so, apart from O2 and O4, the remaining companies do not have a well-defined strategy that guarantees targeted, well-designed, and effective processes that are focused on the ability to integrate innovation and technological development and face challenges and change. However, O3, O4, O7, and O12 are aware of the importance of innovation and plan to guide the functional areas of management toward innovation.

"*Innovation is not a necessity but an imperative for evolution and access to markets"* (O2).

"To innovate we need a fundamental element, the support of consultants and all their advice and proposals to incorporate and change practices and processes. This is what happened with the air quality and gas and odor emission control system and its frequent monitoring, with good results" (O1).

"We have no vocation to innovate, only to acquire innovations" (O11).

"Although the market already values CE, it is not willing to pay for the resulting innovative products" (O6).

"When the innovation and the knowledge obtained is evaluated by the company's managers as being of interest and the potential CE tasks and activities are considered feasible in the current organizational structure, we move towards implementation, establishing objectives, strategies and change plans that communicate to collaborators. In some cases, we experiment on a small scale (as was the case with composting). The interest of the international markets in this topic and in the products that prove to have been produced in accordance with EC is another element that weighs on the decision" (Q10).

Cooperation depends on the will of the managers and the usefulness and value attributed by them. Within the scope of WASP and OASP, all cooperate with each other. Out of this context, when they consider collaboration to be important and even a priority, relationships and collaboration with all interested parties are encouraged. When not, this cooperation does not exist or is limited. The business models of O2, O3, O4, and O9 incorporate protocol cooperation processes with suppliers and other public and private economic agents. Other companies (O1, O5, O6, O7, O8, and O11) collaborate sporadically and occasionally with suppliers in specific actions which are advantageous for both. The O10 and O12 have few synergies with suppliers and other stakeholders.

"In addition to the need for human, material and financial resources, the ability to innovate, change and adopt more sustainable EC solutions depends on the resourcefulness of the manager or decision-maker" (Q6).

"We have a return system for packaging and other materials and components established with each supplier" (O3).

"Many of the decisions taken by the company, regarding the treatment and conservation of water and the use of by-products in heat energy, resulted from cooperation protocols with partners. Used packaging is also recycled by entering a cooperative return system. The grass is also covered and the consumption of fertilizers is reduced and organic waste from the plant nursery is recovered in cooperation with suppliers" (O2).

4.4 Exploitation

In the exploitation dimension, the antecedents identified were the use of new knowledge and experiences as well as proactivity and risk-taking. It is worth highlighting the relevance of the proactivity of the organizations studied in the search for new solutions to their problems, and their flexibility to adjust to changes in the market, technology, policies, institutions, management, and society.

"We have made a structural and organizational change that resulted from the adoption of a strategy where CE practices were contemplated as well as the

involvement of stakeholders, whether in the supply of more ecological materials and components (recycled), or in the collection of plastics, paper and other recyclable materials, as well as reducing water and energy consumption and minimizing surpluses and waste" (Q7).

It was because of their proactivity that they all voluntarily joined the WASP and OASP, seeking to explore new opportunities and solutions and improve agronomic, agro-industrial, and environmental performance practices, aiming to gain in sustainability and competitiveness. The adoption of CE best practices is included in this scope, although there is still a long way to go. All companies consider that there are great CE opportunities though there is still a need for more R&D to find more practical and more economical solutions.

The companies' relatively high aversion to risk leads it to prefer to take small-scale, but safe, CE measures rather than making qualitative and quantitative leaps that imply greater exposure to risk (Q5).

O1 has a high degree of innovation in their treatment of washing water, with a unit that allows its reuse in irrigation and the automation of the industrial process in the extraction of olive oil at low temperature (up to 26 °C), using only physical processes and strict internal criteria and quality control operations at all stages of the production system.

"It was the adoption of best practices, the high degree of self-innovation and internal and external collaboration that led the company to move towards a circular business model. If necessary, the company imports external knowledge for its products and processes and integrates new equipment suggested by suppliers. It is also important to thoroughly review all information and knowledge on an ongoing basis" (Q2).

For O2, cooperation between employees and between functional areas stands out. At O7, self-innovation emerges, with the creation of a hotel for insects that are useful in the biological fight against pests and diseases and the reduction in the consumption of insecticides and nests to promote biodiversity.

Employees at O9 made proposals for improving human resources management and air quality. The adoption of some CE strategies and practices such as adapted and precision irrigation with efficient pumping and off-hours and pressure/flow regulation was suggested by one of the O5 technicians.

"Equipped with cutting-edge technology and machinery and applying optimal processes and procedures in terms of efficiency and air quality, the industrial unit is a reference in the region and has served as a model for other units created later" (Q9).

"The fact that the European Commission has identified the company as a leader in the industry and appointed it as a member of the Technical Secretariat for the study on the carbon footprint of the product, is one of the fundamental elements in the exploration of knowledge and a source of pride for all the collaborators" (Q8).

"Boosting CE at an organizational level involves incorporating CE as an objective in the company's planning and management and strategy. Also, and consequently, for

its operationalization, meaning the development of the recommended CE activities and the valorization of the resulting new products" (Q7).

Having a positive relationship between CE and results—either through their valuation both by market and by consumers who are willing to pay for items resulting from CE processes or through innovation and appreciation by other industries and sectors—was considered essential by most companies for the expansion of circular business models.

5 Discussion

The results highlight the significant contribution provided by ACAP methodology to the analysis of the CE in firms belonging to the wine and olive oil sustainability programs. Also, the findings highlight several antecedents for each absorptive capacity dimension and development potential, exploring their key concepts. It should be noted that the results obtained are in line with those of Marrucci et al. (2022) in a similar study conducted in Italy. The study is in fact by Marrucci et al. (2022) one of the few in the literature that considers ACAP and CE. Most works study CE at a particular stage of the business model or in an aggregated version, without considering the contributions of each dimension and their interconnections.

The first result to be noted is the very close connection between the ACAP dimensions and the antecedents, demonstrating, as a result, the strong bond between all dimensions, which cannot be considered isolated and separated, as shown by Marrucci et al. (2022). Previously in a qualitative approach of case studies, Chalmers and Balan-Vnuk (2013) had already identified the role of the organization in the ability to transform the acquired knowledge into innovation.

In this study, the various factors that contribute to the adoption of CE by WASP and OASP companies, namely ACAP's external and internal antecedents, are in line with the results of Lewin et al. (2011) and Chalmers and Balan-Vnuk (2013). These authors reported knowledge sharing as an internal capability and identified and recognized, as external capabilities, the value of externally generated knowledge and the ability to learn from and with partners, suppliers, customers, competitors, and consultants. These factors were also found in the results of this study. On the other hand, studies on the context of the effects of higher education institutions (see, e.g., Kempton et al, 2021) also highlight the importance of innovation, knowledge transfer, and absorption capacity in institutions (companies and other entities) to make it possible to implement more sustainable production and organization methods. In this framework, the existence of new knowledge only becomes truly useful when it is appropriated by organizations and incorporated into production processes and value creation.

The results also show that the transformation of knowledge fundamentally results from R&D, that this knowledge is supported by research centers and institutions, and that it has high costs. Also Sharma et al. (2021) stated that there are financial barriers to the development of CE and innovation in companies, particularly in SMEs, supporting the need for public policy measures to overcome and encourage this adoption and development. The financial barrier is also highlighted by Marrucci et al. (2021) as an obstacle to the diffusion of LCA and other environmental management and certification systems. Likewise, the authors point out "technology upgradation" as a major prerequisite for the implementation of the CE. Similarly Huynh et al. (2022) state that firms need knowledge and technology to perform the CE actions. To Cicchiello et al. (2023) equity crowdfunding can assist this transition, proposing an array of possibilities for supporting green entrepreneurship.

The work carried out also showed the importance of leadership and managers (top and middle) and human resource management for the transition to an CE. An organizational culture and open communication, conducive to CE and innovation-oriented changes, are perceived as associated with strong leadership and managers' commitment to their employees. The training of workers is also perceived as fundamental to match the firms' skills with CE opportunities. However, the transition to CE and organizational and market changes need to be accelerated (Hazen et al., 2017). For that, it needs more applied research and improvements in firms' dynamic capabilities (Bertassini et al., 2021). To Pham and Tran (2023) building CE solutions takes time and close cooperation with stakeholders. For the authors, the application of the CE business model allows for the whole economy to make better use of resources and energy at a business, local, regional, and world levels so as to operate sustainably. Additionally, the circular supply chain helps trace the origin and control the stages of raw materials, of production, of processing, and of consumption. Simultaneously, it contributes to the development of social infrastructure and enhances the social responsibility of the business community and the economy. These results are also in line with the results obtained in this work.

Adopting CE initiatives and integrating CE business models at the firm level and along the supply chain is key to pursuing business strategies for sustainability (Ethirajan et al., 2020; Rehman Khan et al., 2021; Shahzad et al., 2021). This is also demonstrated by the results of WASP and OASP case studies. Considering the current knowledge-based economy, investing in and adopting new technologies and sustainable practices is valuable for long-term success as well as managing organizational knowledge (Shahzad et al., 2021). To Donner and Vries (2021), business models for a circular bioeconomy depend on various action levels and need radical combined organizational and technological innovations for a most efficient use of agricultural waste and by-products. Specifically for the olive oil sector, Donner et al. (2022) showed that CE and bioeconomy can contribute to transitioning toward more sustainable production and consumption.

Finally, the findings of this work also showed strong aspirations for improved sustainability in the WASP and OASP context. Furthermore, the CE strategies (waste and by-product valorization) foster the creation of innovative practices. These results are in line with Donner et al. (2022) for the olive oil sector.

6 Conclusions

Evolving productive activities toward a paradigm of greater predominance of the CE is a goal that producers, particularly agricultural producers, and society, in general, are confronted with.

This chapter introduces an exploratory study investigating the backgrounds or antecedents of the absorptive capacity (ACAP) theory in successful CE business cases, based on the experiences of the Alentejo programs for wine sustainability and olive oil sustainability. To this end, the investigation followed a methodology for collecting primary and secondary information. For primary information, the study used an exploratory research approach with an in-depth interview tool plus field observation and focus group discussion. For secondary information, the literature review is used to support the background of the study and the construction of the interview guidelines.

This study allows us to conclude that, in all areas of ACAP analysis, the importance of knowledge and, naturally, of the workers who possess it and apply it in their daily activities is fundamental to the construction of an economic paradigm where there are relevant actions aimed at increasing the importance of CE practices, particularly in agricultural production.

Effectively, in the context of the acquisition, the relevance of knowledge of the characteristics associated with the operation of the market, and therefore the importance of competition, stands out: the fact that companies are aware of the importance of new knowledge and innovation, as well as the relevance of the vision and objectives to be achieved by the business organization.

In terms of assimilation, the human resources management and the essential contribution of different types of workers are still more prevalent, which stems from their experience, qualification, flexibility, and motivation to participate in the company's activities, as well as in actions that involve interaction with external entities. In this case, the importance of perceiving the relevance of adopting environmental certification procedures or adopting innovative knowledge in production practices, with a view to gaining better leverage of available resources, minimizing waste, is highlighted.

Regarding transformation, the results obtained highlight the importance of incorporating R&D as well as understanding and assimilating the strategy and cooperation processes. In this domain, interaction with science and higher education institutions is crucial, as these entities enhance access to innovative and differentiated knowledge, with the capacity to improve performance in business organizations. These inter-institutional articulation processes between companies and science institutions reinforce the importance of cooperation, with a view to a more effective creation and sharing of knowledge, that is, underlining the relevance of knowledge co-creation processes.

Regarding the exploitation domain, the use of new knowledge, the ability to implement new experiences, proactivity, and risk-taking were topics considered as decisive. The proactivity of the organizations studied in the search for new solutions to their problems and the flexibility to adjust to changes in the market, technology, policies, institutions, management, and society were identified as very relevant. So, the importance of organizations being able to rely on qualified, competent employees who are knowledgeable about the strategy and mission of business organizations is highlighted.

The results obtained from this study emphasize the importance of human capital for business organizations to be able to respond to the challenges of innovation and value creation in the context of growing competition. The integration into sustainability programs, which identify the main milestones to be achieved for a more economically efficient, socially fairer, and more environmentally balanced resource exploitation performance, results in broader benefits for companies, with the ability to access and integrate innovative knowledge into their context, making it easier to transform it into value creation.

Finally, it should be noted that the study would benefit from a greater number of organizations interviewed and the consultation of secondary sources of information on companies, such as annual and sustainability reports. Nonetheless, the qualitative methodology is indicated in the literature as limiting the risk of social desirability bias; thus, some distortions can be presented in the results as a consequence of different situations. On one hand, all respondents belonged to specific sectors. On the other hand, all the sample consisted of WASP and OASP volunteer members, which presumes an intrinsic motivation to adopt the best sustainable practices. Additionally, the interviewees aimed to provide politically correct answers or responses that could be seen by third parties as favorable from a sustainability point of view.

Ultimately, both the limitations and the findings can be the starting point for future research and provide theoretical and managerial implications with forthcoming propositions. Thus, future studies may not only increase the size of the sample and the number of companies interviewed, but also advance and segment the study of the role of ACAP antecedents in different companies, in terms of dimensions, forms of organization and structures, life cycles, stages of maturity, and availability of resources. Exploring the influence of ACAP in evaluating the performance of sustainable companies could be another line of future research.

A common regulatory framework, public financial measures, new circular business models using innovative technologies, multi-actor collaboration, and increased consumer awareness of the CE and new waste-based products are necessary for more efficient and sustainable use of WASP and OASP resources.

Acknowledgments *This research is supported by national funds through the FCT (Portuguese Foundation for Science and Technology) under the projects UIDB/04007/2020, UIDB/05183/2020, and UIDB/04647/2020. Funding was additionally provided by the FCT– Portuguese Foundation for Science and Technology under the research contract PRT/BD/152273/2021 to Ibrahim Prazeres.*

References

Aranda-Usón, A., Portillo-Tarragona, P., Marìn-Vinuesa, L. M., & Scarpellini, S. (2019). Financial resources for the circular economy: A perspective from businesses. *Sustainability, 11*(3), 888. https://doi.org/10.3390/su11030888

Awan, U., Nauman, S., & Sroufe, R. (2021). Exploring the effect of buyer engagement on green product innovation: Empirical evidence from manufacturers. *Business Strategy and the Environment, 30*(1), 463–477. https://doi.org/10.1002/bse.2631

Barrera, E. A. (2019). Exploring circular economy as an innovative approach through a wine tourism itinerary. MSc thesis, universities of Porto, Rovira I Virgili and Bordeaux. Retrieved January 15, 2023, from https://hdl.handle.net/10216/126700

Berardi, P. C., & Dias, J. M. (2019). How has the wine sector incorporated the premises of circular economy? *Journal of Environmental Science and Engineering B, 8*, 108–117. https://doi.org/10.17265/2162-5263/2019.03.004

Bertassini, A. C., Ometto, A. R., Severengiz, S., & Gerolamo, M. C. (2021). Circular economy and sustainability: The role of organizational behaviour in the transition journey. *Business Strategy and the Environment, 30*(7), 3160–3193. https://doi.org/10.1002/bse.2796

Camisón, C., & Forés, B. (2010). Knowledge absorptive capacity: New insights for its conceptualization and measurement. *Journal of Business Research, 63*, 707–715. https://doi.org/10.1016/j.jbusres.2009.04.022

Carayannis, E. G. (2012). Absorptive capacity and organizational learning. In N. M. Seel (Ed.), *Encyclopedia of the sciences of learning*. Springer.

Centobelli, P., Cerchione, R., Chiaroni, D., del Vecchio, P., & Urbinati, A. (2020). Designing business models in circular economy: A systematic literature review and research agenda. *Business Strategy and the Environment, 29*(4), 1734–1749. https://doi.org/10.1002/bse.2466

Chalmers, D. M., & Balan-Vnuk, E. (2013). Innovating not-for-profit social ventures: Exploring the microfoundations of internal and external absorptive capacity routines. *International Small Business Journal, 31*(7), 785–810. https://doi.org/10.1177/0266242612465630

Chan, R. Y. K., & Ma, K. H. Y. (2020). How and when environmental orientation drives corporate sustainable development in a cross-national buyer–supplier dyad. *Business Strategy and the Environment, 30*(1), 109–121. https://doi.org/10.1002/bse.261

Chebbi, A., Franzetti, A., Duarte Castro, F., Tovar, F. H. G., Tazzari, M., Sbaffoni, S., & vaccari, M. (2021). Potentials of winery and olive oil residues for the production of Rhamnolipids and other biosurfactants: A step towards achieving a circular economy model. *Waste Biomass Valor, 12*, 4733–4743. https://doi.org/10.1007/s12649-020-01315-8

Cicchiello, A., Gatto, A., & Salerno, D. (2023). At the nexus of circular economy, equity crowdfunding and renewable energy sources: Are enterprises from green countries more performant? *Journal of Clear Production, 410*, 138932. https://doi.org/10.1016/j.jclepro.2023.136932

Cohen, W. M., & Levinthal, D. A. (1990). Absorptive capacity: A new perspective on learning and innovation. *Administrative Science Quarterly, 35*(1), 128–152. https://doi.org/10.2307/2393553

Daddi, T., Ceglia, . D., Bianchi, G. & Barcellos, M. D. de (2019). Paradoxical tensions and corporate sustainability: A focus on circular economy business cases. *Corporate Social Responsibility and Environmental Management, 26*(4), 770–780. https://doi.org/10.1002/csr.1719

Donner, M., Erraach, Y., López-i-Gelats, F., Manuel-i-Martin, J., Yatribi, T., Radić, I., & El Hadad-Gauthier, F. (2022). Circular bioeconomy for olive oil waste and by-product valorisation: Actors' strategies and conditions in the Mediterranean area. *Journal of Environmental Management, 321*, 115836. https://doi.org/10.1016/j.jenvman

Donner, M., & Vries, H. (2021). How to innovate business models for a circular bio-economy? *Business Strategy and the Environment, 30*(4), 1932–1947. https://doi.org/10.1002/bse.2725

Ellen Macarthur Foundation. (2017). What is the circular economy? Retrieved January 5, 2023, from https://www.ellenmacarthurfoundation.org/circular-economy/what-is-the-circular-economy

Ethirajan, M., Arasu, T. M., Kandasamy, J., Vimal, K. E. K., Nadeem, S. P., & Kumar, A. (2020). Analysing the risks of adopting circular economy initiatives in manufacturing supply chains. *Business Strategy and the Environment, 30*(1), 204–236. https://doi.org/10.1002/bse.2617

Ghisellini, P., Cialani, C., & Ulgiati, S. (2016). A review on circular economy: The expected transition to a balanced interplay of environmental and economic systems. *Journal of Cleaner Production, 114*, 11–32. https://doi.org/10.1016/j.jclepro.2015.09.007

Geissdoerfer, M., Savaget, P., Bocken, N. M. P., & Erik Jan Hultink, E. J. (2017). The Circular Economy – A new sustainability paradigm?. *Journal of Cleaner Production, 143*, 757–768. https://doi.org/10.1016/j.jclepro.2016.12.048

Hamam, M., Chinnici, G., Di Vita, G., Pappalardo, G., Pecorino, B., Maesano, G., & D'Amico, M. (2021). Circular economy models in agro-food systems: A review. *Sustainability, 13*, 3453. https://doi.org/10.3390/su13063453

Hazen, B. T., Mollenkopf, D. A., & Wang, Y. (2017). Remanufacturing for the circular economy: An examination of consumer switching behavior. *Business Strategy and the Environment, 26*(4), 451–464. https://doi.org/10.1002/bse.1929

Huynh, P. E., Rasmussen, E., & Nenadic, O. (2022). Commercializing circular economy innovations: A taxonomy of academic spin-offs. *Technological Forecasting and Social Change, 185*(C), 122102. https://doi.org/10.1016/j.techfore.2022.122102

Jahan Khan, S., Kaur, P., Jabeen, F., & Dhir, A. (2021). Green process innovation: Where we are and where we are going. *Business Strategy and the Environment, 30*(7), 3273–3296. https://doi.org/10.1002/bse.2802

Kempton, L., Rego, M.C., Alves, L. R., Vallance, P. Serra, M.A., Tewdwr-Jones, M. (2021). Putting Universities in their Place: An evidence-based approach to understanding the contribution of higher education to local and regional development. *Regional Studies Policy Impact Books, 3*(1). Regional Studies Association. ISSN: 2578-711X. ISBN13: 978-1-032-05566 (print); ISBN13: 978-1-003-19815-4 (e-book). https://www.tandfonline.com/toc/rpim20/current

Khan, O., Daddi, T., & Iraldo, F. (2020). Microfoundations of dynamic capabilities: Insights from circular economy business cases. *Business Strategy and the Environment, 29*(3), 1479–1493. https://doi.org/10.1002/bse.2447

Kirchherr, J., Piscicelli, L., Bour, R., Kostense-Smit, E., Muller, J., Huibrechtse-Truijens, A., & Hekkert, M. (2018). Barriers to the circular economy: Evidence from the European Union (EU). *Ecological Economics, 150*, 264–272. https://doi.org/10.1016/j.ecolecon.2018.04.028

Kirchherr, J., Reike, D., & Hekkert, M. (2017). Conceptualizing the circular economy: An analysis of 114 definitions. *Resources, Conservation and Recycling, 127*, 221–232. https://doi.org/10.1016/j.resconrec.2017.09.005

Lewin, A. Y., Massini, S., & Peeters, C. (2011). Microfoundations of internal and external absorptive capacity routines. *Organization Science, 22*(1), 81–98. https://doi.org/10.1287/orsc.1100.0525

Litchfield, R. C., & Gentry, R. J. (2010). Perspective-taking as an organizational capability. *Strategic Organization, 8*(3), 187–205. https://doi.org/10.1177/1476127010374249

Magalhães, B., Fitas da Cruz, V. F., & Ascenço, C. (2017). Relatório sobre boas práticas de utilização eficiente de recursos e de valorização de resíduos e subprodutos. *ALENTEJO CIRCULAR - Promover a Economia Circular nas Explorações Agrícolas e Agroindústrias do Alentejo.* Retrieved February 10, 2023, from https://alentejocircular.uevora.pt/?publicacao=2014

Marrucci, L., Daddi, T., & Iraldo, F. (2019). The integration of circular economy with sustainable consumption and production tools: Systematic review and future research agenda. *Journal of Cleaner Production, 240*, 118268. https://doi.org/10.1016/j.jclepro.2019.118268

Marrucci, L., Iannone, F., Daddi, T., & Iraldo, F. (2022). Antecedents of absorptive capacity in the development of circular economy business models of small and medium enterprises. *Business Strategy and the Environment, 31*(1), 532–544. https://doi.org/10.1002/bse.2908

Marrucci, L., Iraldo, F., & Daddi, T. (2021). Investigating the management challenges of the EU ecolabel through multi-stakeholder surveys. *Inter national Journal of Life Cycle Assessment, 26*, 575–590. https://doi.org/10.1007/s11367-021-01866-5

Murray, A., Skene, K., & Haynes, K. (2017). The circular economy: An inter-disciplinary exploration of the concept and application in a global con-text. *Journal of Business Ethics, 140*(3), 369–380. https://doi.org/10.1007/s10551-015-2693-2

Orazbayeva, B., Plewa, C., Davey, T., & Muros, V. G. (2019). The future of university-business cooperation: Research and practice priorities. *Journal of Engineering and Technology Management, 54*, 67–80. https://doi.org/10.1016/j.jengtecman.2019.10.001

Pacheco, L. M., Alves, M. F. R., & Liboni, L. B. (2018). Green absorptive capacity: A mediation-moderation model of knowledge for innovation. *Business Strategy and the Environment, 27*(8), 1502–1513. https://doi.org/10.1002/bse.2208

Pham, V. K., & Tran, T. T. H. (2023). Circular supply chain: Situation and implementation solutions for agricultural products in Vietnam. *Journal of Trade Science, 11*, 51–63. https://doi.org/10.54404/JTS.2023.11.01.05

Rehman Khan, S. A., Razzaq, A., Yu, Z., & Miller, S. (2021). Industry 4.0 and circular economy practices: A new era business strategies for environ mental sustainability. *Business Strategy and the Environment, 30*(8), 4001–4014. https://doi.org/10.1002/bse.2853

Sales, H., Figueiredo, F., Vaz Patto, M. C., & Nunes, J. (2021). Assessing the environmental sustainability of Portuguese olive growing practices from a life cycle assessment perspective. *Journal of Cleaner Production, 355*, 131692. https://doi.org/10.1016/j.jclepro.2022.131692

Scarpellini, S., Aranda-Usón, A., Aranda, J., Llera, E., & Marco-Fondevila, M. (2012). R&D and eco-innovation: Opportunities for closer collaboration between universities and companies through technology centers. *Clean Technologies and Environmental Policy, 14*, 1047–1058. https://doi.org/10.1007/s10098-012-0514-1

Scarpellini, S., Marín-Vinuesa, L. M., Aranda-Usón, A., & Portillo-Tarragona, P. (2020a). Dynamic capabilities and environmental accounting for the circular economy in businesses. *Sustainability Accounting, Management and Policy Journal, 11*(7), 1129–1158. https://doi.org/10.1108/SAMPJ-04-2019-0150

Scarpellini, S., Ortega-Lapiedra, R., Marco-Fondevila, M., & Aranda-Usón, A. (2017). Human capital in the eco-innovative firms: A case study of eco-innovation projects. *International Journal of Entrepreneurial Behavior and Research, 23*(6), 919–933. https://doi.org/10.1108/IJEBR-07-2017-0219

Scarpellini, S., Portillo-Tarragona, P., Aranda-Usón, A., & Llena-Macarulla, F. (2019). Definition and Measurement of the Circular Economy's Regional Impact. *Journal of Environmental Planning and Management, 62*(13), 2211–2237. https://doi.org/10.1080/09640568.2018.1537974

Scarpellini, S., Valere-Gil, J., Moneva, J. M., & Andreaus, M. (2020b). Environmental management capabilities for a "circular eco-innovation". *Business Strategy and the Environment, 29*(5), 1850–1864. https://doi.org/10.1002/bse.2472

Sehnem, S., Ndubisi, N. O., Preschlak, D., Bernardy, R. J., & Junior, S. S. (2020). Circular economy in the wine chain production: Maturity, challenges, and lessons from an emerging economy perspective. *Production Planning & Control, 31*(11–12), 1014–1034. https://doi.org/10.1080/09537287.2019.1695914

Shahzad, M., Qu, Y., Zafar, A. U., & Appolloni, A. (2021). Does the interaction between the knowledge management process and sustainable development practices boost corporate green innovation? *Business Strategy and the Environment, 30*(8), 4206–4222. https://doi.org/10.1002/bse.2865

Sharma, N. K., Govindan, K., Lai, K. K., Chen, W. K., & Kumar, V. (2021). The transition from linear economy to circular economy for sustainability among SMEs: A study on prospects, impediments and prerequisites. *Business Strategy and the Environment, 30*(4), 1803–1822. https://doi.org/10.1002/bse.2717

Stake, R. E. (2005). Qualitative case studies. In N. K. Denzin & Y. S. Lincoln (Eds.), *Hand book of Qualitative Research* (3rd ed.), 443–466. Sage

Todaro, N. M., Daddi, T., Testa, F., & Iraldo, F. (2020). Organization and management theories in environmental management systems research: A systematic literature review. *Business Strategy and Development, 3*(1), 39–54. https://doi.org/10.1002/bsd2.77

Venâncio, E., Lucas, M.R., & Marta-Costa, A. (2022). Sustainable consumption of Agri-food: Analysis of scientific production. Proceedings of the X Congresso da APDEA/ IV Encontro Lusófono em Economia, Sociologia, Ambiente e Desenvolvimento rural "Territórios, Agriculturas e Agroalimentar: Desafios globais e riscos no século XXI" (p. 56), 14–16 September, ESAC, Coimbra (Portugal). Retrieved January 5, 2023, from https://apdeaesadr.wordpress.com/livro-de-resumos/.

Walton, S., Zhang, A., & O'Kane, C. (2020). Energy eco-innovations for sustainable development: Exploring organizational strategic capabilities through an energy cultures framework. *Business Strategy and the Environment, 29*(3), 812–826. https://doi.org/10.1002/bse.2399

Zahra, S. A., & George, G. (2002). Absorptive capacity: A review, reconceptualization, and extension. *The Academy of Management Review, 27*(2), 185–203. https://doi.org/10.2307/4134351

Circular Economy Business Model—Let's Wine about it: An Exploratory Research Applied to Portuguese Wineries

Sandrina B. Moreira, Luísa Cagica Carvalho, and Ricardo Travassos

Abstract Sustainability requires businesses to adopt a more holistic approach, rethinking how products are designed, manufactured, and distributed and finding ways to reduce waste and increase resource efficiency.

The wine sector can benefit significantly from adopting circular business models as they offer opportunities to reduce waste, increase resource efficiency, and create value from by-products.

This chapter aims to discuss the following questions: Are companies in the wine sector adopting circular economy practices? Are the managers engaged in a circular economy strategy? What are companies' main difficulties when switching to circular economy business models?

This exploratory study is a case study. The qualitative methodology applied semi-structured interviews in two wineries and one wine association.

The results suggested a scarcity in the adoption of circular economy models. Generally, the managers are not engaged in sustaining the transition to a circular approach on the assumption of maximum efficiency. Finally, the study also reveals the perception by critical informants of wineries that higher education institutions can make an essential contribution to the transition from a linear to a circular economy. Therefore, the creation of idea-sharing networks with them can be a catalyst for new developments.

Keywords Sustainable development · Circular economy · Wineries · Business model

S. B. Moreira (✉) · L. C. Carvalho · R. Travassos
Instituto Politécnico de Setúbal, Setúbal, Portugal
e-mail: sandrina.moreira@esce.ips.pt; luisa.c.carvalho@esce.ips.pt

C. Rego et al. (eds.), *Entrepreneurship, Technological Change and Circular Economy for a Green Transition*, Studies on Entrepreneurship, Structural Change and Industrial Dynamics, https://doi.org/10.1007/978-3-031-48079-9_12

1 Introduction

The circular economy is becoming increasingly important because it offers a new approach to economic development that is sustainable, regenerative, and inclusive. The traditional linear economy model, based on the principles of "take-make-dispose," is not sustainable in the long term and leads to environmental degradation, resource depletion, and social inequality. The circular economy model aims to keep resources in use for as long as possible, extract maximum value from them, and minimize waste and pollution. Here are some reasons why the circular economy is essential:

Environmental sustainability: The circular economy aims to minimize waste, reduce pollution, and preserve natural resources. Keeping materials and resources in use for as long as possible reduces the need for new resource extraction and minimizes the environmental impact of economic activities.

Economic benefits: The circular economy can create new economic opportunities and value by promoting resource efficiency, innovation, and new business models. Businesses can reduce costs, increase productivity, and create new revenue streams by optimizing resource use.

Social benefits: The circular economy can promote social inclusion and equality by creating new job opportunities, reducing community waste and pollution, and improving access to affordable goods and services.

Climate change mitigation: The circular economy can mitigate climate change by reducing greenhouse gas emissions through resource efficiency, renewable energy use, and waste reduction.

Resource security: The circular economy can improve resource security by reducing dependence on finite resources and promoting the use of renewable resources.

Circular business models can offer significant benefits for traditional companies, including cost savings, revenue generation, environmental performance, reputation enhancement, and meeting customer demand. By embracing circularity, traditional companies can remain competitive and relevant in a rapidly changing business landscape.

By adopting circular business models, the wine sector can reduce waste, optimize resource use, and create economic value while minimizing environmental impact. These models can also help the sector meet consumer demand for environmentally sustainable products and services.

The wine sector represents a significant part of wealth creation in Portugal and plays a crucial role in creating stable links in the environments where it operates. It can present necessary premises regarding the relations between economic, social, and environmental aspects and introduce new business models, namely the circular economy business model, reusing the waste generated throughout the production chain so that it can be the inputs for new products and thus reuse all the resources maintaining all their value in a continuous cycle.

It is essential to create mechanisms, techniques, and methodologies to reverse the situation we see today, using new technologies and new business models in favor of sustainable development in order to guarantee the prosperity of future generations and, above all, preserve the environment because that is what we need to continue to develop our activities and continue our development as a society.

This chapter aims to provide insights into the circular business model in a traditional sector, such as the wine sector, that still needs to be studied. In order to shed light on the issue, this chapter is divided into the following sections. Section 2 is related to the theoretical framework of the issues where the urgency and importance of environmental sustainability practices are addressed and the methodologies that can be implemented to help companies in this chapter. Section 3 grounds the methodological options for the research in question and explains the techniques used for data collection. Section 4 presents the main results obtained, and the chapter ends with final considerations.

2 Theoretical Framework

2.1 Sustainable Development

The concept of sustainable development has mostly stayed the same over the years. Today we live in a global context where countries have become closer, commercial transactions have become more profound and complex, and connections between individuals from different societies have increased. Therefore, we can assume that globalization has underlying political and socio-economic aspects, cultural orientation, and economic and technological factors that bring societies together and develop them and push as a whole toward similar goals (Mensah, 2019; Parjanadze, 2009). In this sense, the concept of sustainable development is also similar.

The Brundtland report in 1987 states that sustainable development is based on three assumptions: economic sustainability, respect and preservation of the environment and its limits, and social cohesion and equity (Moreira & Crespo, 2012; Holden et al., 2017; Madeira & Silveira, 2020). However, in addition to these pillars, we must also consider the relationships between them: inclusive growth, green growth, green society, and inclusive development (Gupta & Vegelin, 2016; Fig. 1).

In an inclusive way we can talk about the Sustainable Development Goals (SDGs), which end up translating the relationships present in the figure above into 17 concrete goals as shown in Table 1.

Fig. 1 Sustainable development relations, adapted from Gupta and Vegelin (2016)

Table 1 Sustainable development goals, adapted from Fonseca et al. (2020)

SDGs	Description
1. Poverty elimination	Ending poverty
2. Eradicate hunger	Eliminating hunger, increasing food nutrition and promoting sustainable agriculture
3. Health and well-being	Ensuring access to quality health care globally
4. Quality instruction	Promoting equity and inclusive instruction and promoting lifetime learning opportunities
5. Gender equality	Achieving gender equality and empowering all women and girls
6. Drinking water and sanitation	Ensuring everyone has access to clean water and sanitation
7. Clean and affordable energy	Ensuring clean, reliable, modern, and efficient energy for all
8. Sustainable work and economic growth	Inclusive and sustainable promotion of economic growth and promotion of work with good conditions
9. Industry, innovation, and infrastructure	Creating strong infrastructures, promoting inclusive and sustainable industry and focusing on innovation
10. Reduce inequalities	Reducing inequalities within and between countries
11. Sustainable communities and cities	Making cities and communities inclusive, safe, secure, resilient, and sustainable
12. Sustainable production and consumption	Ensuring sustainable patterns of consumption and production
13. Climate protection	Urgent action to fight against climate change and its impacts
14. Marine life protection	Sustainable use of ocean resources for sustainable development

Table 1 (continued)

SDGs	Description
15. Protection of earth life	Protecting, restoring, and promoting the sustainable use of resources and ecosystems
16. Peace, justice, and effective institutions	Promoting peace and providing justice for all, based on capable and inclusive institutions
17. Partnerships to achieve the objectives	Revitalizing and strengthening partnerships for sustainable development

2.2 *Circular Economy*

In addition to the SDGs mentioned above, knowing the strategies created to achieve these goals is imperative. Thus, European agencies have tried to safeguard the interests of Earth and future generations. Based on this assumption, we see, at the European level, the application of laws that aim to mitigate the negative impacts of economic activities on the planet, such as the Climate Law or, more importantly, the Green Deal.

2.2.1 European Green Deal

According to the European Commission (2021, p. 2), the Green Deal aims to create "(...) a new growth strategy that aims to transform the EU into a fair and prosperous society, with a modern, resource-efficient, and competitive economy where there are no net emissions of greenhouse gases in 2050 and where economic growth is decoupled from resource use." Therefore, the agreement aims to preserve and enhance the existing capital and heritage of the European Union (EU) and protect the well-being and health of societies resulting from environmental impacts (European Commission, 2021).

The objectives inscribed in the Green Agreement are based on the premise that the economy is in a symbiotic relationship with ecology, thus allowing the protection of the environment in the perspective of an investment opportunity that results in greater efficiency of the production chain and the entire logistics chain, as well as the emergence of new and more profitable products, access to subsidies to invest in continuous innovation, the creation of more and new jobs and, above all, the emergence of green technologies.

Reinforced by the Green Deal is the pursuit of the circular economy, which is increasingly vital concerning the goal of sustainable development as a whole, but even more critical of economic prosperity and respect for the environment (Kirchherr et al., 2017; Ossewaarde & Ossewaarde-Lowtoo, 2020).

2.2.2 Circular Economy Goals

The circular economy aims to achieve a closed loop of material flow by reducing inputs, reusing and recycling products and waste in order to achieve a high level of quality of life by increasing resource efficiency (Prieto-Sandoval et al., 2018), while reducing energy and waste consumption, generating more jobs, and reducing subsequent pollution (Stahel, 2016). A circular economy as an economic system changes the concept of "end-of-life" by reducing or reusing, recycling, or recovering materials in production and distribution, and consumption throughout processes (Kirchherr et al., 2017).

According to Morseletto (2020), there are five goals for the circular economy:

The goal of increasing resource efficiency;
The goal of reducing waste and pollutant emissions;
Eco-design goals;
The goal of recovering once wasted resources;
Recycling goals.

In this sense, the 10R's policy, as stated by Morseletto (2020), represents a sustained way to achieve the assumptions of the circular economy (Table 2).

Table 2 10R's, adapted from Morseletto (2020)

Groups	Strategies	Objectives
Smart use of the product and the production	R0—refuse	Refuse products that didn't offer value and can harm the public health and the environment
	R1—rethinking	Rethinking the consumption
	R2—reduce	Reduce consumption, giving more longevity to existing products
Product life increase	R3—reuse	Finding new functions for obsolete products
	R4—repair	Repairing products that have defects
	R5—refurbish	Upgrade obsolete products
	R6—rebuild	Using parts from other products to produce a new one with similar functions
	R7—regain	Use obsolete/damaged parts to produce a new one with different functions
Careful application of materials	R8—Recycle	Processing materials to obtain products with similar or higher quality than when they were new
	R9—recover	Using incineration processes to generate energy

2.2.3 Advantages, Criticisms, and Challenges

The circular economy, even though it has noble and rational assumptions about consumption for better utilization and use of products, can have inherent disadvantages (Reike et al., 2018).

In terms of advantages, we can point to the reuse of material resources and waste reduction, reduced exploitation of natural resources, increased competitiveness and innovation, promotion of corporate image, promotion of new business models that create wealth, increased quality and value of products at lower prices (Morseletto, 2020; European Parliament, 2021); these can translate into three primary levels (Table 3):

In terms of criticism, we have the so-called rebound effect, where the secondary production process uses and exploits more significant amounts of natural resources than the primary production process, thus obtaining a negative result in the cost equation (environmental and economic), i.e., costs associated with the primary production process + costs associated with the secondary production process (Zink & Geyer, 2017). In order to avoid the rebound effect, the market segment should be

Table 3 Advantages of the circular economy strategy, adapted from Ellen MacArthur Foundation ANBI (2017)

Economic level:
* Reduction of raw material prices
* Decrease of price volatility and supply chain risks
* Business opportunities opening up (primary, secondary, and tertiary sectors)
* Reduction of negative externalities (ecological footprint)
* More robust and resilient economic systems
Organizational level:
* Reinforced synergies of reverse logistics companies
* Expansion of the collaborative economy
* Specialization of companies in remanufactured materials/components (innovation)
* New funding opportunities (e.g., community funds)
* Adding greater solidity to competitive advantages
* Improved relationship with stakeholders
* Increased product simplicity and more predictable life cycles (eco-design)
Consumers level:
* Reduced product costs due to planned obsolescence
* Larger variety of product choice
* Products with multiple functionalities

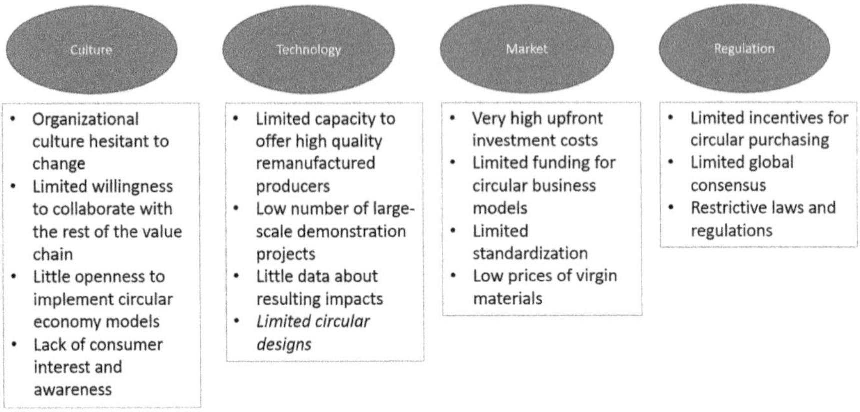

Fig. 2 Challenges of implementing circular economy strategies, adapted from Kirchherr et al. (2018)

chosen with criteria considering the product's characteristics. The secondary production chain cannot significantly impact the final prices of the products besides the focus on marketing (Zink & Geyer, 2017).

Another criticism pointed out to the circular economy is known as greenwashing. According to Ferrón Vílchez (2017) and Todaro et al. (2019), this effect is directly related to the intentions behind using strategies such as the circular economy. The focus is centered on the gain in business (due to the external corporate image) that leads to entry into new markets, access to better resources and know-how, and thus capitalizing the business. That is, there is thinking at the micro level instead of a macro perspective, where only the increase in profit matters, and there is a bias between the ends that these sustainable strategies aim for and the ends that economic agents perceive with their use (Ferrón Vílchez, 2017; Todaro et al., 2019).

Houston et al. (2018) also mention that another critical barrier lies in how the circular economy is perceived in many sectors. That is, the circular economy is often, and almost exclusively, seen as the re-entry of waste into a virtuous system, ultimately ignoring possible profitable and sustainable businesses within the circular economy theme (Houston et al., 2018).

There are also other obstacles, but this time associated with the difficulties of implementing a business model based on a circular economy. Kirchherr et al. (2018) point out that the main difficulties are present in four main axes: culture (either organizational or societal), technology, market, and laws/legal regulations (Fig. 2).

2.2.4 Green Technologies

The use of green technologies is a central part of sustainable development and is also, in an incisive way, possible to capitalize on all the proposed goals at the sustainable development level and thus mitigate the negative impacts of the traditional economic model (Lv et al., 2020; Oláh et al., 2020); that is, green technologies aim to impact the environment and society positively, seeking the satisfaction of consumers' wants and needs and reducing the negative impacts on the planet (Wicki & Hansen, 2019).

According to Du and Li (2019), the development and implementation of green technologies will be able to reduce the pollution generated in the production processes (greater efficiency and effectiveness in the use of resources), also allowing a quick way to provide reports to the strategic decision makers when possible, deviations from what was initially intended and expected are perceived (Oláh et al., 2020). In summary, the adoption of green technologies provides optimization in the entire value chain, reducing waste, improving the development and quality of products, increasing the efficiency level of the production chain, and in the efficient allocation of resources (Li et al., 2020).

2.2.5 Circular Economy Business Models

The growing concern about making the entire economic-financial ecosystem more sustainable means that companies are almost obliged to create or implement sustainable business models that enable companies to leverage their sustainability and the emergence of competitive advantages (Geissdoerfer et al., 2018; Mungai et al., 2020). That said, companies must be more awake and able to introduce sustainable business models based on innovation (Yang & Kang, 2020).

According to Geissdoerfe et al. (2018), these sustainable business models should pay attention to nine main strategies: (1) maximizing resource utilization and greater efficiency; (2) striving for closed resource cycles/loops; (3) introducing processes for renewable materials; (4) delivering functionality rather than proprietary products; (5) adopting a model of utility; (6) encouraging a sufficiency thinking; (7) reusing materials, keeping in mind creating value for society and the environment, (8) inclusive value creation; and (9) developing massively sustainable solutions.

This search for "new paths" results in innovations that lead companies to new innovation-based business models (Evans et al., 2017) that, when related to the theme of sustainability, aims to target sustainable developments with less impact on the environment and society and value creation for the transactional and contextual environment (Geissdoerfer et al., 2018).

A circular economy business model, based on its innovation, thus presupposes that four main strategies drive it: (1) cycles: resources are recycled within the process through reuse, renewal, and recycling; (2) time extension: the phase of "product use" should be extended as much as possible, and there should be efforts at the level of marketing, maintenance, and repair of these same products; (3) intensify use: the same product should be used intensively and if possible by several people—shared economy; and (4) dematerialization: the usefulness of the product should come from the replacement of hardware by software services and solutions (Geissdoerfer et al., 2020) (Fig. 3).

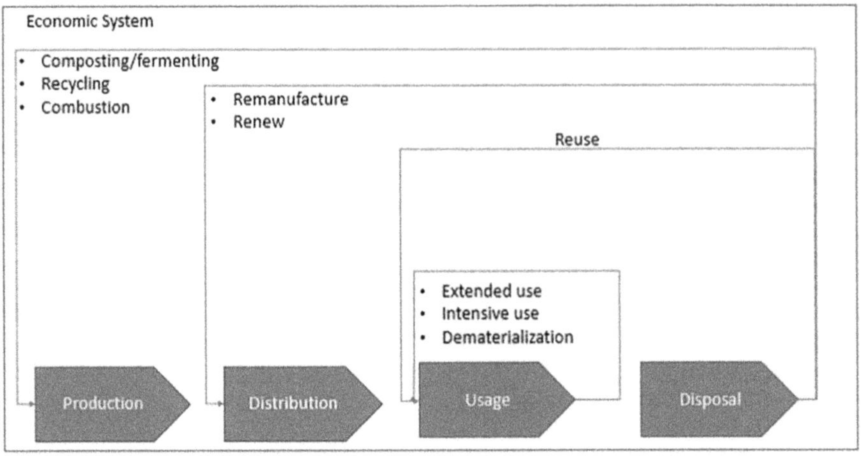

Fig. 3 Circular economy business model, adapted from Geissdoerfer et al. (2020)

The Ellen MacArthur Foundation and McKinsey Center for Business and Environment (2015) have identified six main lines of action in this regard:

1. Regenerate: switching to renewable materials and energy sources and returning to nature organized and thoughtfully.
2. Share: proceed to share products, extending their life cycle.
3. Optimize production chains to achieve zero waste.
4. Circular: opt for closed resource loops and thus maintain their value.
5. Dematerialize: transform physical products into virtual ones.
6. Moreover, replace old materials with new sustainable ones.

Thus, a given organization can adopt a circular economy model that, according to Esposito et al. (2018) and Lacy and Rutqvist (2015), can be of five types: Circular supply chain, Recovery and recycling, Product life extension, Sharing platform, and Product as a Service (PaaS).

Circular Supply Chain The circular supply chain involves the efforts of a given organization supplying materials/products and components to other organizations or itself, where they only use sustainable material sources (e.g., through recycling), being a sustainable, predictable, long-term, high-quality, reasonably priced alternative to supplying goods, making their customers more secure, whether in terms of supply chain risk reduction, pricing, or regulations and thus establishing fruitful long-term relationships (Lacy & Rutqvist, 2015).

Holistically, the circular supply chain "(…) forth the use of fully renewable, recyclable, or biodegradable inputs as substitutes for linear ones" (Esposito et al., 2018, p. 11).

Recovery and Recycling This circular economy business model is mainly based on the maximum exploitation of all by-products and waste streams in both directions, i.e., through recycling and upcycling (converting something dated into something current and valuable), in a bidirectional supply chain perspective where recycled products are supplied. End-of-life products are recovered, coming back into the system, and so on (Lacy & Rutqvist, 2015).

The benefits associated with the recovery and recycling circular economy business model are as follows:

Reducing the cost of quality and waste management;
Revenue from the sale of unwanted materials;
Production of positive externalities due to lower demand for virgin raw materials;
Sustainable waste "disposal";
More incisive exchange of know-how and synergies between the parties translates into product developments at the level of functionality and design to improve their recyclability.

Reduction in the cost associated with materials due to the exchange of primary resources for secondary ones (depending on the sectors) (Lacy & Rutqvist, 2015).

Product Life Extension According to Lacy and Rutqvist (2015), this business model is based on extending the product life cycle, generating revenue through longevity and not short product life cycles. Thus, characteristics such as durability, quality, and functionality are the most valued, and updates and services can be implemented throughout the product life cycle to keep the life cycle as long as possible, where six main activities are identified: durable and high-quality construction, product restoration, buying/selling/buying back products, updating, refueling and repair.

Sharing Platform The sharing platform works by renting, sharing, swapping, lending, or exchanging products, where the platform holder does not offer a product per se. Instead, it generates a revenue stream by matching supply with demand for finite resources, thus charging a percentage fee for all transactions. This leads to lower demand for new materials and lower product production (Lacy & Rutqvist, 2015).

PaaS (Product as a Service) In PaaS, companies retain product ownership and offered them to various individuals. They are responsible for the design, use, maintenance, reuse, remanufacturing, and recycling of this same product and, therefore, the individual is purchasing the purpose of the product rather than the product itself, where the value of the performance and purpose is above the value of its ownership (Lacy & Rutqvist, 2015). The product-as-a-service model can take on four typologies (Esposito et al., 2018):

- Pay per use: customers pay according to the use of the product (e.g., hours of use).
- Leasing: establishment of a contractual relationship for the exclusive use of a particular product during a specific period (long term);

– Rental: customers pay royalties for using a product for a limited time (short term).
– Performance agreement: when customers buy this service, in addition to the product, they also buy a usage quality guarantee, where the owning organization commits to a certain level of usage quality.

2.2.6 Wine Sector

Brief Description of the Sector According to the I.V.V (2021a), the wine sector is very important in Portugal's economic panorama, being very relevant in terms of exports. As shown in Table 4, Portugal has been increasing the volume of exported hectoliters of wine.

According to Fig. 4, we see that even in spite of the negative impact of pandemic COVID-19 in the world economy, Portugal reached by the end of 2020 a value of 858 million resulting from wine exports.

According to Table 5, the number of hectares (Ha) used for wine production has remained stable. There was a more pronounced decrease in the year 2015, which could mean the replanting of new vineyards to replace the older ones.

From Fig. 5, the top 5 countries in importing Portuguese wine are France, the United Kingdom (UK), the United States of America (USA), Brazil and Canada, and in that set, about 416 million Euros were generated (48.5% of total revenue).

Introduction of Circular Economy Models and Strategies in the Wine Sector As the wine sector is of great importance to the Portuguese panorama, both economically (wealth creation) and socially (direct and indirect employment), as well as a standard bearer for the preservation and protection of the environment, it is essential to infer how the economy can act in order to produce positive externalities.

The entire production chain in the wine industry (planting, maintenance, harvesting, winemaking) generates waste. It can be harnessed in a circular economy since 20%–30% of the production chain is waste (stalk, grape marc, yeast, wine lees, and saturated residual waters from the production process) (Ncube et al., 2021).

During the pruning season, the residues left by this activity are commonly abandoned or burned in large extensions, significantly increasing greenhouse gas emissions into the atmosphere (Nunes et al., 2021). Few studies in the area prove that if the residues left by the pruning activity are transformed into bio-coal, the production and emission of pollution are avoided. At the same time, it can be used as another alternative energy source (instead of using another type of more polluting coal, e.g., in the bread industry during baking). This material can also be used as a soil regenerator due to its nutrient retention characteristics in the soil, or it can also be used as a CO2 trap due to its high level of capture of pollution elements using pyrolysis techniques in the production of this bio-coal (Nunes et al., 2021).

As mentioned above, quite a few residues are produced during the production of wine that can be used for various purposes, such as fertilizing compounds, supplements, animal feed, biofuels, fuel additives, or even chemical platforms. So the

Table 4 Wine exports 2010–2021 (hectoliters), I.V.V. (2021a)

Intra+Extra EU	Export												Delta
	2010	2011	2012	2013	2014	2015	2016	2017	2018	2019	2020	2021	2021/202
Jan-out	2,156,394	2,479,852	2,757,682	2,562,664	2,354,802	2,325,344	2,272,297	2,434,812	2,482,621	2,475,655	2,605,127	2,729,849	4.80%

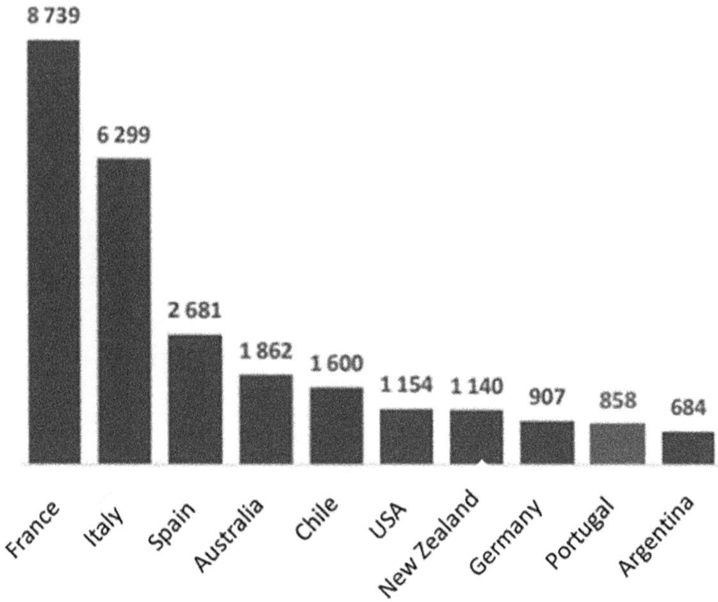

Fig. 4 Value of wine exports 2020 in million €, International Trade Statistics. Adapted from I.V.V. (2021a)

Table 5 Evolution of the total vineyard area in Portugal (Ha), adapted from I.V.V. (2021b)

Evolution of total vineyard area—Portugal (Ha)					
	2015	2016	2017	2018	2019
Total—mainland	198,683	190,467	188,881	187,562	189,988
Total—self-governing regions	2761	2752	2751	2760	2755
Total	201,444	193,219	191,632	190,322	192,743

wine industry can be exploited, as a consequence, as a bio-refinery (Ncube et al., 2021). Thus, the grape marc can be used for the extraction of oil that can be used in the cosmetic and personal care industry (by obtaining biosurfactants), the bioethanol resulting from its fermentation can be used for incorporation in animal feed or as organic fertilizer, the stem can be processed in order to extract its cellulose fibers for incorporation in plastics making them more "environmentally friendly," and the red wine grape skin recovered from pomace production can be processed to produce by-products that prevent glycol-oxidative stress associated with type 2 diabetes (Chebbi et al.,2021; Ncube et al., 2021).

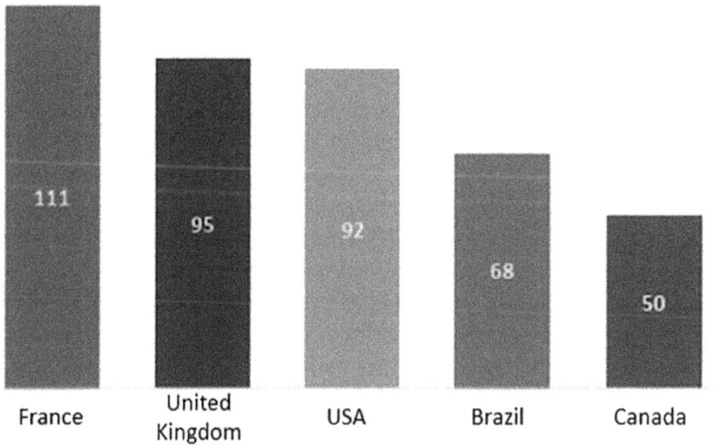

Fig. 5 Top 5 export destinations 2020 in million €, I.V.V. (2021a)

3 Research Methods & Methodology

3.1 Methodological Options

As the first stage, the information collected on the issue under study invariably falls through the literature review to authors recognized in the field and journals specific to the industry area.

To better study the phenomenon of circular economy in the wine sector, and given the goals of the present research, a qualitative methodology/approach was chosen.

Given the little existing information on circular economy business models in the wine sector, this qualitative research assumed a constructivist character where it is intended to build through the representation of typical cases (Stake, 2012), i.e., a technical procedure of case studies was adopted where the "(...) focus is on contemporary phenomena embedded in some real-life context" (Prodanov & Freitas, 2013, p. 128). Thus, the case study method allows the researcher to understand complex social phenomena, retaining the characteristics of real life as it is demonstrated (Yin, 2010).

As for the data collection technique, we used the interview survey. The interview was applied to three entities linked to the wine sector: one small and one medium-sized company, the first with national representation and the second with both national and international representation, as they have more resources and can be used as a comparison for an eventual more incredible/better adoption of circular economy models, as well as to understand their point of view, particularly about barriers and opportunities; the third entity interviewed is a winegrowers' association, intending to study sustainable circular economy business models in the wine sector in a global manner. Therefore, two specific points of view regarding each company (with different structural characteristics) are covered and contrasted with a global and impartial point of view of a winegrowers' association.

The interview guide used in this research was based on a previously validated questionnaire survey, and its content was adapted for the companies and the association described above since they have different objectives, which culminated in the production of two sets of questions, one more directed to the companies and the other to the association.

On the other hand, improvements and changes were made, following a structure adapted in light of the literature review and by researching information about the companies that allowed us to confirm some questions and add others.

The interviews were conducted between April 15 and May 20, 2022. They consisted of three blocks of questions. The first and second blocks characterized the respondent and the company. The third block aimed to characterize the adoption of circular economy business models, the use and reuse of resources in companies, and to understand the associated impacts that respondents expect for the company, society, and the environment. This third block also includes the main difficulties in implementing circular economy projects and what higher education institutions/other national and international private and public associations can do to help organizations in the transition phase.

4 Results

4.1 Characterization of the Wine Companies and Association and the Respective Interviewees

The methodological approach selected for the present study (semi-structured interview) allowed the interviewees to express their opinion about the theme in question in an accessible way while still following identified guidelines.

The table below characterizes the interviewees through their socio-demographic characterization and proceeds to the exact characterization concerning the companies and associations. Regarding the interviewees, information is presented regarding gender, organizational function, seniority in the company and the sector (winemaking), and academic qualifications. Regarding the companies and the association, it focuses on seniority, the number of employees, and turnover (Table 6).

The average age of the interviewees in the current companies/association is 16 years, as well as the average age in terms of seniority in the wine sector (16 years). All the interviewees hold at least one mid-level position (viticulture technician,

Table 6 Characterization of the interviewees, companies, and association, author's elaboration

	Gender	Organization role	Seniority	Seniority in the sector	Educational level	Foundation date	No of workers	Business volume
Association	Female	Viticulture technic	18 years	18 years	Degree in agricultural engineering	1984 (38 years)	7	No KPI
Company 1	Female	Quality manager	23 years	23 years	Degree in food engineering	1958 (64 years)	90	22 million €
Company 2	Female	Managing partner	7 years	7 years	Degree in business management in marketing	1950 (72 years)	12	2 million €

director of the quality and exploitation department, and managing partner), reinforcing the possible solidity and guarantee of good results despite the small number of interviewees. In addition to their positions, the number of years they have been working at the winery/association and in this specific sector also ensures the credibility of the data collected.

In turn, the companies in the study have an average age of 68 years, and the association has fewer years of seniority (38 years). The companies in the study also have in their structure, on average, about 51 workers, with a predominance of 91.8% of the workforce for the company that has national and international representation, and the association has seven workers. It is important to emphasize that this number of workers refers to the workers with permanent contracts with the companies because, due to the characteristics of the product in question, in the summer, during the grape harvest, and in the spring, with pruning and delimbing, the labor force is continually reinforced, being safeguarded through the provision of temporary workers. Finally, the average turnover is 12 million euros, and the association needs indicators that measure turnover because its objectives are limited exclusively to technical support to associated wine companies.

4.2 Business Plan for Sustainability Objectives and the Internal Projects/Practices at the Awareness Level

When the interviewees were questioned about the sustainability objectives in the business plan in several time goals (short/medium/long term) and what would result in internal projects/practices that the company could have in order to raise awareness for the reduction of resource consumption, both (from companies 1 and 2) said that there is already a practical demonstration toward the sustainability theme, even highlighting the investment in Wastewater Treatment Plants (WWTP), water recirculation and recycling systems and solar panels. The interviewee from company two also showed that the company is also very focused on the use of recycled green glass in bottled wine bottles, the reuse of cardboard boxes in order to decrease the plastic consumption, and the improvement of cargo stowage and packaging of products so that fewer trips and fewer resources are used before, during and after transport even though, due to the products in question, it is still necessary to use plastic films to accommodate the pallets so that no moisture is created inside the packages which, in turn, may damage the wine.

A measure also adopted by company 2 is the issue of selling wine in bulk. It is a new bet of the company that intends to reduce the use of cardboard packages and plastic inserts in the packages, being the customers responsible for taking their packages to fill with wine, being the primary purpose of the winery is the reduction of the use of packages. On this assumption, the interviewee states that the zero-waste policy is something the company relies on to obtain "(…) the maximum production efficiency, even to exploit the economic profitability better."

However, from a more global point of view, the interviewee from the Association of Grape Growers states that, at this time, farmers are still doing the essentials, within the assumption of integrated production, with particular attention to the best allocation of water resources, care of the soil, and treatments for crops. Still, he states that "(...) in the theoretical context, all wineries respond positively when it comes to these measures, however in the practical context, I have some doubts," also often a result of each company's capital for investments in sustainability.

It is thus verified that with awareness of environmental sustainability and the business plans themselves. There is already some movement, and changes in mentality are already denoted. However, it also depends on compliance with standards, mandatory sector regulations, and waste reduction for maximum economic efficiency. Thus, despite all the measures taken and included in the business plans of the company/Adega, the primary focus is on efficiency for better profitability, the consequence being the improvement of its economic sustainability, which corroborates what is written in the literature about the phenomenon of "greenwashing."

4.3 *ISO14001 Certification*

ISO14001 is a standard that aggregates a set of requirements to allow effective and sustainable management of the business activities of each company, having as basic assumptions the environmental protection and the reduction of the ecological footprint.

Both interviewed companies state that they are not certified by ISO14001, despite being certified by other standards, namely ISO9001 as an imposition of the market and the sector in which they operate to ensure minimum standards of food safety and the HACCP standard (Hazard Analysis and Critical Control Point). The interviewee from Company 1 states that the company is not certified by ISO14001 but follows all the rules and directives required by law.

The interviewee from Company 2 also states that despite not being certified by this standard due to the high bureaucracy involved, the company intends to be certified by it in short/medium term.

The interviewee related to the association also mentions that the large wineries already have a wide range of certifications, namely the exporting ones, due to market impositions.

Replacement of internal materials/machinery and processes with new efficient, sustainable, or computerized/technological ones.

New technologies and new generations of machines and equipment allow companies to produce more efficiently and robustly. They are also more sustainable, eventually transferring all the valence to an organization's internal processes.

When the interviewees were questioned about this theme, it was verified that both companies opted to modernize all the work material (equipment) for new, more efficient ones and even use electric equipment instead of fossil fuel. This corroborates the literature regarding the efficiency that more modern machines and equipment (green technologies) can offer to an industrial production such as the winery, thus obtaining greater efficiency and causing a lower environmental impact.

Despite the testimony of the individuals from the interviewed companies, it is vital to highlight the testimony of the interviewee from the association who states that "(…) in general I do not see this region as being at all geared towards innovation, despite the improvement in some processes, but I do see them concerned with new production methods, new technologies," objecting, from a concise point of view, that technologies are an essential point in the pursuit of the search for environmental sustainability and that farmers are aware of this issue. However, this innovation becomes more prolonged, resulting in most wine companies needing more financial capabilities to invest in new technologies.

4.4 Department/Structure and/or Person in Charge of Product Innovation Specialized in Transforming Waste into New Products

Regarding this sub-dimension of analysis, all interviewees (companies and associations) demonstrated that no department or even human resources are directly allocated to an innovation structure that transforms waste into new products from a circular economy perspective. If in company one, it is stated that the non-existence of a department and person in charge of innovation is due to the existing limitation by the rules and standards imposed on the sector, company two does not have one because the organizational culture itself understands that it is not necessary. As stated by the interviewee, "the organizational culture is family-based, and we want to keep it as small as possible." Only one person is responsible for innovation at the wine product's level, the winemaker. The interviewee in the association states that the inexistence of a specialized department is mainly because there are no "(…) means to invest." We thus verify three different points of view. However, they all corroborate with the literature about the barriers to implementing circular economy business models. First of all, we talk about the company's organizational culture (as stated by the interviewee from company 2), the castrating laws and regulations (as stated in company 1), and the fact that the costs involved are high. The companies/Adegas do not have the capacity for this investment (as stated by the interviewee from the association).

4.5 Waste Treatment

As for treating the residues of the winemaking process, both companies have some processes. According to Company 1, the residues from pruning or harvesting are deposited in the field and the vine stalks, thus serving as a natural fertilizer. If the quantity is large, they are incinerated. In contrast, the stalks are sent to distilleries that will use them to create brandy. In company two, the same thing happens, except that the brandy from the waste goes back into the company for the production of muscatel, in a circular economy perspective. Still, the interviewee in company 2 mentions that it is difficult to find better solutions despite the concern with waste treatment because the processes still need to be robust, and their treatment is costly.

However, when asked about this issue, the association was forceful in stating that there is no waste treatment at all; besides the masses and dregs that are required to be sold, there is no analysis throughout the process in order to ensure that the waste is being treated appropriately and following the assumptions of environmental sustainability. In the words of the interviewee, "(…) they just want to get rid of the waste." Once again, there are extreme points between what companies do and the opinion of an institution with a more globalized and independent point of view.

Thus, we again put into perspective that there is a gap between what companies/waste collectors do and think they are doing right and what they are doing. We may be facing another case of only doing this waste management because they are required by law, without careful control over what will happen to them in the rest of the process, as is also said by the interviewee of company two who states that it is challenging to give a better treatment to waste.

4.6 The Company/Association Has Practices Aimed at Creating Synergies Between Itself and the Surrounding Communities

In this point of analysis, both interviews conducted with the companies/Adegas denote that there is no demand for a relationship between the company and the community around it as a way to attract new ideas. Both interviewees refer to it as something other than what is sought. The interviewee in company two even states that the strategic vertex needs to perceive these synergies as necessary for the business.

On the contrary, in the interview conducted with the association, it perceives that the creation of networks of synergies is very positive and that the dynamics with institutions like the Polytechnic Institute of Setubal, which have great dynamism, are beneficial. However, it is not easy to verify the result of these dynamics in the practical and labor context of the wineries because there needs to be more data.

According to the literature, from the start, we are not witnessing a constant idea and assumption of sustainability, as confronted in the first question, because to achieve sustainable development, there is necessary to balance three dynamics (society, environment, and economic growth). There is a contradiction between what was said above and this question. We can see in this question that these companies/Madeiras are more concerned with the economic growth aspect, neglecting the advantages that inclusive growth could bring to the winery.

4.7 *The Strategic Top Encourages Awareness of Sustainable Practices as a Way to Ensure the Well-Being of Stakeholders*

The answers to this sub-dimension were unanimous, where all said that the strategic top is concerned with its stakeholders. In this question, the interviewee from company 1 mentioned that the main concern is at the level of training both the workers and the grape-producing partners (their raw material suppliers) because "(…) the greater our efficiency, the lower our waste and the more profitable will be the production," focusing here the main activity. Regarding company 2, the interviewee explains that the focus is on raising awareness of more sustainable practices among its employees, and there is no particular concern with suppliers in terms of environmental sustainability; the focus in the choice is to ensure good relationships with suppliers based on focusing exclusively on the economic component.

As for the interviewee from the association, this also corroborates what had been said previously. That is, the overall perception by the association is that the strategic top tends to focus on strengthening the training in the sustainability component to its workers and other stakeholders.

The main focus of wineries is on the training and qualification of their workers, in a perspective of greater efficiency so that profitability is higher. There are some measures, such as the distribution of reusable bottles to reduce the consumption of plastic, the awareness for the reduction of water consumption, and the adoption of recycling behaviors (company 2), although only the minimum is done. Not all stakeholders are involved, and even suppliers are not the target of awareness-raising actions or, when they are, it is with the perspective of providing a better product and not in the reduction of the ecological footprint, at the same time that the intention is to improve the product.

4.8 *Awareness of the Strategic Top for Circular Economy Models*

According to the interviewee, the strategic top in company 1 needs to be made aware of circular economy models. There is no such openness due to the significant investment that must be made to accommodate such paradigm shifts. At the level of company 2, the strategic top seems awake to the theme, stating to the interviewee that "there have been efforts to give new life to this waste resulting from wine." However, once again, it is perceived that the association has a different view, stating to the interviewee that "companies say yes, but in a practical context, it does not seem to happen. The very concept of the circular economy must be understood and partially understood by wineries." This statement is essential to understand, as it may point to a real gap between what companies do and what they understand as a circular economy. Therefore, possible resources could be allocated but nothing at the moment.

4.9 Company Integration in Circular Economy Projects

The interviewee in company one states that the winery has never been involved in any circular economy project and that she does not intend, shortly, to be involved in a project of this type. In contrast, company two has already been involved in a circular economy project, where she participated in a study about the winemaking process. However, they are open to entering new projects of this type. As for the association, it is always involved in projects of this nature, as this is where it gets the necessary know-how to be able to provide technical support to its members, so the last one was the "Red Wine," which intended to make a survey of the waste resulting from the winemaking process and to transform it into new products (cosmetics, agricultural applications, biostimulants, and creation of algae with the CO2 produced in the wine fermentation process). The interviewee from the association also states that, although the new managers have a different vision concerning environmental sustainability, this has not yet been translated into the practical field, and, therefore, the largest wineries have been developing a greater awareness of this issue, but have not achieved objective results.

In sum, there is still no awareness of the top management regarding circular economy models since the "method works," as evidenced by the interview of company 1. This shows that the top management is still very little concerned about changing their current business model (linear) and moving to the circular business model, which is reinforced by the fact that both wineries state that they are not inserted in any circular economy project at present or in the future, being focused only on production and resources that are allocated to production. Both interviewees perceive, nevertheless, that it might be beneficial for companies to participate in these projects.

4.10 Barriers to Embrace Circular Economy Models

The barriers mentioned by the companies were economic-financial (high initial investment), organizational culture, and know-how for company 1. Know-how, technology, human resources, economic-financial (the lack of state support for the transition), and the high state bureaucracy are some of the barriers mentioned by the interviewee for company 2.

From the association's point of view and according to the interviewee, the main barriers observed are "cultural barrier (farmer's way of thinking), market barrier (impositions of the consumer market), and economic-financial barrier (organizations with little capital for investment and that only try to guarantee the organization's survival)."

Also concerning the barriers to the adoption of circular economy models, which were mentioned by the interviewees from the association and company 1, is the fact that there is increasing market pressure for products to have an increasingly sustainable origin. However, then there is no openness to pay for these products that usually entail more costs (higher price) when compared to similar linearly produced products. This is also referred to in the literature, corroborating it, namely in the lack of awareness on the part of consumers about the higher price.

4.11 Contribution of Higher Education Institutions to the Integration of Circular Economy Models in Companies

All interviewees acknowledge that higher education institutions can contribute to the transition from a linear to a circular economy. According to the interviewee from Company 1, higher education institutions can be "important allies in the transition from theoretical knowledge to the field." In contrast, the interviewee from Company 2 adds that "there should be an exchange of know-how and ideas and that contact with various agents can bring together experiences and contacts that can help the company become more sustainable." However, although they want to move toward the transition to a circular economy, the focus will invariably have to be on wine production to ensure the company's survival, so they must refrain from investing resources in the search for this knowledge. The interviewee from Company 2 also says that the existence of state support would help immensely in the transition, as wineries, due to their characteristics, are not available to move alone in projects of this nature and size where the success is very uncertain, as well as the debureaucratization of projects so that they are faster to implement when there is the opportunity.

The association recognizes that higher education institutions and, more specifically, the Polytechnic Institute of Setúbal (IPS) have led to significant advances in several areas, including sustainability, where it is "recognized the merit of the association for the IPS." Once again, the idea remains that state support can provide and contribute significantly so that change can occur, exceptionally when consumers and the rest of the market are increasingly demanding, and companies must adapt.

5 Conclusions

The latest manifestations of the planet, indicating that we are at a frantic pace toward its destruction, have precipitated the acquisition of a new awareness about sustainable practices, how to make these sustainable practices "core" in companies, and how to monetize them. Innovation enables fundamental advances to be made in technology and business models to guide, in a sustained and robust way, all stakeholders toward more excellent resource rationalization and decarbonization. A linear model ("make, use, dispose of") can no longer be the norm in order to preserve "our home." Still, circular economy business models have advantages, disadvantages, and criticisms inherent to any business model. However, resilience is needed to overcome these disadvantages and barriers to meet sustainability goals and survive.

Given the knowledge obtained through reading specialized literature in the area, it was found that there is still much work to be done concerning environmental sustainability and, more specifically, CE business models in the wine sector that can contribute to the development of the sector. Therefore, much can still be done, especially when we talk about a predominantly agricultural and industrial activity that exploits various natural resources for its activity.

The results of this research show that there is still a long way to go, which may present itself as a critical area of study to be intensively explored in the coming years, not only for its importance at the economic-social and environmental level but also for the possibilities it presents in the field of circular economy business models and products to explore.

Accordingly, several questions were raised, mainly whether circular economy policies/practices already exist in these companies, what motivations lead top management to adopt or not these policies, and what impact entities like the Polytechnic Institute of Setúbal could have to help and support companies in this transition.

Both companies show some attention regarding the business plan's sustainability objectives and the existing internal projects/practices. Both interviewees mention that the companies have RAST for water reuse, recycling systems, solar panels investment to create renewable energy, also some recycled materials use (such as green glass), and a more significant concern with the stowage of the product when it is placed on pallets to be sent to its destination, using less plastic wrap, with plastic of lower microns, and thus spending much less plastic. Specifically, in Company 2, there is also an innovation: selling wine in bulk, where consumers bring their packages.

In general, one says that companies are very focused on maximum efficiency for better exploitation of economic profitability. Still, and from an independent point of view, the interviewee from the association mentions that farmers are still doing the basics in terms of environmental sustainability practices, but that they show a concern with the use of water resources, care with the soil use and in the treatments applied to crops. The interviewee also mentions that, although wineries claim to be environmentally sustainable in theory, they do not do it in practice due to their complex economic possibilities.

Regarding ISO14001, no winery is certified by it, and only Winery 2 states that in the short term, it intends to be certified by it, but not before due to the high bureaucratic burden it entails. Still, both interviewees state that their respective companies comply with all legal requirements and standards for protecting nature and the environment.

About new technologies and what they can offer in favor of sustainability, the interviewees were asked whether their wineries replace their machinery and equipment and internal processes with more efficient ones, to which both interviewees replied that whenever there is the opportunity to acquire new equipment, they opt for new more efficient and even electric ones, to avoid the consumption of fossil fuels when possible. The interviewees only mentioned equipment/materials, which shows that they have remained the same in their internal practices. The interviewee from the association corroborates this statement by saying that. However, the region is geared toward something other than innovation. There is a concern with acquiring new production methods and technologies because they are more efficient and effective.

As for the existence of a department and resources directed to the transformation of waste into new products, all interviewees stated that there is no structure or resources allocated to it, and we are also told by the interviewee in company one that there is no such department due to castrating standards in the sector, while the interviewee from company two states that the company culture is the main responsible since it is intended to be small. The interviewee from the association goes further by saying that it is also tricky due to the need for more capacity and means to invest. Thus, there is no structure to transform waste from one production process into inputs for another. This shows that these wineries may need to prepare to move toward circular economy business models.

Regarding waste treatment, it is almost null. There is no concern with waste treatment, stating the interviewee from the association stated that wineries want to "get rid of them." As for the interviewees of the companies, they state that part of the waste produced in the field is placed there to serve as biofertilizer and, in turn, the pulp and dregs are sent to distilleries where they are transformed into brandy, which precisely in company two is used for the production of Muscat. An important aspect is a statement made by the interviewee of company two regarding the difficulty in finding better solutions for the waste because the processes could be more robust. The treatment is costly, which may also mean that there is an excellent unawareness on the part of wineries about what can be done with the "waste" and that they do not do it because it is expensive, and they do not want to do it.

In the triangle of sustainability relationships, one of the assumptions is society and what companies can do together with their surroundings. Thus, regarding creating synergies between companies and the communities around them, both state that there are none and that it needs to be sought and understood by top management (they do not perceive it as necessary for business development). On the other hand, the interviewee from the association mentions that creating synergies adds fundamental advances. So, until now, we can say that in the triangle of sustainability assumptions, despite what the companies tell us, they are still focused only on the economic aspect, and the environmental aspect is only a positive externality (a consequence of the search for better efficiency).

The top management is also concerned with its stakeholders. However, in a merely formative sense and geared toward greater production efficiency, the interviewee of company two also states that there is some effort in raising awareness of workers regarding sustainable practices. Also highlighted is that the suppliers of these wineries are chosen taking into account purely economic assumptions, which indicates once again that the main focus is on the economic aspect of sustainability.

Regarding the awareness of top management for circular economy models, it is unanimous in all interviews that him/her is not at all aware or concerned with implementing these business models, which is also reflected in the integration of these companies in circular economy projects. However, the association has been present in projects of this nature, the latest being Red Wine.

Even so, the interviewees were asked about the main barriers to the non-adoption of circular economy models, and it was said that the main barriers are the following: economic and financial (due to high initial investments), know-how, technology, human resources, and the associated high bureaucracy. The interviewee from Company 2 also states that the market still needs to prepare to support the extra costs of sustainable products, even though there is pressure from the market to be more sustainable. As such, these barriers also help explain why managers do not see circular economy business models as a central and critical part of the future success of the business and the sector, which may have a lot to offer.

Finally, concerning the contribution of higher education institutions and other entities to assist the integration of circular economy models in companies, the response is also unanimous, with the interviewees saying that these entities can make an essential contribution in the search and sharing of knowledge and ideas that can be put into practice by wineries and thus provide them with better sustainable practices, particularly in terms of environmental sustainability.

To sum up, companies are already implementing some practices that, at first glance, are concrete actions toward environmental sustainability. However, these practices are more focused on the economic aspect, with the environmental aspect left behind and being only a consequence of the measures applied; at the social level, the wineries need to be made aware and focused. Suppose at the level of environmental awareness. It has already been seen that it is only a consequence concerning more sustainable business models (such as the circular economy business model). In that case, these are entirely put aside and are not seen as important as the current methods have worked, and therefore, there is no reason to change, which can also be explained due to existing barriers. However, they recognize that the creation of idea-sharing networks with higher education institutions could be a catalyst for new developments.

There is, therefore, a long way to go. This sector is underused theoretically and practically (according to the results of this research). It can be a focus for studies aiming to implement circular economy business models, but it has yet to be considered fundamental. Decision makers often need to understand it in wineries. This research pioneered the area, and much can still be investigated. Namely, future studies can be developed around projects that can be applied in wineries to take advantage of the waste generated as inputs for other production processes and what can be transformed.

In terms of the limitations of the research carried out, we highlight mainly the difficulty in reaching the interviewees since, at the time of the study, we were still under contingencies related to COVID-19, which made time management a little complicated and made it impossible for the authors to closely assess the entire production system and thus better understand what had been said during the interviews.

References

Chebbi, A., Franzetti, A., Duarte Castro, F., Gomez Tovar, F. H., Tazzari, M., Sbaffoni, S., & Vaccari, M. (2021). Potentials of winery and olive oil residues for the production of Rhamnolipids and other biosurfactants: A step towards achieving a circular economy model. *Waste and Biomass Valorization, 12*(8), 4733–4743. https://doi.org/10.1007/s12649-020-01315-8

Du, K., & Li, J. (2019). Towards a green world: How do green technology innovations affect total-factor carbon productivity. *Energy Policy, 131*(December 2018), 240–250. https://doi.org/10.1016/j.enpol.2019.04.033

Ellen MacArthur Foundation & McKinsey Center for Business and Environment. (2015). Growth within: A circular economy vision for a competitive Europe. Ellen MacArthur Foundation, 100.

Ellen MacArthur Foundation ANBI. (2017). What is the circular economy? Retrieved January 22, 2022, from https://archive.ellenmacarthurfoundation.org/circulareconomy/what-is-the-circular-economy

Esposito, M., Tse, T., & Soufani, K. (2018). Introducing a circular economy: New thinking with new managerial and policy implications. *California Management Review, 60*(3), 5–19. https://doi.org/10.1177/0008125618764691

European Commission. (2021, December 18). The European Green Deal. https://ec.europa.eu/info/sites/default/files/european-green-dealcommunication_en.pdf

European Parliament. (2021, February 12). Circular economy: definition, importance and benefits. Retrieved May 7, 2021, from https://www.europarl.europa.eu/news/pt/headlines/economy/20151201STO05603/economia-circular-definicao-importancia-e-beneficios

Evans, S., Vladimirova, D., Holgado, M., Van Fossen, K., Yang, M., Silva, E. A., & Barlow, C. Y. (2017). Business model innovation for sustainability: Towards a unified perspective for creation of sustainable business models. *Business Strategy and the Environment, 26*(5), 597–608. https://doi.org/10.1002/bse.1939

Ferrón Vílchez, V. (2017). The dark side of ISO 14001: The symbolic environmental behavior. *European Research on Management and Business Economics, 23*(1), 33–39. https://doi.org/10.1016/j.iedeen.2016.09.002

Fonseca, L. M., Domingues, J. P., & Dima, A. M. (2020). Mapping the sustainable development goals relationships. *Sustainability (Switzerland), 12*(8), 3359. https://doi.org/10.3390/SU12083359

Geissdoerfer, M., Pieroni, M. P. P., Pigosso, D. C. A., & Soufani, K. (2020). Circular business models: A review. *Journal of Cleaner Production, 277*, 123741. https://doi.org/10.1016/j.jclepro.2020.123741

Geissdoerfer, M., Vladimirova, D., & Evans, S. (2018). Sustainable business model innovation: A review. *Journal of Cleaner Production, 198*, 401–416. https://doi.org/10.1016/j.jclepro.2018.06.240

Gupta, J., & Vegelin, C. (2016). Sustainable development goals and inclusive development. *International Environmental Agreements: Politics, Law and Economics, 16*(3), 433–448. https://doi.org/10.1007/s10784-016-9323-z

Holden, E., Linnerud, K., & Banister, D. (2017). The imperatives of sustainable development. *Sustainable Development, 25*(3), 213–226. https://doi.org/10.1002/sd.1647

Houston, J., Casazza, E., Briguglio, M., & Spiteri, J. (2018). Stakeholder views report: Enablers and barriers to a circular economy, 730378, 1–30.

I.V.V. (2021a, November 24). Exports. Retrieved January 08, 2022, from https://www.ivv.gov.pt/np4/9840.html

I.V.V. (2021b, March 8). Vineyard Statistics. Retrieved January 8, 2022, from Instituto da Vinha e do Vinho: https://www.ivv.gov.pt/np4/35/

Kirchherr, J., Piscicelli, L., Bour, R., Kostense-Smit, E., Muller, J., Huibrechtse-Truijens, A., & Hekkert, M. (2018). Barriers to the circular economy: Evidence from the European Union (EU). *Ecological Economics, 150*(April), 264–272. https://doi.org/10.1016/j.ecolecon.2018.04.028

Kirchherr, J., Reike, D., & Hekkert, M. (2017). Conceptualizing the circular economy: An analysis of 114 definitions. *Resources, Conservation and Recycling, 127*(April), 221–232. https://doi.org/10.1016/j.resconrec.2017.09.005

Lacy, P., & Rutqvist, J. (2015). *Waste to wealth: The circular economy advantage* (1st ed.). Palgrave Macmillan. https://doi.org/10.1057/9781137530707

Li, Y., Dai, J., & Cui, L. (2020). The impact of digital technologies on economic and environmental performance in the context of industry 4.0: A moderated mediation model. *International Journal of Production Economics, 229*(May 2019), 107777. https://doi.org/10.1016/j.ijpe.2020.107777

Lv, S., Liao, T., Xi, F., Zhang, J., Zuo, X., & Cao, X. (2020). Resources, Conservation & Recycling Classifying green technologies for sustainable innovation and investment. *Resources, Conservation & Recycling, 153*(November 2019), 104580. https://doi.org/10.1016/j.resconrec.2019.104580

Madeira, S., & Silveira, L. (2020). Sustainable development goals: Decent work and economic growth within the context of sustainable tourism. Atheorethical, practical and critical analysis. *Cadernos de Geografia, 41*, 31–41.

Mensah, J. (2019). Sustainable development: Meaning, history, principles, pillars, and implications for human action: Literature review. *Cogent Social Sciences, 5*(1), 1653531. https://doi.org/10.1080/23311886.2019.1653531

Moreira, S., & Crespo, N. (2012). Development economics-from traditional approaches to new concepts of development. *Journal of Economics, 38*(2), 25–50.

Morseletto, P. (2020). Targets for a circular economy. *Resources, Conservation and Recycling, 153*(October 2019), 104553. https://doi.org/10.1016/j.resconrec.2019.104553

Mungai, E. M., Ndiritu, S. W., & Rajwani, T. (2020). Do voluntary environmental management systems improve environmental performance? Evidence from waste management by Kenyan firms. *Journal of Cleaner Production, 265*, 121636. https://doi.org/10.1016/j.jclepro.2020.121636

Ncube, A., Fiorentino, G., Colella, M., & Ulgiati, S. (2021). Upgrading wineries to biorefineries within a circular economy perspective: An Italian case study. *Science of the Total Environment, 775*, 145809. https://doi.org/10.1016/j.scitotenv.2021.145809

Nunes, L. J. R., Rodrigues, A. M., Matias, J. C. O., Ferraz, A. I., & Rodrigues, A. C. (2021). Production of biochar from vine pruning: Waste recovery in the wine industry. *Agriculture (Switzerland), 11*(6), 1–15. https://doi.org/10.3390/agriculture11060489

Oláh, J., Aburumman, N., Popp, J., Khan, M. A., Haddad, H., & Kitukutha, N. (2020). Impact of industry 4.0 on environmental sustainability. *Sustainability (Switzerland), 12*(11), 1–21. https://doi.org/10.3390/su12114674

Ossewaarde, M., & Ossewaarde-Lowtoo, R. (2020). The EU'S green deal: A third alternative to green growth and degrowth? *Sustainability (Switzerland), 12*(23), 1–15. https://doi.org/10.3390/su12239825

Parjanadze, N. (2009). Globalisation theories and their effect on education. *International Black Sea University Scientific Journal, 3*(3), 77–88. http://search.proquest.com/docview/865523946?accountid=14771

Prieto-Sandoval, V., Jaca, C., & Ormazabal, M. (2018). Towards a consensus on the circular economy. *Journal of Cleaner Production, 179*, 605–615. https://doi.org/10.1016/j.jclepro.2017.12.224

Prodanov, C. C., & Freitas, E. C. (2013). *Metodologia do trabalho científico: métodos e técnicas da pesquisa e do trabalho acadêmico* (2nd ed.). Universidade Freevale.

Reike, D., Vermeulen, W. J. V., & Witjes, S. (2018). The circular economy: New or refurbished as CE 3.0? – exploring controversies in the conceptualization of the circular economy through a focus on history and resource value retention options. *Resources, Conservation and Recycling, 135*(February), 246–264. https://doi.org/10.1016/j.resconrec.2017.08.027

Stahel, R. W. (2016). Circular economy. *Nature, 531*, 435–438. https://doi.org/10.1038/531435a

Stake, R. E. (2012). *A arte da investigação com estudos de caso* (3rd ed.). Fundação Calouste Gulbenkian.

Todaro, N. M., Testa, F., Daddi, T., & Iraldo, F. (2019). Antecedents of environmental management system internalization: Assessing managerial interpretations and cognitive framings of sustainability issues. *Journal of Environmental Management, 247*(June), 804–815. https://doi.org/10.1016/j.jenvman.2019.06.106

Wicki, S., & Hansen, E. G. (2019). Green technology innovation: Anatomy of exploration processes from a learning perspective. *Business Strategy and the Environment, 28*(6), 1–19. https://doi.org/10.1002/bse.2295

Yang, M. G., & Kang, M. (2020). An integrated framework of mimetic pressures, quality and environmental management, and firm performances. *Production Planning and Control, 31*(9), 709–722. https://doi.org/10.1080/09537287.2019.1681533

Yin, R. K. (2010). *Case study: Planning and methods* (4th ed.). BOOKMAN.

Zink, T., & Geyer, R. (2017). Circular economy rebound. *Journal of Industrial Ecology, 21*(3), 593–602. https://doi.org/10.1111/jiec.12545

Towards Circular Economy: Barriers, Drivers, Challenges and Potentialities in Angola's Agribusiness

Antonino Kamutali, Pedro Henriques, Maria Raquel Lucas, and Ana Marta-Costa

Abstract Agribusiness has a huge impact on the three dimensions of sustainability and it also plays a very important role in humanity's livelihood by providing food safety and food security, fibre, fuel and animal feed, as well as being a sector that moves the economy and the agricultural policies in the world.

In Angola, the contribution of agribusiness to GDP is still very low, but increasing from 0.03% in 2019 to 6.3% in 2021. Angola has the potential to become one of the leading countries in agribusiness in the region, due to the availability and quality of resources (human and natural) and edaphoclimatic conditions (weather, water and soils). In order to meet its internal food needs, Angola needs a modern agribusiness sector. Such implies incorporating innovation and new production technologies and other technological changes and/or technological adequacy. This chapter aims to analyse the situation of agribusiness in the Angolan provinces of Cuanza Sul, Huambo, Bié and Luanda, namely, the barriers, potentialities and challenges to its

A. Kamutali (✉)
Faculty of Agricultural Sciences, Universidade José Eduardo dos Santos, Huambo, Angola
and CEFAGE- Center forAdvanced Studies in Management and Economics, Universidade de Évora, Évora, Portugal
e-mail: ninokamutali@gmail.com

P. Henriques
Department of Economics and MED - Mediterranean Institute for Agriculture, Environment and Development, Universidade de Évora, Évora, Portugal
e-mail: pdamiao@uevora.pt

M. R. Lucas
Department of Management and CEFAGE- Center for Advanced Studies in Management and Economics, Universidade de Évora, Évora, Portugal
e-mail: mrlucas@uevora.pt

A. Marta-Costa
Department of Economy, Sociology and Management, and CETRAD - Centre for Transdisciplinary Development Studies, Universidade de Trás-os-Montes e Alto Douro, Vila Real, Portugal
e-mail: amarta@utad.pt

C. Rego et al. (eds.), *Entrepreneurship, Technological Change and Circular Economy for a Green Transition*, Studies on Entrepreneurship, Structural Change and Industrial Dynamics, https://doi.org/10.1007/978-3-031-48079-9_13

development from a perspective of circular economy and sustainability. The findings reveal that Angolan agribusiness has a weak performance, but that it has the potential to grow, in terms of size, quantity, quality and level of sustainability. The poor performance is fundamentally due to the use of inappropriate agronomic practices, including the scarce use of improved technologies and the low technical capacity of the players along the entire value chain.

Keywords Agriculture production · Focus group · Performance · Stakeholders · Supply chain · Sustainability · Technology

1 Introduction

Agribusiness has a huge impact on the three dimensions of sustainability, and it plays a very important role in humanity's livelihood by providing food safety and food security, fibre, fuel and animal feed, as well as being a sector that moves the economy and the agricultural policies in the world. According to the United Nations Conference on Trade and Development (UNCTAD, 2022), agribusiness contributes with 65% of the employment and 75% of the internal trade in Africa. The continent boasts a growth rate of 5% per year, which could result in its Gross Domestic Product (GDP) tripling by 2030. Growing at this rate, Africa's GDP could surpass Asia's by 2050 (UNCTAD, 2022). Due to its direct links with the rest of the economy, agribusiness could prove to be the necessary catalyst to boost this level of growth in Africa.

In Angola, the contribution of agribusiness to the GDP is still very low, but it increased from 0.03% in 2019 to 6.3% in 2021 (MEP, 2021). Angola has the potential to become one of the leading countries in agribusiness in ~~the~~ Central Africa, due to the availability and quality of resources (human and natural), as well as its edaphoclimatic conditions (weather, water and soils). The arable land occupies about 28.2% of the country, of which only 17% is cultivated, with family farming being responsible for 91.6% of the sown area and the business sector accounting for 8.4% (MINAGRIF, 2020). The agribusiness sector in Angola is becoming very relevant. At social level, it employs 46% of the workforce and represents the main source of income for 90% of the 8.5 million Angolans who live in rural areas (MINAGRIF, 2020). With few economic alternatives outside ~~of~~ agriculture, women make up for more than 70% of the active agricultural workforce and dominate food production and informal marketing channels, in addition to being responsible for household food security (MINAGRIF, 2020).

Despite being a major oil exporting country, rich in diamonds and other mineral resources, boasting a per capita GDP of over US$3500, poverty remains prevalent, with a third of its population living on or below the poverty line of US$ 1.90 per day, of which 64% are in rural areas (INE, 2020). Additionally, a large part of the population still does not have access to basic services, especially in rural areas, and life expectancy is 52.3 years old, making it the 149th country in the United Nations Human Development Index (BM, 2018), out of a total of 188 countries.

In several years of economic diversification based on import substitution and large-scale government-sponsored projects, the country has solely reached modest results and a significant number of the goods consumed continue to be imported (MEP, 2021). Moreover, Angola's private sector is small and dual, with a large number of microenterprises, largely informal, and key sectors dominated by non-competitive state-owned or government-affiliated companies. Few new and better jobs were created, particularly for the young population below 24 years old, which represents 65% of the total population (MINAGRIF, 2020).

Based on the objectives of sustainable development and in order to meet its internal food requests, Angola needs a modern agribusiness sector. In this context, the chapter aims to analyse the Agribusiness situation in the Angolan provinces of Cuanza Sul, Huambo, Bié and Luanda, namely, the barriers, potentialities and challenges to its development from the perspective of circular economy and sustainability. This work is integrated in the "Accelerating Inclusive and Sustainable Agribusiness Investment in the Economic Growth Corridors" (Agro-PRODESI), a project resulting from an initiative by the Ministry of Economy and Planning (MEP) of the Angolan government, which was implemented in partnership with the Food and Agriculture Organization of the United Nations (FAO). The project aims to contribute to an increase in national production by fostering prioritized value chains and connecting market actors and small farmers along the Economic Growth Corridors.

According to Sehen et al. (2019), company performance supported by sustainable operations is essential for the creation of circular economy based production systems and investment in industrial synergies can be seen as a strategic tool for economic development. Such will require an integration of subjects such as economic diversification, innovation and human resources qualification, which can represent technological changes and advances, essential to finding permanent solutions to the economic, social and environmental challenges of agribusiness. This involves technology changes in the equipment, environmental control of agricultural practices, use and control of herbicides and pesticides, animal production, welfare, processing, logistics, management processes and marketing (Weick, 2001), aimed towards the circular economy principles.

Enhancing Angola's agriculture performance can increase foreign market competitors, provide a response to the country's food challenges and potentially improve the competitiveness of the agribusiness' participation in the trade balance (Cristóvão et al., 2022). Additionally, a new developed model based on stronger private sector participation can be promoted to increase food production while simultaneously consuming less resources. In this context, the circular economy approach gains significant attention and highlights the urgent need for its adoption to preserve resources at all levels and contribute to ~~the~~ sustainability (Velenturf & Purnell, 2021).

2 Background

Agribusiness is considered one of the drivers of the economy of many countries, as it generates a high number of jobs which are directly reflected on the income of the workers. Concerning its performance, Costa (2006) exposes the importance of agribusiness in the global sphere, which surpasses the average productive capacity of the industrial sector, due to its dynamics and participation in the economy.

The term agribusiness was created by Davis and Goldberg (1957) and refers to the Commodity system approach or approach by the system of product—raw material—commodity. For Freitas et al. (2007), the term agribusiness can be associated with the principle of supply chain, presenting activities and/or operations which relate to each other.

From an economic point of view, agribusiness is a conjunction of businesses linked to agriculture and livestock, which represents rural producers, ~~constituted~~ in the form of individuals, ~~the~~ farmers, and legal entities, i.e. companies. Other businesses, in turn, are represented by the commerce and industry that supply inputs for farming production, which is the case, for example, of fertilizer, manufacturers, pesticide manufacturers and equipment professionals. Lastly, there are the businesses of purchase, transport, processing and sale of agricultural products to the final consumer, which include slaughterhouses, textile and footwear industries, packers, supermarkets and food distributors (Barros et al., 2020).

Thus, agriculture is no longer restricted to the physical barriers of the property, but it is continuously and increasingly dependent on inputs coming from outside the farm. Moreover, its production strategies are closely related to the consumer market, involving farmers in the process, who seek constant reconciliation between quantities and prices (Kamutali et al., 2020).

All the concepts do not display agribusiness as a set of isolated elements, but instead, as a supply chain as a whole, where the interdependence of the economic agents takes the products to the final consumer, starting at the input producer. Thus, in a short space of time, an activity that used to predominate in subsistence became ~~considered of~~ a great flow of products, services and capital, capable of moving ~~the~~ major economies in the world (Barcellos et al., 2016).

Buainain et al. (2013, 2014) add that the agribusiness sector presents a peculiar characteristic, which is the fact that it is composed of factors such as: seasonality, perishability and heterogeneity. The context in which it operates directs the actions of the companies in this branch so that they may continuously and durably innovate and have the ability to keep up with the dynamism of the market. The agribusiness sector is constantly undergoing changes ~~in order~~ to increase its productivity and competitiveness and to meet the increasingly challenging consumer demand. This results in a stimulus for innovations that can support the speed of these transformations (Rajapathirana & Hui, 2018).

Although agribusiness is a largely solid sector, from the economic and social points of view, it has its limitations and barriers, which require great efforts from its players in order to remain competitive. At the producer sphere, for example, there are several bottlenecks that directly affect the system, such as government policies, seasonality and local economy (Sehen et al., 2019). Additionally, the current system development has been considered harmful to the ecological system (Amui et al., 2016), contributing to the depletion and waste of natural resources, pollution and damage of the planet (Yong, 2007; Marta-Costa et al., 2012).

In this context, the optimization and the efficient use of resources has been strongly encouraged. To this end, the transition from the original economic model towards circular economy models is playing a key role (Atanasovska et al., 2022).

Circular economy assumes the circulation of material, capital and labour and their supply-demand connexion as endogenous factors, which promote economic growth, while the environmental costs and scarcity of natural resources are internalized among the total costs of the socioeconomic activities. This theoretical principle, in addition to encouraging resource saving and mitigating pollution, guides those involved towards resource sustainability (Yong, 2007). However, the transitioning process to the circular economy requires several trade-offs, which may generate tensions among the stakeholders involved in the process (Pitkanen et al., 2016; Sehen et al., 2019).

According to Burgo-Bencomo et al. (2019), three phases are needed to implement circular farm management models. At the production planning level, the food demand and possible surpluses should be known to determine the required area of land, variety of products and plan the sowing. The organization of the production includes all of the actions which seek the correct development of the crop and compatibility with the preservation of ecosystems, while implying the circularity of materials, energy and labour. The third phase is dedicated to the production system application, which includes planting, harvesting, determining crop yields and their natural integration with the environment. At this stage, it is essential to optimize the use of nutrients according to the needs of the crops or to minimize the use of pesticides, not only to avoid polluting conditions and adverse effects on biodiversity, but also to control and regulate the process.

The implementation of these actions is in line with Ekins (2019), who highlights the reduction of the depletion of scarce natural resources, the decrease of environmental damage with the use of virgin materials, and the abatement of the resulting pollution as inherent goals of the circular economy. The author also associates two aspects to this concept, although he attributes less importance to them, which are the flow of materials through an economy and the reflection on the economic conditions that can give rise to this flow.

Thus, the circular economy model emphasizes the shift to ecologically sound products and renewable energy (Kopnina & Blewitt, 2015), being a strategic concept based on the reduction, reuse, recovery and recycling of materials and energy towards achieving sustainable production and consumption processes, through ensuring the reduction of negative externalities (Górecki et al., 2019). By coordinating production and consumption systems in closed circuits, a reorganized and more efficient economic model is promoted (Hofmann, 2019).

To implement circular economy initiatives in firms, different strategies have been pursued by developed and developing countries (Sehen et al., 2019). In the first group, they tend to be bottom up and stimulate reverse and closed-loop logistic mechanisms for the company's sustainability, supported in a structured institutional framework (Ghisellini et al., 2016; Haddad-Sisakht & Ryan, 2018; Sehen et al., 2019; Korhonen et al., 2018). From the perspective of an emerging economy, these initiatives follow a top-down pattern and they are very slow and dominated by the effects of waste on the environment (Ghisellini et al., 2016; Rótolo et al., 2022; Sehen et al., 2019).

This is the particular context of Africa, where circular economy methods help improve food security and agricultural productivity. In this geographical area, the use of regenerative agricultural practices and post-harvest handling techniques stand out, which allow the adoption of closed nutrient cycles through the use of their coproducts and organic waste as inputs (GRID-Arendal, 2021). In this perspective, the GRID-Arendal (2021) document identifies the reconfiguration of resource use, labour and economies as promoting systemic, inclusive and operational approaches at various scales.

3 Methodology

3.1 The Study Location

Angola is located on the west coast of Africa, below the equator, bordering Namibia to the south, Congo to the north and Zambia to the east. The country is divided between an arid coastal strip stretching from Namibia to Luanda, a humid inland central plateau, a dry savanna in the southern and south eastern interior, and tropical rainforest in the north and Cabinda enclave.

The territory has a surface area of 1,245,700 km^2 and is the home to communities whom in bygone centuries formed nations (Kingdoms of Congo, Matamba and Benguela), with their own social structures and organizations. The evolution of history determined the borders known today, which means that in the eighteen provinces that make up the current administrative division, live seven indigenous cultural communities, from which other subgroups emerge (about a hundred) that are distinguished by their socio-linguistic and cultural particularities, namely, the Bacongo, Ambumbu, Ovimbundu, Tchokwe, Ovinganguela, Ociwambo and Khoisan.

The 2014 population census data estimates that the province of Luanda is the most populous, concentrating 27% of the country's population. This province is followed by the provinces of Huíla 10%, Benguela and Huambo with 8%, Cuanza Sul 7% and Bié and Uíge with 6%. These 7 provinces concentrate 72% of the total resident population of the country (INE, 2020).

In terms of climate, two distinct seasons occur: the rainy season—humid and hot, which runs from September to April, occurring earlier or later in some regions; and the "cacimbo" season—dry and cold, which usually runs from May to September. Given the size of the country, there is a variety of climates from region to region, on the coast the average annual rainfall is less than 600 mm, but the province of Cabinda, to the north, sometimes reaches values in the order of 800 mm, while the province of Namibe, on the south coast, reaches only 50 mm.

In the pre-independence period, the agrarian production structure was organized into two socioeconomic types, "the family or traditional sector" and the "business or employer" sector. The family sector was responsible for most of the marketed production of the main crops in Angola, corn (88%), cassava (100%), beans (94%), peanuts (100%) and potatoes (71%) (Pacheco, 2003). Up to the present day, there have been no major changes, as family farming in Angola constitutes an economic activity which occupies a large part of the population and is responsible for the production of about 81% of its cereals, 92% of its roots and tubers, 89% of Angolan legumes and oilseeds, 85% of meat and 30% of fish (MINAGRIF, 2020). Family farming is still the sector ~~the sector~~ that supplies the rural and urban markets of the main cities.

3.2 Data Collection

This work is integrated in the project Agro-PRODESI, namely, in the axe dedicated to the AgroProdesi's Territorial Capacity Building Program "Agro-PRODESI Territorial", which foresees the key actor's building capacity in agribusiness development, within the prioritized value chains, promotion and facilitation of sustainable investments in agriculture, allowing them to participate effectively and competently in the implementation of the territorial investment strategy for the agricultural sector.

The Territorial Agro-PRODESI initiative was implemented during the period between May and December ~~of~~ 2021, in all 18 provinces of Angola. It was during this training sessions period that information was collected through focus groups, in 4 provinces.

The distinct points of view of the different value chain participants were explored, in light of the focus of this research, which aims to understand and deepen the challenges and potentialities of agribusiness in Angola. All the participants expressed themselves freely on aspects specifically related to the context of their localities. The qualitative approach was selected, because, according to Guerra (2006), qualitative research allows for an understanding of the participants' perspectives (individuals or small groups of people) on the phenomena that surrounds them, as well as a greater insight on the way in which the participants themselves subjectively perceive their reality, through a deepening of their experiences, opinions and meanings, without worrying about numerical representativeness, statistical generalizations and linear relationships of cause and effect.

The information was obtained through focus groups, which are recommended by several authors (Plummer-D'Amato, 2008; Nyumba et al., 2018), based on the participants' discourse and interaction, and assuming the construction of knowledge in spaces of intersubjectivity. In other words, it is assumed that, through the focus group, similarities and differences, constructions and deconstructions, inherent to the intersubjectivity of its participants, are individualized.

Different actors of the value chain, who were previously invited by MEP, participated in these focus groups, ranging from representatives of public institutions, academics, NGOs, civil society organizations, individual producers and consumers, cooperatives, input sellers and service providers. In each province and according to their role in the value chain, the following seven thematic focus groups were formed: technical assistance (public, NGOs, academia), input sellers, producers (individual and cooperatives), processors (mills and breweries), transporters, sellers (supermarkets and markets) and consumers (buyers).

Within the focus groups, the work methodology was based on tools that allowed the participants to value their experiences, work practices and knowledge, combining a theoretical and practical approach which promoted dialogue, based on concrete work experiences and approaches to agribusiness. Each focus group identified the main challenges and pointed out possible solutions for an improved agribusiness value chain, in terms of linkages among different value chain actors. Afterwards, the focus groups were brought together and all the work done separately was discussed in plenary.

Participation by province by gender and by actors is presented in Table 1. Producers were in the majority in all the provinces, followed by consumers. The province of Cuanza Sul had the highest number of participants followed by Bié, while Luanda had the lowest participation. In all provinces, the majority of participants were men.

Table 1 Participation by value chain actors, sex and province

		Angola Province				
		Huambo	Cuanza Sul	Bié	Luanda	Total
Value chain actors	Technical support	8	6	8	4	26
	Input and equipment salesmen	6	5	5	4	20
	Farmers	14	22	20	17	73
	Transformers	4	3	4	3	14
	Transporters	8	4	3	4	19
	Sellers	8	10	7	4	29
	Consumers	10	12	13	8	43
Sex	Women	20	22	18	14	74
	Men	38	40	42	30	150
Total		58	62	60	44	224

Source: Author's elaboration

4 Results and Discussion

The results of the research with the focus groups are presented in Table 2. They are separated by the seven value chain actors, followed by a discussion focused on the barriers in Angolan agriculture and the application of circular economy models.

Technical assistance is mostly provided to producers free of charge, by public institutions and non-governmental organizations (NGOs). Among the public institutions, the Institute for Agrarian Development (IDA) stands out. Although free, it is not sufficient in quantity and quality to meet the needs of producers.

Table 2 Barriers identified in each focus group

Value chain actors	Barriers
Technical support	• Precariousness of technicians' contracts. • Insufficient extension services. • Low-skilled labour force. • Insufficient technical staff.
Input and equipment salesmen	• External dependency. • High purchase prices. • High customs tariffs. • Weak financial capacity of family producers. • Weak financial capacity to cover the country.
Farmers	• Few technical experts. • Excessive bureaucracy for land legalization. • Low productivity. • Poor utilization of water resources. • High prices and short supply of agricultural inputs and services. • Difficulty of commercial banking in dealing with the agricultural sector. • High prices and low supply of agricultural mechanization services. • Poor protection against pests and diseases.
Transformers	• Insufficiency of electrical infrastructure. • Insufficiency and high prices of raw materials for transformation. • Insufficient storage infrastructure. • Scarce sector funding. • Insufficient skilled labour.
Transporters	• Degraded secondary and tertiary roads. • Difficulty in acquiring spare parts. • Difficult access to the production areas. • Poor technical condition of the vehicles.
Sellers	• Lack of market information. • Competition of imported products versus local products. • Informality of the market.
Consumers	• High product prices. • Little information on the products.

Source: Authors' elaboration

In light of the challenges, there are a couple of initiatives that can be pointed out as possible solutions to mitigate these challenges, such as the opening of a public competition for entry to the public technical assistance services, the stimulation of the emergence of public–private partnerships for technical assistance, the training of specialized technicians with priority for agricultural mechanics, agricultural machinery operators, specialists in grains, tubers, legumes and specialists in irrigation and drainage, and finally, the strengthening/emergence of technical assistance services in the private sector.

Regarding agricultural input supply, the stimulus for the emergence, in the country, of factories of agricultural fertilizers, pesticides, agricultural machinery and equipment as well as seed producers and multipliers are pointed out as measures to mitigate foreign dependence. On the other hand, while the factories do not exist, the subsidy of inputs and the exemption of customs duties on agricultural inputs can contribute to those actors having a margin that allows them to open input store representations at the level of municipalities and communes, which would allow an easier access to producers.

With respect to farmers, land preparation is done mainly by hand (in ridges) or with the use of animal traction (shallow, in furrows), with the incorporation of plant remains from the past agricultural season in the soil. The main fertilizer used in Angola is the compost N-P-K, in different doses and in many cases without soil analysis. The cover fertilization is done with nitrogen fertilizers (urea and ammonium sulphate). The use of organic fertilizers (manure) is mainly verified in families with animals (cattle, chickens). This practice, combined with the incorporation of plant remains, provides the soil with the appropriate conditions for biochemical transformations to occur and good circular economy practices to exist.

The slowness of the land legalization process is pointed out as one of the main challenges, since it limits producers' access to financing. An improvement in the process of granting land titles can be achieved by reducing the number of public agents participating in the process, or by concentrating the services in a sort of one-stop shop. Producers have difficulties with financing, which, considering the high rates practiced by commercial banks as well as the excessively short grace period, makes it difficult for them to solve problems related to the acquisition of inputs, services and technical assistance, which contribute to low productivity.

The processors in the chain were found more as service providers than producers of their product brands. Extending the electric grid to the production and transformation zones, access to financing and the creation of short-term university courses or technical schools are some of the measures pointed out to overcome the identified challenges.

As far as transporters are concerned, roads and transportation vehicles are considered the main challenges. It is worth noting that among these challenges, the poor quality of the roads was the most referenced.

The results obtained from sellers revealed that there is a high degree of informality in agri-food markets, little information on prices and market operation, and the existence of many imported products in the market, which are believed to be produced at low cost and apparently with higher quality than those produced locally.

For consumers, the main challenges are high prices and information. Consumers say that agricultural production in Angola is quite diversified, but insufficient to supply the market, and many of the products consumed are imported. In relation to their origin, consumers do not always have information and do not seek to have it. However, for products such as honey, vegetables, fruits and meat, consumers tend to choose local products, believing them to be natural.

According to the results of the focus groups and based on the authors' knowledge of the local context, it was possible to construct Table 3, which identifies the opportunities and weaknesses of Angolan agribusiness.

4.1 Towards Circular Economy

An integrated vision of the development of the Angolan agribusiness sector should include policy proposals to overcome some of the barriers and take advantage of the opportunities and strengths of the current situation, while gradually incorporating the fundamentals of the agri-food circular economy.

Regarding the policy proposals, Table 4 shows the measures for different areas: land, input supply, agricultural credit and insurance, agricultural production, agricultural distribution, product innovation and agroindustry, technical support, professional training, research, infrastructure, linkages and gender equity.

A global vision on circular economy in the Angolan agribusiness sector will require strong agri-food circular policies and partnerships between public,

Table 3 Policy proposals for the agribusiness in Angola

Opportunities
• Possibility of establishing partnerships and/or productive alliances to find niche markets of high added value; • Strong competition and very strong competition with some imported products; • Formal market concentration in some (few) operators; • Concentration of distributors and retailers with high negotiating power; • Need to create the identifying and differentiating brand of the products and to have a good/better relationship with buyers and suppliers.
Strengths
• The roads that connect the municipal headquarters with the capitals of the provinces are passable, this facilitates logistics; • Existence of research and teaching institutions: Institute of Veterinary Research, Institute of Agricultural Research, Faculty of Veterinary Medicine and Faculty of agricultural sciences, technical agricultural institutes facilitate the emergence of innovation; • Established hydrographic network: There is a vast hydrographic network in the region, which is a factor of potentiality for the agricultural sector; • Human resources: There are many people trained in the agricultural field, who can contribute to the development of agriculture; • Various actors working in the rural environment: There are several NGOs working in the rural environment.

Source: Authors' elaboration

Table 4 Policy proposals for different areas

Area	Policy proposals
Land	• Celerity in granting the land titles by local governments;.
Input supply	• Subsidy of production factors and tax exemption to its importers; • Stimulate the emergence of additional seed multipliers; • Stimulate the emergence of new mechanization operators; • Review, in the long term, strategies for the acquisition of fertilizers and other inputs (installation of factories);
Agricultural credit and insurance	• Stimulate the emergence of specialized banks that meet the needs of the agriculture; • Guarantee access to technical assistance, bank loans, training and markets; • Adjust the grace period for agricultural activity; • Stimulate the creation of agricultural insurance;
Agricultural production	• Improve operational efficiency: Reduce production and transaction costs; stimulate the business sector with subsidized credits; • Establish an outflow strategy for field products;
Product innovation and agroindustry	• Improve operational efficiency: Reduce production and transaction costs; stimulate the business sector with subsidized credits; • Establish an outflow strategy for field products; • Invest more in the processed products market (establish partnership with processing companies to expand the markets and at the same time make the current production profitable and dynamic); • Create/strengthen product brands, expanding to different products; • Bet in innovative products (with high added value and low competition, in order to increase margins);
Agricultural distribution	• Pursue improvements in distribution, seeking greater intervention from the industrial sector;
Research	• Increase the connection with universities and research centres for studies and partnerships that find solutions to the concrete problems identified;
Technical support	• Financial support for soil analysis laboratories; • Encourage the private sector to join the activity of soil analysis as a business;
Professional training	• Training actions for tractor drivers and agricultural mechanics; • Establish/strengthen farmers' capacity in production techniques, with high quality standards;
Infrastructure	• Establish a strategy for the improvement of secondary and tertiary access roads;
Linkages	• Promote linkages between the agricultural sector and other sectors, as well as between rural areas and cities; • Improve coordination of all actors operating in agribusiness; • Dynamize the relationship with suppliers and distributors; • Improve communication (with suppliers and consumers); • Establish/strengthen the relationship/partnership and communication among actors in the value chain;

(continued)

Table 4 (continued)

Area	Policy proposals
Gender equity	• Ensure women's access to productive resources equal to those of men; • Need to invest in women empowerment programs, with emphasis on socio-cultural and educational areas, among others.

Source: Authors' elaboration

cooperative and private sectors in order to reduce food loss, improve livelihoods, keep materials in use to create new value opportunities, regenerate natural systems for better societal and environmental outcomes and enable the roles of design and technology, which can empower food designers to support healthy diets (Jeffries, 2023). The agri-food circular policies should be based on the principles of multiple R's such as reduce, reuse, refuse, rethink, repair and recycle (Boon & Anuga, 2020).

An overall view of the Angolan agribusiness sector shows two main groups of activities: to supply farmers with agricultural inputs (machinery, fertilizers and chemicals) and to supply consumers with agricultural products, which is dependent on farmers (storage, processing and distribution). The first group of activities concerns mainly imported goods and it is unforeseen that this situation will change in the next few years. The second group of activities starts at farm level and ends at the final consumer, and thus, it should be subject to special attention as means of inducing the principles of circular economy.

The production challenge for farmers in Angola is to close the gap between current and potential yields, in order to ensure that farmers are able to provide food for their families and for consumers in the rural and urban markets. To achieve this core goal, the development process should incorporate measures towards circular economy. In this context, there are four strategies that can facilitate the development and implementation of circular models in agriculture (Velasco-Muñoz et al., 2021): (i) narrowing resource circuits, in order to maximize resource utilization; (ii) slowing resource circuits, in order to extend the useful life of products within the agri-food system; (iii) closing resource circuits, based on the reuse and recycling of agricultural materials; and (iv) regenerating resource flows, which groups all actions that promote the conservation and improvement of natural capital.

Although no studies on circular economy in Angolan agriculture were found, in general terms, the intervenient in the agricultural sector must look carefully at each agricultural activity and identify the main material inputs and outputs, in order to propose measures and business models towards circular economy.

The implementation of cleaner production is a core element to circular economy at enterprise level. In order to raise the efficiency of resource utilization and reduce or avoid emissions, there are three key tasks for an enterprise-level cleaner production, which are: closed material flows that are technologically and economically reasonable, a step-by-step full use of energy and a maximum reduction in the use of toxic materials (Yong, 2007). A recurring criticism levelled at the circular economy is that it ignores much of the established knowledge. It neglects, in particular, the thermodynamic principle that matter cannot be created or destroyed; meaning that whatever the resources used, they cannot be destroyed but only converted and dissipated, ending up somewhere in the environmental system (Giampietro & Funtowicz, 2020).

Regarding agriculture and its main activities, crop and livestock, Table 5 summarizes the main material inputs and outputs used.

A large number of studies analyse agricultural practices to improve material efficiency in crops, for instance in the fertilizers used. An important aspect of the circular economy for the agricultural sector, which has been widely considered, is to enhance phosphorus efficiency and recovery.

Regarding land use, specific solutions and techniques for soil improvement, which mitigate aspects of climate change and, at the same time, deal with food security solutions, are urgent and necessary. The application of circular economy principles to land use can make agricultural production more sustainable and efficient (Toop et al., 2017).

The inputs and outputs of materials and energy used in agriculture constitute a wide range of alternatives, depending on crop or animal agricultural activities. Depending on the agricultural system, the more complex the system is, the more information is needed to improve it. The main resources used in animal production are water, energy, animal feed, vitamins, medicines, and in crop production, seeds and plants, fertilizers, chemicals and water, among others. The processes and cultural practices carried out in order to improve the production and quality of the products can include alternatives towards circular economy principles, such as mulching, irrigation techniques, pruning, pest control and the use of crop residues and animal waste as crop fertilizers. Thus, the level of resource use depends on the characteristics of the crop or livestock and on their needs at each stage of crop or animal development.

In order to improve the circularity of the agricultural processes and cultural practices, a greater level of knowledge is needed on the possibilities of reusing the different materials and the circular treatment of permanent and semi-permanent farm elements, such as greenhouse structures, irrigation systems, ventilation and heating systems, etc. This knowledge isn't always available, though it provides an opportunity to develop circular models based on the management of technical agriculture

Table 5 Inputs and outputs for a circular crop and livestock activities

	Crops	**Livestock**
Inputs	Carbon dioxide Oxygen Fertilizers Chemicals Energy (human, animal and oil) Water (rain and irrigation)	Oxygen On-farm feed Off-farm feed Natural pastures and grasses Water Medicines Energy (human, animal and oil)
Outputs	Harvested materials Crop residues	Animal products Manure Gaz emissions

Source: Adapted from Arru et al. (2022)

materials. Managing these farm elements is particularly relevant to developing integrated value chains which include more than the links of a single production process, as they interconnect different processes and sectors. All this must be complemented with other approaches, in a holistic manner, which implies the simultaneous application of the different circular economy principles, at the different stages of the involved value chains' life cycles and considering all stakeholders (Mangers et al., 2021).

The role of energy, generated from waste from agricultural activities, which addresses the circular economy and closes the loop, deserves to be highlighted, although it is not new. For many years and in many contexts, agriculture has been using the loop closing perspective for materials and waste. However, with the advancement of the circular economy and the closing loops knowledge, new methods for further efficiency in generating energy from agricultural waste have been emerging in recent years. The most advantageous of these is biogas. Thus, using the new biodigester technologies, the generation of energy of different types (electrical, thermal, mechanical) adds value to the agricultural system, making it stronger, more competitive, sustainable and circular (Barros, 2020). Moreover, biomass, such as wood, forest and crop residues, can also be used to generate electricity. Biomass is a significant resource in an agricultural circular economy in terms of material flow and energy supply, and choosing appropriate raw materials can be a challenge if resource offer is scarce when compared to demand (Velasco-Muñoz et al., 2022). Although such practices are not often identified in Angola, it is not denied that the future will pass through them.

The installation of biodigester systems can be linked to sustainable development and consequently to bioenergy. In economic terms, the biodigester can generate electricity which can be used both on-farm and as a product to be sold to supply the electricity grid. Ultimately, biomass, animal fat, agro-industrial waste, manure and food waste contribute to the environmental aspect of sustainability.

Reducing food losses could address two of the biggest challenges affecting African countries today, namely food security and the livelihoods of smallholder farmers. Due to a variety of factors—including the quality of and access to storage containers, poor transport systems, and lack of refrigeration—food losses in Africa remain significant. For example, Nigeria, the second largest tomato producer in the continent, loses up to 50% of its tomatoes after harvest. Improving roads and other infrastructure will be key in addressing this issue, however with an almost USD 50 billion yearly infrastructure gap, between what is needed and what is available, other options need to be explored (Savolainen et al., 2020). In addressing these interconnected issues, there is an opportunity for African countries to follow a different trajectory and establish a food system that can help tackle some of the biggest food security and economic challenges the continent is facing (Savolainen et al., 2020).

With regard to transport, which is a major contributor to greenhouse gases, urgent changes are needed to reduce emissions. A new business model based on clean energy is required, hence, in this particular case, circular transport starts at the fuel source (GRID-Arendal, 2021). Angola can play a key role in the mobility of the southern region of Africa. To do so, it will need to rehabilitate its main transportation infrastructures, namely, the rehabilitation of secondary and tertiary roads and interconnect the main transportation networks, roads, railways and sea ports, with the objective of shortening the transportation circuits of products and increasing their efficiency.

In the case of the transformation actors, industrial factories can also be improved, specially their electrical and water infrastructures, allowing for important gains in water and energy savings. Today, many transformation units function based on generator groups with high fuel expenditures. On the other hand, the reuse and recycle of the waste produced is also an important challenge for the near future.

In the sellers and consumer field, the use of plastic packaging is constant and has been increasing with the existent business models. Concerning this matter, the main challenge for Angola is the reduction, the reuse and the recycling of plastic packaging. When travelling in urban and rural areas of Angola and many others countries, the presence of plastic bags is common at the roadsides and proves a disturbing view of our shared development model.

The limitations of the established knowledge also extend to how specific circular economy organizations understand consumption. Advocates of circular economy and circular business models have been found to adopt, many times, a simplified understanding of consumption, reducing it to purchase and recycling (Casson & Welch, 2021).

The lack of consumer interest in green alternatives is a common problem. However, unlike, for example, the substitution of fossil fuels for biofuels, the circular economy requires a radical reformulation of the role of the consumer—from consumer to user (Lazarevic & Valve, 2017). Therefore, replacing traditional ownership with dematerialized services may not always be feasible or appealing to consumers (Hobson, 2019).

Circular consumption confronts consumers with choices and trade-offs that are difficult to resolve, while the basic technological orientation of the circular economy approach and its ecomodernist idea of gradually adapting the current production system to material resource limitations tend to sideline, temporality and spatiality, in which consumption occurs (Holmes et al., 2021).

The implementation of circular models in the agricultural sector involves a series of challenges, such as the imposition of regulatory limitations, the need to optimize reverse logistics chains, the geographical dispersion of the companies, system boundaries and the leakage of materials, the ignorance and lack of acceptance among consumers, the technological limitations, the lack of certainty and incentives with respect to investments (Velasco-Muñoz et al., 2021).

Therefore, the economic (investments, costs and revenues), technical (inputs and technology), social (equity, job creation and income), cultural (traditions and indigenous knowledge) and environmental aspects (ecosystems preservation and diversity of human and living beings) appear as demands for future research, along with new circular business models which are expected to be designed to facilitate business management. In terms of tools and indicators to measure the circular economy in rural properties, further elaboration is needed in order to facilitate the management of the material and energy which circulates inside and outside the organizations (Barros et al., 2020).

Human societies have been using with greater or lesser success, depending on the contexts, the reduce, reuse and recycle R's. The next challenge for all societies is to expand the use of these three R's, as well as to implement innovation production technologies for goods that are easily repair, to rethink our lifestyles and the way products are designed or made, and to have the courage not to buy and not to use materials and products that are unsustainable.

5 Concluding Remarks

This exploratory and descriptive research sought to analyse the reality of agribusiness in Angola, focusing on the analysis of the value chain, namely technical assistance, input sellers, producers, processors, transporters, vendors and consumers.

The results reveal that Angolan agribusiness is underperforming but has the potential to grow. The poor performance of Angolan agribusiness is mainly due to the use of inadequate agronomic practices, among which the scarce use of improved technologies, with emphasis on the insufficient use of improved seeds, fertilizers, plant protection and irrigation technology, which contribute to low productivity. However, most agricultural inputs and technologies are imported and most small and medium-sized agribusinesses cannot access them due to their excessive costs. Weak support infrastructure and little market information also contribute to poor performance. These deficiencies are further compounded by low technical capacity throughout the value chain, which starts with insufficient technical assistance institutions, deficient agricultural research and low funding dedicated to technical assistance and research.

Aware of this situation, the government has developed actions aimed at restoring growth, with the implementation of policies, programs and projects aimed at diversifying the economy and replacing imports of various products. The government recognizes the fundamental role that agribusiness plays, and in this context, has made it a priority to put the country's natural resources at the service of food security, to generate foreign exchange by promoting agribusiness and to become competitive at the regional scene.

However, there seems to be no doubt that the development of agribusiness in Angola will necessarily go through an intense process of technological innovation and management of value chain, but such innovation requires research structures to generate the knowledge needed to make the various value chains sustainable and competitive.

A global vision on circular economy in the Angolan agribusiness sector will require strong agri-food circular policies with partnerships between public, cooperative and private sectors, in order to reduce food loss, improve livelihoods, keep materials in use to create new value opportunities, regenerate natural systems for better societal and environmental outcomes and enable the roles of design and technology, which can empower food designers to support healthy diets (Jeffries, 2023). The agri-food circular policies should be based on the principles of multiple R's such as reduce, reuse, refuse, rethink, repair and recycle (Boon & Anuga, 2020).

An overall view of the Angolan agribusiness sector shows two main groups of activities: the supply of agricultural inputs to farmers (machinery, fertilizers and chemicals) and the farmers' supply of agricultural products to consumers (storage, processing and distribution). The first group of activities are mainly imported goods and are unexpected that this situation will change in the next years. The second group of activities, starting at farm level and ending at the final consumer, should be subject to special attention as to induce the principles of circular economy.

In this context, the necessary changes to be implemented in a sector with serious weaknesses, susceptible to few innovations in its history, characteristic of developing economies, make it potentially able to benefit from the most up-to-date knowledge (subscribed to the circular economy and sustainability theme), allowing productive systems to be adapted to recent circumstances and with greater gains at various levels. This is undoubtedly a great opportunity for the advancement of the agricultural sector, with positive impacts on its value chains.

We are possibly at a unique moment for the development of the Angolan agro-industrial sector, since it is possible to begin to integrate the principles of circular economy. The process is complex, requires commitment, innovation and know-how capacity from agricultural services, and in particular from extension services, input suppliers, research institutes, cooperatives, farmers, transport, distribution chains and consumers. Everyone is tasked with looking at their resources and processes, and start applying circular economy principles such as reduce, reuse, refuse, rethink, repair and recycle, in a step-by-step process to reach the main objective, the circularity of the agri-food chain of Angola.

The path is long, so the sooner it starts, the better it will be for the sustainability of the Angolan agro-industrial sector.

Acknowledgements This research is supported by the Food and Agriculture Organization of the United Nations (FAO), Angola offices through contract-index number 3089160 to Antonino Kamutali. It is also supported by national funds through the FCT (Portuguese Foundation for Science and Technology) under the projects UIDB/04011/2020, UIDB/04007/2020 and UIDB/05183/2020. Funding was additionally provided by the FCT– Portuguese Foundation for Science and Technology under the research contract PRT/BD/152098/2021 to Antonino Kamutali.

References

Amui, L. B. L., Jabbour, C. J. C., Jabbour, A. B. L. S., & Kannan, D. (2016). Sustainability as a dynamic organizational capability: A systematic review and a future agenda toward a sustainable transition. *Journal of Cleaner Production, 142*(1), 308–322. https://doi.org/10.1016/j.jclepro.2016.07.103

Arru, B., Furesi, R., Pulina, P., Sau, P., & Madau, F. A. (2022). The circular economy in the Agri-food system: A performance measurement of European countries. *Economia agro-alimentare, 2*(2), 1–35.

Atanasovska, I., Choudhary, S., Koh, L., Ketikidis, P. H., & Solomon, A. (2022). Research gaps and future directions on social value stemming from circular economy practices in Agri-food industrial parks: Insights from a systematic literature review. *Journal of Cleaner Production, 354*, 131753. https://doi.org/10.1016/j.jclepro.2022.131753

Barcellos, J., Oliveira, T. E., & Marques, C. S. S. (2016). Apontamentos estratégicos sobre a bovinocultura de corte Brasileira. *Archivos Latinoamericanos deProducción Animal, 24*(4). *(6) (PDF) factors that affect the response of rural beef cattle property to technical assistance.* Retrieved April 08, 2023, from https://www.researchgate.net/publication/360326416_Factors_that_affect_the_response_of_rural_beef_cattle_property_to_technical_assistance

Barros, M. V., Salvador, R., de Francisco, A. C., & Piekarski, C. M. (2020). Mapping of research lines on circular economy practices in agriculture: From waste to energy. *Renewable and Sustainable Energy Reviews, 131*, 109958. https://doi.org/10.1016/j.rser.2020.109958

BM – Banco Mundial. (2018). Projeto de Desenvolvimento da Agricultura Comercial em Angola. Relatório no: PAD2029, Luanda, Angola.

Boon, E. K., & Anuga, S. W. (2020). Circular economy and its relevance for improving food and nutrition security in sub-Saharan Africa: The case of Ghana. *Materials Circular Economy, 2*(5), 1–14.

Buainain, A. M., Alves, E., Silveira, J. M., & Navarro, Z. (2013). Sete teses sobre o mundo rural brasileiro. *Revista de Política Agrícola, 22*(21), 105–121. Brasília.

Buainain, A. M., Contini, E., & Viera, P. A. (2014). *Revolução da gestão ou gestão da revolução na agropecuária*. Embrapa, Documento interno para discussão.

Burgo-Bencomo, O. B., Gaitán, V. S., Yanez, J. S., Zambrano, A. A. M., Castellanos, G. P., & Estrada, J. A. H. (2019). La Economia circular una al-ternativa sostenible para el desarrollo de la agricul-tura. *Espacios, 40*(13), 2–6. http://www.revistaespacios.com/a19v40n13/a19v40n13p02.pdf

Casson, C., & Welch, D. (2021). Histories and futures of circular economy. In R. B. Swain & S. Sweet (Eds.), *Sustainable consumption and production* (Vol. 2, pp. 35–54). Springer Books. https://doi.org/10.1007/978-3-030-55285-5_3

Costa, E. J. (2006). Gestão de riscos no agronegócio. *AgroAnalysis, 26*(2), 37–37.

Cristóvão, G. S. F., de Medeiros, G. A., Mancini, S. D., & Pessela, J. D. (2022). Index for the assessment of municipal solid waste management in Angola: A case study in Kuito, bié province. *African Geographical Review, 41*(4), 545–557.

Davis, J. H., & Goldberg, R. A. (1957). *Concept of agribusiness*. Division of Research, Graduate School of Business Administration, Harvard University.

Ekins, P., Domenech, T., Drummond, P., Bleischwitz, R., Hughes, N., & Lotti, L. (2019). The circular economy: What, why, how and where. Background paper for an OECD/EC workshop on 5 July 2019 within the workshop series "Managing environmental and energy transitions for regions and cities", Paris.

Freitas, C. A. D., Bacha, C. J. C., & Fossatti, D. M. (2007). Avaliação do desenvolvimento do setor agropecuário no Brasil: período de 1970 a 2000. *Economia e Sociedade, 16*, 111–124.

Ghisellini, P., Cialani, C., & Ulgiati, S. (2016). A review on circular economy: The expected transition to a balanced interplay of environmental and economic systems. *Journal of Cleaner Production, 114*, 11–32.

Giampietro, M., & Funtowicz, S. O. (2020). From elite folk science to the policy legend of the circular economy. *Environmental Sceince & Policy, 109*, 64–72. https://doi.org/10.1016/j.envsci.2020.04.012

Górecki, J., Núñez-Cacho, P., Corpas-Iglesias, F. A., & Molina, V. (2019). How to convince players in construction market? Strategies for effective implementation of circular economy in construction sector. *Cogent Engineering, 6*(1), 1690760. https://doi.org/10.1080/23311916.2019.1690760

GRID-Arendal. (2021). *Circular economy on the African continent: Perspectives and potential* (p. 48). GRID-Arendal.

Guerra, I. C. (2006). Pesquisa qualitativa e análise de conteúdo: sentidos e formas de uso. Estoril, Principia Editora

Haddad-Sisakht, A., & Ryan, S. M. (2018). Closed-loop supply chain network design with multiple transportation modes under stochastic demand and uncertain carbon tax. *International Journal of Production Economics, 195*, 118–131. https://doi.org/10.1016/j.ijpe.2017.09.009

Hobson, K. (2019). Small stories of closing loops': Social circularity and the everyday circular economy. *Climatic Change, 169*, 99–116. https://doi.org/10.1007/s10584-019-02480-z

Hofmann, F. (2019). Circular business models: Business approach as driver or obstructer of sustainability transitions? *Journal of Cleaner Production, 224*, 361–374. https://doi.org/10.1016/j.jclepro.2019.03.115

Holmes, H., Wieser, H., & Kasmire, J. (2021). Critical approaches to circular economy research: Time, space and evolution. In R. B. Swain & S. Sweet (Eds.), *Sustainable consumption and production, volume II: Circular economy and beyond* (pp. 55–74). Springer International Publishing. https://doi.org/10.1007/978-3-030-55285-5

INE – Instituto Nacional de Estatística. (2020). *Relatório da Pobreza para Angola*. Luanda.

Jeffries. (2023). ESD-DEI Report. Building on our sustainability strategy. Retrieve from: https://www.jefferies.com/CMSFiles/Jefferies.com/Files/Sustainability/Jefferies-2023-ESG-DEI-Report.pdf

Kamutali, A., Henriques, P. D. D. S., Lucas, M. R., & Marta-Costa, A. (2020). Agricultura familiar no huambo: Uma abordagem da sustentabilidade económica e social para superar a pobreza. In *Economía empresarial: aportaciones científicas para su desarrollo, libro de resumenes del XXII Seminario Hispano Luso de Economía Empresarial. 5 and 6 of November of 2020 Facultade de CC Empresariais e Turismo Campus de Ourense Universidade de Vigo* (p. 117). Universidade de Vigo.

Kopnina, H., & Blewitt, J. (2015). *Sustainable business: Key issues*. Routledge.

Korhonen, J., Honkasalo, A., & Seppälä, J. (2018). Circular economy: The concept and its limitations. *Ecological Economics, 143*, 37–46. https://doi.org/10.1016/j.ecolecon.2017.06.041

Lazarevic, D., & Valve, H. (2017). Narrating expectations for the circular economy: Towards a common and contested European transition. *Energy Research & Social Science, 31*, 60–69. https://doi.org/10.1016/j.erss.2017.05.006

Mangers, J., Minoufekr, M., Plapper, P., & Kolla, S. (2021). An innovative strategy allowing a holistic system change towards circular economy within supply-chains. *Energies, 14*(14), 4375. https://doi.org/10.3390/en14144375

Marta-Costa, A., Torres-Manso, F., & Tibério, L. (2012). Sustainability diagnosis of an agroforestry system. *Regional Science Inquiry Journal, IV, 2*, 111–124.

MEP – Ministério de Economia e Planeamento. (2021). *Projecto de Aceleração do investimento inclusivo e sustentável do agro-negócio nos Corredores de Crescimento Económico: Programa de fortalecimento de capacidades em sistemas agroalimentares e desenvolvimento de cadeias de valor*. Luanda.

MINAGRIF – Ministério da Agricultura e Florestas. (2020). *Projeto de Desenvolvimento da Agricultura Comercial*. Luanda.

Nyumba, T. O., Wilson, K., Derrick, C. J., & Mukherjee, N. (2018). The use of focus group discussion methodology: Insights from two decades of application in conservation. *Methods in Ecology and Evolution, 9*(1), 20–32.

Pacheco, F. (2003). *A Reforma da Terra em Angola: Força motriz ou obstáculo para o desenvolvimento democrático* (p. 28). Colóquio Internacional sobre Reforma Agrária na África Austral.

Pitkanen, K., Antikainen, R., Droste, N., Loiseau, E., Saikku, L., Aissani, L., Hansjürgens, B., Kuikman, P. J., Leskinen, P., & Thomsen, M. (2016). What can be learned from practical cases of green economy? Estudies from five European countries. *Journal of Cleaner Production, 139*(666), e 676. https://doi.org/10.1016/j.jclepro.2016.08.071

Plummer-D'Amato, P. (2008). Focus group methodology part 1: Considerations for design. *International Journal of Therapy and Rehabilitation, 15*(2), 69–73.

Rajapathirana, R. J., & Hui, Y. (2018). Relationship between innovation capability, innovation type, and firm performance. *Journal of Innovation & Knowledge, 3*(1), 44–55.

Rótolo, G. C., Vassilo, C., Rodriguez, A. A., Magnano, L., Milo Vaccaro, M., Civit, B. M., Covacevich, M. S., Arena, A. P., & Ulgiati, S. (2022). Perception and awareness of circular economy options within sectors related to agriculture in Argentina. *Journal of Cleaner Production, 373*, 133805. https://doi.org/10.1016/j.jclepro.2022.133805

Savolainen, V., Clottey, V. A., Doubi, B. T. S., Konan, J. L., Quain, M., Bezeng, B. S., Logah, V., Wireko-Kena, A., Osekre, E. A., Atuah, L., Angui, C. M. V., Ameka, G., Turkson, B., Boatemaa, A., Anankware, J. P., Boafo, H. A., Dwarko, D. A., & Collins, C. M. (2020). Systems thinking creates opportunities for a circular economy and sustainable palm agriculture in Africa. *Current Research in Environmental Sustainability, 1*, 31–34. https://doi.org/10.1016/j.crsust.2020.05.001

Sehen, S., Jabbour, C. J. C., Pereira, S. C. F., & Jabbour, A. B. L. S. (2019). Improving sustainable supply chains performance through operational excellence: Circular economy approach. *Resources, Conservation & Recycling, 149*, 236–248. https://doi.org/10.1016/j.resconrec.2019.05.021

Toop, T. A., Ward, S., Oldfield, T., Hull, M., Kirby, M. E., & Theodorou, M. K. (2017). AgroCycle–developing a circular economy in agriculture. *Energy Procedia, 123*, 76–80. https://doi.org/10.1016/j.egypro.2017.07.269

UNCTAD. (2022). *Trade and development report 2022* (Development prospects in a fractured world: Global disorder and regional respondes. UNCTAD/TDR/2022). United Nations. https://unctad.org/publication/trade-and-development-report-2022

Velasco-Muñoz, J. F., Aznar-Sánchez, J. A., López-Felices, B., & Román-Sánchez, I. M. (2022). Circular economy in agriculture. An analysis of the state of research based on the life cycle. *Sustainable Production and Consumption, 34*, 257–270. https://doi.org/10.1016/j.spc.2022.09.017

Velasco-Muñoz, J. F., Mendoza, J. M. F., Aznar-Sánchez, J. A., & Gallego-Schmid, A. (2021). Circular economy implementation in the agricultural sector: Definition, strategies and indicators. *Resources, Conservation and Recycling, 170*, 105618. https://doi.org/10.1016/j.resconrec.2021.105618

Velenturf, A. P., & Purnell, P. (2021). Principles for a sustainable circular economy. *Sustainable Production and Consumption, 27*, 1437–1457. https://doi.org/10.1016/j.spc.2021.02.018

Weick, C. W. (2001). Agribusiness technology in 2010: Directions and challenges. *Technology in Society, 23*(1), 59–72.

Yong, R. (2007). The circular economy in China. *Journal of Material Cycles and Waste Management, 9*, 121–129. https://doi.org/10.1007/s10163-007-0183-z

Reputational Leadership, Family Business and Sustainability in the Creation of Economic and Social Value. The Case of the Company "Agroparque Esperanza"

Juan Luis Tato Jiménez, Ramsés Cabrera Gala, and Maria Victoria Carrillo Durán

Abstract This chapter proposes a research study based on a case methodology examining the Agroparque Esperanza company, an example of entrepreneurship in the circular economy (CE), within the Mexican scope.

Currently, society demands companies to be involved in the creation of a fairer and more sustainable economy, and hence the rise of the so-called circular economy. Agroparque Esperanza is a clear example of sustainable entrepreneurship, both because it is a family business with a particular leadership style and because it is an agricultural company that focuses its activity on producing and processing apples under a sustainable context.

This is an innovative case of value creation due to the regional environment in which the business project is developed and the business model it follows. Its business model stands out because of its contribution to the creation of economic value and social value, thanks to promoting a family business, a particular leadership style under a sustainable context.

On the one hand, Agroparque Esperanza creates economic value thanks to its different business model as a circular Economy company. It is a family business where some members are in charge of the management of the company and with a particular entrepreneurship "gene" that helps to innovate in a sustainable context.

J. L. Tato Jiménez (✉)
Dpto. de Dirección de Empresas y Sociología, Universidad de Extremadura Avda, Badajoz, Spain
e-mail: jltato@unex.es

R. Cabrera Gala
Benemérita Universidad Autónoma de Puebla CRS, Puebla, Mexico
e-mail: ramses.cabrera@correo.buap.mx

M. V. Carrillo Durán
Dpto. De Información y Comunicación, Universidad de Extremadura Plazuela Ibn Marwan, Badajoz, Spain
e-mail: vicduran@unex.es

C. Rego et al. (eds.), *Entrepreneurship, Technological Change and Circular Economy for a Green Transition*, Studies on Entrepreneurship, Structural Change and Industrial Dynamics, https://doi.org/10.1007/978-3-031-48079-9_14

This situation leads to the company to use reputation leadership and reputation family as a strategic factor in the business model.

On the other hand, Agroparque Esperanza is a company that is supported by the creation of social value. Agroparque Esperanza provides solutions through local actions in a particular Mexican region. Thus, it solves some of the important challenges facing sustainable companies, and it provides a business model that is replicable and viable as a contribution to the United Nations 2030 Agenda on Sustainable Development.

Keywords Circular economy · Reputational leadership

1 Introduction

The challenges that societies face today show the need to evolve towards the circular economy (CE) where the 5 R's predominate—rethink, redesign, reduce, reuse, and recycle. Furthermore, CE is an economic model considered an option for the linear economy with relevant repercussions for a large set of recent industrial activities (Bîrgovan et al., 2022). In this same sense, CE provides a growing industrial focus on sustainability, which could be the global engine of green growth (Welfens et al., 2017) being an opportunity for entrepreneurship in some regional contexts particularly.

For this reason, it seems essential to understand and know more about how to proceed in this transition, paying attention to cultural, technological, and social change and knowing the difficulties and critical issues for the success of this transformation. To succeed in this change process, a new leadership style and management of the leader's reputation are required to guide toward new business models based on the CE and associated innovations.

Thus, the link between CE and companies is becoming closer and more frequent, in which different productive activities are also incorporated (see, e.g., D'Adamo et al., 2019; Esbeih et al., 2021; Becerra et al., 2020; Farhana et al., 2022; Anzules-Falcones et al., 2021), especially if family businesses are considered. Family businesses in Mexico result from shared values, history, and identity that provide a source of great pride for the family and strength for the family business. These values align with society and include relevant aspects such as environmental care and social economy. In this sense, for Kazancoglu et al. (2021), the ethical development of these types of firms promotes improvement in their social responsibility and environmental management systems in the environment of the new technological industry.

In addition, family businesses support their total or partial management in the leadership of one of its members; however, they owe it to their stakeholders; therefore, the leader's reputation and the family business play a strategic role. Hence, in this point, the role of the founder stands out, both in entrepreneurial development and when passing the baton to the next generation (Jamil et al., 2021). Among other important aspects, such as motivation or personality traits, leadership's role stands out since it is considered a very relevant quality for developing entrepreneurship and business sustainability (Purba & Tan, 2018). We must understand this leadership both in the company and in the family. The importance of this leadership is given by the fact that the failure of the family business begins with the generational change (Kirsipuu, 2012), turning the generational transition into a critical challenge for its survival.

If the survival of the family business is desired, entrepreneurial and business skills must be transferred to the family in the same way ownership of the business is transferred. The knowledge and learning derived from the involvement of family members in the business is the basis for creating family business value (Barros-Contreras et al., 2020). Families contribute to creating new ideas, processes, products, and services (Barroso et al., 2016) and innovation (Rondi et al., 2019) because of their potential capabilities to recognize, assimilate, and exploit the knowledge of each member.

For this reason, entrepreneurial families have been considered in relevant contexts to study entrepreneurial behaviors (Zehrer & Leiß, 2019). Entrepreneurship results from social interactions of the family, firm, and non-family managers, conditioned by family cultural patterns (Arz, 2019). We are interested in knowing the family values of a family business whose activities protect the environment, what entrepreneurial spirit they develop, and what role the leader plays in all this.

For Hall et al. (2004), the reputation of leaders influences the trust that stakeholders have in them. Additionally, for these authors, how stakeholders perceive the leader's actual behavior is moderated by motives and intentions that stakeholders attribute to the leader's behavior so that the attitudes, decisions, and actions surrounding the CEOs could influence his/her reputation, perceptions of his/her leadership style, and even corporate reputation.

Additionally, in the case of Mexico, CE and sustainability are in an initial phase; hence, it is necessary to design an action plan where the government, organizations, and civil society are committed to the conversion to these models (Córdova-Preciado et al., 2021). In other words, there is a need to deepen the scientific study of CE and its application in the organizational context in Mexico.

In this chapter is analyzed the case of the company Agroparque Esperanza, the leader in the Mexican market's central region and located in Esperanza Puebla. The economic activity of this family-owned company is based on a comprehensive portfolio of services that includes a collection of materials as in-plant collection, as well as dismantling of machinery and complete factories, destruction of obsolete and defective products, brokerage to foundries, steel mills and paper mills, and industrial bazaar offerings. This Mexican business started in the circular economy as a family company, and it is still owned and managed by two generations of the same family through innovation, entrepreneurship, and adaptability to survive through time.

Other similar case studies are the case of "La Hormiga Verde" (Tato-Jiménez et al., 2021) and "La Hormiga Verde as a case of social marketing to improve the quality of life through recycling waste electrical and electronic equipment (WEEE)"

(Carrillo-Durán et al., 2021) that follow a similar pattern of analysis. In the case of a Special Employment Center in the circular economy, the originality of the business model of this company as well as the efficiency and positive results in the circular economy and for all its stakeholders is evident.

We can also mention the work of Dressler (2023) who analyzes several cases of wine producing companies that follow sustainable policies and strategies due to changeable demanding dynamics of markets nowadays. In this paper, the analysis consists of two parts, one that explores the concept of dynamic capabilities in the context of social, environmental, and economic sustainability and another that compares business models in order to verify that the efforts made in sustainability should be based on strategic principles.

For its part, the work of Konovalov (2022) studies the entrepreneurial spirit of an Australian family that owns the company Natural Evolution. The work deals with issues of circular economy, eco-innovation, organizational change, and business collaboration, reaching the conclusion of the importance of the entrepreneurial spirit of the family. Finally, and in the same line and with similar results, one can refer to the work of Islas-Moreno et al. (2023). This is a multi-case study in the Jalisco state (México) regarding some business families of the agriculture sector. According to the authors and after the analysis of different categories: "strategies actions," "entrepreneurship stages," and "business family types," there is the necessity of cultivating both strategies and the entrepreneurship spirit in the family; this entrepreneurship spirit in the family leads the organization to balance the speed of the entrepreneurship and the products innovation, measure the risk, and accumulate knowledge and capabilities generation by generation.

According to this, the objectives for this work are:

To describe the business model of this company as an example of entrepreneurship and that operates in the CE as a family business and with a particular reputational leadership style.

To identify and explain the economic value of the company and how it gets it in a regional context and with a particular business model.

To determine the social value of a company that works in the CE in a particular context, and how it impacts entrepreneurship.

2 Preliminary Aspects: Agroparque Esperanza as a Mexican Company

Agroparque Esperanza is defined as a company whose economic activity includes the production, industrialization, and commercialization of agricultural products. This company is responsible for the transformation of agricultural products into processed products for food consumption. This transformation includes the processes of quality selection, classification (by size), packaging, and storage of agricultural production.

The company is a family business founded in 2006, belonging to the agri-food sector and located in the area of Serdán – Esperanza (Mexico). It has the support, as a partner, of the Philippine company AlasGroup with a clear sustainable character, which has contributed with agro-industrial developments to the production, industrialization, and commercialization of agricultural products, adding value to them.

Agroparque Esperanza is the largest agro-park in Mexico. It currently has a staff of more than 300 employees, both permanent and temporary, all of them from nearby towns, which classifies it as a large company in Mexico. It consists of 52 hectares of Gala variety table apples, 2.5 hectares of micro and macro raspberry tunnels, a fruit processing plant, and by-products, such as juices and spirits produced under the Nietos de Asturias and Zacatlán brand.

The company has a wide range of liquors, distillates, juices, waters, soft drinks, ciders, jams, and apple vinegar. In addition to the well-known apples, they also harvest raspberries, broccoli, etc. Finally, we must point out that they are constantly innovating, improving, and proposing new products aimed at different audiences and markets.

At the time of its foundation, more than 65 thousand trees, apple trees, were planted with the purpose of taking advantage of the type of soil and climatic conditions of the region, which indicates its purpose of becoming a company with a leadership vocation that, together with its sustainable and social character, makes it a reference for the creation of economic wealth for its region.

Agroparque Esperanza can be considered a company belonging to the so-called circular economy due to its waste and rainwater recovery processes, as well as its concern for the environment, sustainability, and the rich heritage and social wealth of its region. The company's success, according to its CEO, lies in its excellent value chain, which makes it unique and different from its competitors.

The vision of the company consists of: "To be a market-oriented agro-industrial business system (Agroparque), consolidated and recognized nationally and internationally for the creation of products and services, ensuring the satisfaction and well-being of our customers, suppliers and collaborators."

This vision is materialized in the following mission: "We are a company dedicated to the production, transportation and commercialization of innovative and superior quality products and services in the agri-food sector, minimizing the use of energy through the use of technologies and maximizing the creation of value, committed to environmental care and social responsibility."

Agroparque Esperanza is a company whose main activity (agricultural production) is seasonal and develops a concentric diversification strategy. Its activity is organized around three main lines of business:

- The agricultural division, which carries out the production process through four agroclusters (Agrocluster Esperanza, Agrocluster Tlachichuca, Agrocluster San Andrés, and Agrocluster Guadalupe Victoria).
- The agro-industrial division, which develops the industrialization of apples that are planted and harvested in the agricultural division for the manufacture of juices and beverages, whose maximum exponent is the Jugos y Bebidas Poblanos

SA de CV project supported by the purchase of an old factory for the traditional production of fermented beverages and cider called Manzanita Zacatlán in Zacatlán Puebla. The factory produces alcoholic and non-alcoholic beverages, as well as other gourmet products.
- The tourism division, which carries out different activities:
  - Guided visits to the apple fields and techno-ecological macro-tunnels to enjoy and participate in the apple blossoming, tour of the production area, barrels, distillation, apple packing, and milling, production, and bottling of beverages and liquors. These visits of approximately 2 h also include the museums of the apple, cider, railroad, pre-Hispanic cultures, and renewable energies, as well as more than 20 robotic dinosaurs with noises and movements. All these activities make it a great attraction for all audiences and help to familiarize them with its products.
  - A wind farm. Within this division, Agroparque Esperanza includes the creation and operation of a wind farm in its apple orchards that produces energy free of carbon footprint and collaborates with ecology and sustainability. In addition, they have built two rainwater catchment ponds for irrigation of the productive field.
  - A cider and railroad museum about the history of cider in Mexico and the Pomares (apple plantations) which is closely related to the founding of the city of Puebla.
  - Finally, they also have a number of stores throughout Mexico where you can purchase their products, such as ciders, liqueurs and creams, soft drinks, drinking teas, juices, bottled water, and preserves.

3 Methodology: The Case Method

3.1 Sample Size and Technique

This chapter is based on a qualitative case method. According to Yin (1994), case study research is an empirical form of research that is used to analyze a contemporary phenomenon within its real context, especially when the limits between the phenomenon (in this case, new business models) and the context being studied (family businesses) need to be exemplified and analyzed. Thus, a case study tries to review a reality in which there are many variables of interest beyond the directly observable data. In this case, and within the possible types of design for case studies according to their typology (Yin, 2009, p. 46), the "type of design for a single case" was chosen. The case is considered to be a single unit of analysis that is original and relevant enough to be representative of the context in which the work is being approached. Agroparque Esperanza is located in the Puebla State of Mexico, and it is a different company, because it has an agricultural business unit, an agro-industrial business unit, and a tourism business unit, in which it carries out activities related to the circular economy, such as reforestation, recycling, and reuse. It also has a wind farm, which is the first registered agro-park located in central Mexico, allowing it to generate clean electricity without leaving a footprint on nature and promoting the use of all natural resources. Finally, Agroparque Esperanza is a family business, and for these reasons, it is relevant as a case study in the new ideas and business models of the family business.

In addition, within case studies, Stake (2005) states that there are several types depending on their purposes. In our research, we opted for the "intrinsic type of case study," which must have its own specificities and a value in itself. In this type of research, the case is studied in order to describe its uniqueness and because it is itself of interest to exemplify a specific reality within the study context. These premises are fulfilled by Agroparque Esperanza because it is a company representative of the problem that it illustrates and presents solutions about wasting management, reforestation, and recycling. Moreover, the case itself is interesting to know examples within the family business in the Mexican context.

To develop the study, a qualitative research design was implemented in which the information collection methods were aimed at filling the gap that exists in social science research on these issues (Rubio-Mozos et al., 2019). In this sense, one researcher was responsible for collecting data, who observed, interviewed, and conducted the sessions with the top management in the company. The sample consisted of five managers and the CEO of Agroparque Esperanza, who participated in in-depth semi-structured interviews.

The focus was on explaining how the firm creates economic value, social value, and sustainability value based on the understanding of the economic keys that maximize the impact of the firm—in addition to determine the implications of leadership on the personal and corporate reputation. Complementarily, the social scope was explained, due to the relevance of the circular economy for Agroparque Esperanza. Finally, it analyzed the family scope, in which the role implications of the family in the economic, social, and sustainability scope were studied.

3.2 Participants

The interviewees were Mexican citizens with an age range of between 33 and 63 years. The mean and the standard deviation were 44.5 and 9.7 years, respectively. On the other hand, the interviewees were the CEO and five managers-supervisors (the top management positions) that cover the business units of the firm. The interviews were conducted entirely by an online forms program, and they were conducted in the month of March during the period from 03/13/2023 to 03/16/2023.

3.3 Data Collection

The data collection was after the approval of informed consent. It is important to stress that the research was conducted online and the answers were transcribed in a word processor. Following, to guarantee the quality of the research the interviewees had the right to leave the forms when they desired and for each interviewee a code was assigned, ranging from AGE1 to AGE5 for the top management and C1 for the CEO's case. The coding was performed by researchers and they employed citations to integrate the findings.

In this same sense, the semi-structured interview followed a pre-established question guide. The first guide was applied to the CEO and managers-supervisors nonetheless; additional questions were asked to obtain precise information from the interviewees (Ryen, 2013).

3.4 Measurement

The question guide used for CEO and managers-supervisors considered the following blocks:

The origin of the business idea and the difficulties encountered in creating the company.

The business model: Scope of the firm's market, description of its services, results and viability of the project, human resources policy, awareness, and applications of CE and approach as a socially responsible company.

The implications as a family business: The family role in business, entrepreneurship, and social economy.

The implications of leadership and reputation: The style in which the CEO leads the firm and the effects in the organizational behavior, his/her personal reputation, and corporate reputation.

Questions were designed and classified according to the existing literature: in the case of business model Iturrioz et al. (2022); Santos et al. (2021) and in the case of family business Ratten (2021); Bierl and Kammerlander (2019).

The questions that analyzed the implications of leadership and reputation were based on Hall et al. (2004), Zinko et al. (2016), and Cabrera-Gala et al. (2023). Finally, the documental review of secondary sources through the analysis of contents generated by the company was based on its website and files provided by the human resources department.

4 The Analysis Model

Starting from the study areas described in Table 1, the information obtained was carried out based on an analysis model prepared for this purpose (Fig. 1) defining key areas of the Agroparque Esperanza business model (Fig. 2). Following Rubio-Mozos et al. (2019), this analysis model addresses the role that Agroparque Esperanza has to play to advance toward the United Nations Agenda (2030) as a family business in the circular economy and in the context of entrepreneurship.

In overall terms, the analysis model is based on three fundamental aspects (business model's impact in regional economic, reputational leadership, and environmental care) to guarantee the business project, forming a unit among themselves despite being apparently different spheres. All these dimensions are important and depend on each other, not being separable at the strategic level of the company but rather at the analytical level to study the organization and its business.

The economic efficiency of the company depends on its impact as a family business following a particular reputational leadership style, since it can make its turnover grow. In turn, the creation of social value must be analyzed through the way in which the company carries out its activity environmental care in the circular economy area and in a particular region. Regarding the creation of economic value, the business activity must be efficient to have a positive effect in this area.

In this way, as another important consideration, the two dimensions (economic and social dimensions) will be interrelated in their activity and must be developed in three temporal action planes—the strategic plane (where the paths to follow are established and the long-term approach is delimited), the tactical plane (where the way to achieve the short- and medium-term goals is defined), and the operational or application plane (where the specific action policies are established).

Table 1 Areas of analysis and value creation

Areas of analysis	Value creation
Economic scope: Study the economic keys of the company that help to maximize the impact attending to the leadership model and the reputational scope of it	Creation of **economic value** through the way of defining and developing its activity within a different business model, strongly supported by a particular style of leadership in the context of a family business
Social scope: Describe the role that the company plays in promoting circular economy and regional development	Creation of **social value** because of being a circular economy business and the effects on personal-corporate reputation and regional value
Sustainability scope: Analyze the importance of the environment in the firm, and the role implications of the family and entrepreneurship in the economic and social scope	Creation of **sustainability value** by the role of the family members in the firm and also by the development of economic-social value

Source: The authors

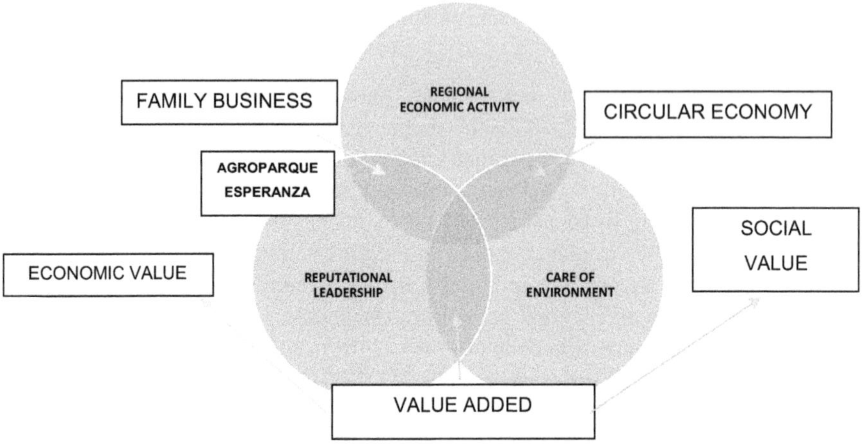

Fig. 1 Value areas of the business model. Source: Author's elaboration

Next, we shall define the areas of value creation and the aspects on which it is convenient to focus in each of these areas and on which each area depends, in order to obtain an X-ray of the company.

This model (Table 2) summarizes the key aspects and has been adapted by the authors based on the contribution by Osterwalder and Pigneur (2009).

After conducting the interviews and gathering the necessary documentation, we shall proceed to present the results obtained for the different areas of the Agroparque Esperanza business model established in Table 2.

5 Discussion and Results

In the following section, the interview data will be analyzed and discussed in relation to the aims of this research by examining the business model in a family business, in addition to analyzing the reputational leadership style and its implications and exploring the economic value and the social value added and their effects.

5.1 Economic Value Added

To carry out the study we started from the methodology of Osterwalder and Pigneur (2009) and adapted it to our case, describing key aspects of the company to create economic value. Therefore, we have divided the economic activity into the following sections: the family business and the entrepreneurial family, the leadership style, strategic partners, key resources and processes, customers and distribution, and value proposition.

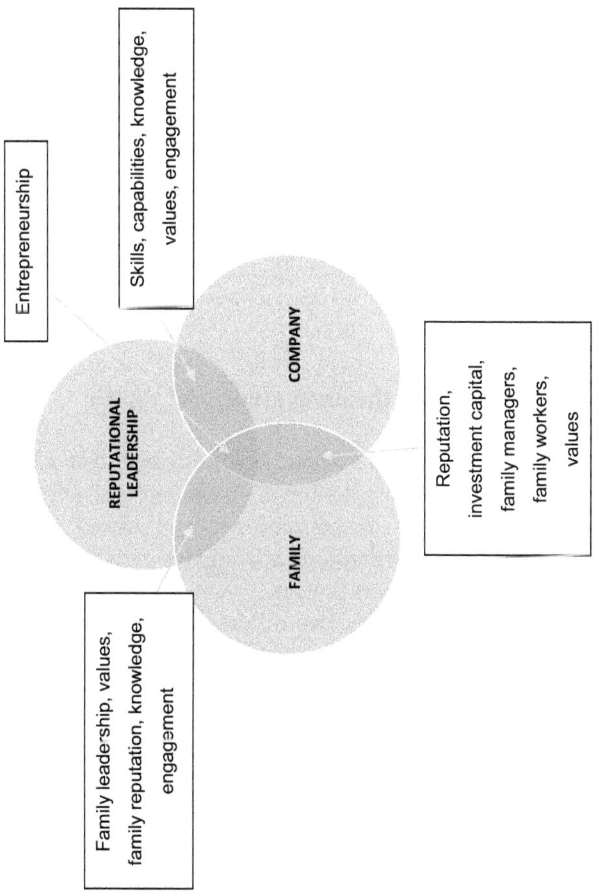

Fig. 2 Agroparque Esperanza business model. Source: Author's elaboration

Table 2 Analysis model and creation of new companies

Creation of economic value	Creation of social value
Economic activity of the business as a family business	Environmental care activity in a particular region
Analyze the key resources and activities that define the competitive capacity	Define the environmental value proposition that is developed through the activity carried out in a region
Assess the company's capacity for growth and the possibility of replicating the business model	Know how environmental care is done within the circular economy
Study the management of operations and logistics	Delimit the scope of the environmental need at geographical, quantitative, and qualitative levels
Know the leadership style and the role of family reputation	Explain how to collaborate in environmental care with the agents involved
Analyze the economic profitability (know how they obtain profitability)	Analyze alliances and associations with social agents and institutions and public or private companies, keys to the activity
Family values, innovation, and entrepreneurship	Assess the company and family capacity for the environment and economic innovation in new lines of entrepreneurship

Source: The authors based on and Pigneur (2015)

5.1.1 The Family Business and the Entrepreneurial Family

Agroparque Esperanza is a family business, firstly because that is how its CEO defines it, secondly because of its distribution of ownership, and, thirdly, because its organizational chart includes up to five family members, two of whom are in the management of the company. It is, therefore, a family business that is incorporating its second generation into the business project.

The family and the company, being two different organizations, in the case of Agroparque Esperanza and as it is usual in first-generation family businesses, falls on the same person, the CEO, who is responsible for the decisions of both family and business.

In this way, we can assure a total transfer between the values of family and business, with honesty and hard work, for the greater purpose of enhancing the legacy of the family and, likewise, the State of Puebla.

Other values they claim are commitment, innovation, teamwork, and care for the environment. They also point out responsibility, creativity, and solidarity as personal values. All of them make up a great cocktail capable of making any company competitive.

On the other hand, the capacity for entrepreneurship is intrinsic to them. This desire for entrepreneurship has led them to undertake new businesses such as the opening of the hotel "Nietos de Asturias," which makes clear mention of their roots and the growth of the business through the purchase of industrial buildings, and the creation of industrial parks, agricultural parks, agrotourism, etc.

5.1.2 Reputational Leadership Style and its Implications

According to Agroparque Esperanza CEO, the first implication about leadership focuses on the establishment of positive working relationships and serves as a platform for building personal relationships, which can promote the development of reputation, particularly, the social type. In this line, the most outstanding way of relating with the collaborators is in terms of labor issues. In this area, they achieved recognition and awards thanks to the work of everyone, in which all collaborators had to get involved and contribute ideas, improvements, and proposals.

Therefore, social interaction is very important because the firm considers family and teamwork as something relevant promoting gatherings or celebrations. The interaction occurs when new projects are carried out, and socially when religious festivities take place where all members participate.

The second implication was that leadership effectively applied to job performance develops a positive reputation, specifically with a focus on task reputation. Also, it can enhance corporate reputation, especially through social and environmental responsibility practices. A manner to unfold all the above mentioned is by the prizes, awards, and corporate recognition.

In this sense, the CEO tries to involve people in new topics and projects and allows them to acquire experience and new knowledge bringing new ideas, sharing information and knowledge so that the project is exactly what they are looking for, and giving feedback on the pending projects or advances in all the activities that are carried out in the company. The managers understand that the experience of the CEO has allowed Agroparque Esperanza to become a successful company, with projects that benefit everyone, in addition to being a socially responsible and ecologically sustainable organization.

The third implication found was that ethical consistency, teamwork, and co-responsibility in new challenges favor the development of a reputation for integrity.

Agroparque Esperanza seems to be integrated by a large group of companies and all collaborators support each other to solve difficult situations. In addition, the CEO undertakes new projects for the benefit of all and faces problems with the best attitude, giving solutions, thinking of the well-being of everyone, and always having an adequate and professional way of directing every project. The leadership style is supported in an ethical way to solve any situation of the company or problem that happens in the company.

The leadership style is based on collaborative work, which seeks to achieve challenging goals within the parameters of ethics, professionalism, social responsibility, and care for the environment. At the same time, it is sustained by having high levels of performance and establishing interpersonal relationships through labor relations. Considering the discussion above, the CEO develops a positive and growing reputational leadership, which in turn can favor the corporate reputation.

5.1.3 Strategic Partners

Due to the growth of the company, replacing imports of 200 million kilos of table apples per year[1], Agroparque Esperanza counts on key partners for its development, such as the AlasGroup Company, which has provided production and control techniques.

As for other key partners, we can mention other local apple suppliers that would lead them to increase their potential production area by 6700 hectares belonging to some 3500 local producers[2]. The relationship with them is based on reputation, achieved through years of experience and family reputation.

5.1.4 Key Resources and Processes

In terms of Agroparque Esperanza's key resources, we should point out that its productive infrastructure consists of fertile cropland with controlled irrigation, techno-ecological macro-tunnels for raspberry production, a fruit processing plant, and a plant for manufacturing metal structures and greenhouses.

The company has advanced equipment and technology to continue improving day by day with personalized advice and quality care and trained personnel in an excellent working environment, paying special attention to attracting and retaining highly qualified personnel.

5.1.5 Customers and Distribution

The main market for apple is the Chihuahua market and among the main customers are Walmart, Prissa, and Gran Bodega. The ways to reach them are face-to-face visits. They also carry out communication campaigns through promotions and surveys that provide them with qualitative and quantitative information.

The company has two brands of beverages: Nietos de Asturias and Manzanita Zacatlán; these products are sold in the region and neighboring states, since the products are mainly marketed in Zacatlán, Puebla, Orizaba, and Córdoba Veracruz and then distributed to other places.

5.2 Social-Sustainable Value Added

Agroparque Esperanza promotes sustainable social development through its commitment to regional development and the recovery of local traditions and culture. Its intention is clear with respect to its concern for its immediate surroundings. This clearly indicates its vocation to provide value to its region and, therefore, to Mexico generating employment, offering a dignified life to workers, and offering a producing municipality.

The company understands social value as taking advantage of all opportunities to add economic value to its environment in two lines: regional development and sustainability.

First, the social value is understood as regional development. On the one hand, they do this not only by using local suppliers, with their agricultural productive activity, indirectly benefitting more than 3500 producing families, but also when other resources are needed. In this line, local materials and labor have been used in the construction of its warehouses, as well as in the manufacture of the extensive line of products produced in its processing plant.

On the other hand, they have also contributed to improving and preserving the historical and cultural heritage of their area through the recovery of the tradition of artisanal cider production and the opening of the Cider and Railway Museum, committed to the rescue of the cider culture and the history of the apple.

Second, the social value is understood as the commitment with the environment and sustainability. On the one hand, this can be seen in its collaboration with the Instituto Tecnológico de Ciudad Serdán and the Universidad Hispana de Ciudad Serdán, with whom it works on technological exchange projects with teachers and internship students. On the other hand, it has a wind farm that generates clean electricity and promotes the use of all natural resources. In addition, its managers boast of their good agricultural production practices, with which all the ground apple returns to the field or soil via compost through leachates.

Finally, it should be noted that sustainability is not a management fad for this company; rather, they intend to continue with their good practices to be a sustainable company and even adopt new processes in this area. A good proof of this is that Agroparque Esperanza develops sustainability policies assisted by AlasGroup, making investments to consolidate a logistics platform to market products, minimize the use of energy, and maximize the creation of value in the production of healthy food, through investments in technology, increasing productivity and competitiveness.

6 Conclusions

To conclude, Agroparque Esperanza solves some of the important challenges facing sustainable companies, and it provides a business model that is replicable and viable as a contribution to the United Nations 2030 Agenda on Sustainable Development.

In accordance with Dressler (2023), it has been possible to verify the importance of combining product quality, innovation, leadership, and entrepreneurship.

In this sense, the company's value proposition, in the entrepreneurial context, is based on a clear commitment to product and service quality, traditions, care for the environment, and sustainability. The rootedness with the local environment also comes from the fact that it is a family business, which implies a double commitment, as a company and as a family with the local community. The leadership style also has clear traits of a family business that helps to differentiate the company and boosting its corporate reputation.

Regarding the economic value, it is based on quality of the product; all products comply with the regulations, which guarantees quality and needs of the local consumer, paying special attention to flavor, good presentation and high nutritional value, working on the quality of raw materials, continuous improvement with state-of-the-art technology, deliveries and commitment to distributors, and high commitment to customer service, as well as with the region and its traditions. Regarding the care for traditions, it can be seen in the production of cider, whose main objective is the historical and patrimonial recovery of the drink, using artisanal methods to rescue regional manners. In this line, and in the context of entrepreneurship, the company plans are focused on cultural tourism.

On the other hand, in the line of social and sustainable value, the company is particularly original in the way they are aligning regional development and sustainability, supporting the traditions and economy of the region. To this end, they are working on improving the standardized and child labor-free processes, and on acquiring more technical knowledge taking care of the environment in the CE implementing hygienic processes, pesticide-free planting, and harvesting systems.

In this sense, the care of their local suppliers also gives them a competitive advantage. This concern for their region constitutes a true alliance with their suppliers backed by their good reputation. This reputation is based on values that support economic and environmental sustainability and is intensified by the reputation of the family, as well as its entrepreneurial projects.

As regards sustainability, it should be said that in addition to being a value for this company, it is also a necessity since the markets are increasingly demanding in terms of sustainability, such as care for water, soil, energy, among others. These results are similar to those of the study by Tato et al. (2021).

Therefore, in line with Tato et al. (2021) and Carrillo et al. (2021), we can point out the need to use sustainability and circular economy strategies and policies to be competitive.

Finally, in line with the results of the work by Konovalov (2022) and Islas-Moreno et al. (2023), in our analysis we have verified the importance of the entrepreneurial capacity of a family based on strengthening the family entrepreneurial spirit.

Regarding the weaknesses of Agroparque Esperanza, first, it highlights the need to improve production processes to increase the productivity of its crops and incorporate new technologies. For this, it counts on AlasGroup as a trusted partner with a relationship based on confidence. Reputational leadership contributes to it. On the other hand, the family will also need new knowledge and skills in different aspects of the business: production, innovation, management, etc., which the new generations must learn. Likewise, the company must be able to attract new talent to maintain the pace of growth and innovation. At Agroparque Esperanza, sustainability and care for the environment must be an attractive value for this talent.

To conclude, this model can be replicated in different contexts for big agricultural companies in different areas and contexts due to its success based on these important and replicable premises:

First, taking advantage of being a family business, with a particular leadership style and values, second, aligning social and sustainable values with the regional development, and third, creating new entrepreneurial projects in new lines of business (e.g., tourism) clearly connected to social values, and always under the same premise: boosting the reputation through the commitment to a close leadership style as a family business, to product and service quality, to local customers, to workers, and to the regional context and its environmental and economic preservation.

Acknowledgments The authors acknowledge Agroparque Esperanza for its support.

Research funded by the Junta de Extremadura (Consejería de Economía, Ciencia y Agenda Digital) and by the European Union "European Regional Development Fund. A way of shaping Europe."

References

Anzules-Falcones, W., Diaz-Marquez, A. M., Padilla, L., Hernan-Hidalgo, D., & Sanchez-Grisales, D. (2021). Foresight for small and medium enterprises in the context of the circular economy. *Foresight and STI Governance, 15*(1), 86–96. https://doi.org/10.17323/2500-2597.2021.1.86.96

Arz, C. (2019). Bridging the micro-macro gap: A multi-layer culture framework for understanding entrepreneurial orientation in family firms. *Journal of Family Business Strategy, 10*(3), 100287. https://doi.org/10.1016/j.jfbs.2019.04.006

Barros-Contreras, I., Basco, R., Martín-Cruz, N., & Hernangómez, J. (2020). Strategic management in family business. The missing concept of the familiness learning mechanism. *Journal of Family Business Management, 12*(1), 67–89. https://doi.org/10.1108/jfbm-10-2019-0066

Barroso, A., Sanguino, R., & Banegil, T. (2016). An empirical study about knowledge transfer entrepreneurial orientation and performance in family firms European J. of. *International Management, 10*(5), 534. https://doi.org/10.1504/EJIM.2016.078790

Becerra, L., Carenzo, S., & Juarez, P. (2020). When circular economy meets inclusive development. Insights from urban recycling and rural water access in Argentina. *Sustainability, 12*(23), 9809. https://doi.org/10.3390/su12239809

Bierl, P., & Kammerlander, N. (2019). Family equity as a transgenerational mechanism for entrepreneurial families. *Journal of Family Business Management*. ahead-of-print. https://doi.org/10.1108/JFBM-09-2018-0043

Bîrgovan, A. L., Vatca, S. D., Bacali, L., Szilagyi, A., Lakatos, E. S., Cioca, L. I., & Ciobanu, G. (2022). Enabling the circular economy transition in organizations: A moderated mediation model. *International Journal of Environmental Research and Public Health, 19*(2), 677. https://doi.org/10.3390/ijerph19020677

Cabrera-Gala, R., Carrillo Durán, M., Sánchez Baltasar, L., & Olivera Pérez, E. (2023). La construcción de la reputación personal de los colaboradores en las organizaciones. AACINI – Revista Internacional De Ingeniería Industrial, (6). http://www3.fi.mdp.edu.ar/otec/revista/index.php/AACINI-RIII/article/view/58

Carrillo-Duran, M. V., Garcia-Garcia, M., & Tato-Jimenez, J. L. (2021). La Hormiga Verde as a case of social marketing to improve the quality of life through recycling waste electrical and electronic equipment (WEEE). In M. M. Galan-Ladero & R. G. Rivera (Eds.), *Applied social marketing and quality of life. Applying quality of life research*. Springer. https://doi.org/10.1007/978-3-030-83286-5_15

Córdova Preciado, M. L., Salgado Beltrán, L., & Bravo Díaz, B. (2021). Economía circular y su situación en México. *Indiciales, 1*(1), 25–37. https://doi.org/10.52906/ind.v1i1.7

D'Adamo, I., Falcone, P. M., Gastaldi, M., & Morone, P. (2019). A social analysis of the olive oil sector: The role of family business. *Resources, 8*(3), 151. https://doi.org/10.3390/resources8030151
Dressler, M. (2023). Sustainable business model design: A multi-case approach exploring generic strategies and dynamic capabilities on the example of German Wine Estates. *Sustainability., 15*, 3880. https://doi.org/10.3390/su15053880
Esbeih, K. N., Molina-Moreno, V., Núñez-Cacho, P., & Silva-Santos, B. (2021). Transition to the circular economy in the fashion industry: The case of the Inditex family business. *Sustainability, 13*(18), 10202. https://doi.org/10.3390/su131810202
Farhana, K., Kadirgama, K., Mahamude, A. S., & Mica, M. T. (2022). Energy consumption, environmental impact, and implementation of renewable energy resources in global textile industries: An overview towards circularity and sustainability. *Materials Circular Economy, 4*(1), 15. https://doi.org/10.1007/s42824-022-00059-1
Hall, A. T., Blass, F. R., Ferris, G. R., & Massengale, R. (2004). Leader reputation and accountability in organizations: Implications for dysfunctional leader behavior. *The Leadership Quarterly, 15*(4), 515–536. https://doi.org/10.1016/j.leaqua.2004.05.005
Islas-Moreno, A., Rodríguez, M., Santoyo, H., Aguilar-Gallegos, N., Martínez-González, E., & Morris, W. (2023). Understanding the process of agricultural entrepreneurship: Perspective from strategic movements and entrepreneurial families. *Journal of Agribusiness in Developing and Emerging Economies, 13*(2), 323–341. https://doi.org/10.1108/JADEE-08-2021-0202
Iturrioz-Landart, C., Aragón-Amonarriz, C., & Cabrera-Suárez, M. K. (2022). Family social capital as a driver to leverage challenged transgenerational entrepreneurship. *Journal of Family Business Management, 13*(3), 762–779. ahead-of-print No. ahead-of-print. https://doi.org/10.1108/JFBM-02-2022-0015
Jamil, M., Md Fadzil, A. F., Waqar, A., & Yaacoob, M. R. (2021). A systematic review of entrepreneurial family: Perspectives and theoretical. *Journal of Contemporary Issues in Business and Government, 27*(2), 4858–4871.
Kazancoglu, Y., Sezer, M. D., Ozkan-Ozen, Y. D., Mangla, S. K., & Kumar, A. (2021). Industry 4.0 impacts on responsible environmental and societal management in the family business. *Technological Forecasting and Social Change, 173*, 121108. https://doi.org/10.1016/j.techfore.2021.121108
Kirsipuu, M. (2012). Sustainability of rural family enterprises. *Discussions on Estonian Economic Policy: Theory and Practice of Economic Policy, 20*(1).
Konovalov, E. (2022). Agricultural innovation in a remote setting. Natural evolution company in far North Queensland. In Q. Alam & R. Grose (Eds.), *Regional businesses in a changing global economy. The Australian experience* (1st ed.). Routledge. ISBN 9781032188720.
Osterwalder, A., & Pigneur, Y. (2009). *Business Model Generation. A handbook for visionaries, games changers, and challengers*. John Wiley & Sons Limited.
Purba, J. T., & Tan, J. D. (2018). Owners 'integrity, customers' relations, and focused attitude as strategies for family business sustainability and growth: An empirical study. *Journal of Economics, 21*, 187–198.
Ratten, V. (2021). Family business and sport entrepreneurship. *Journal of Family Business Management, 11*(3), 300–312. https://doi.org/10.1108/JFBM-05-2020-0041
Rondi, E., De Massis, A., & Kotlar, J. (2019). Unlocking innovation potential: A typology of family business innovation postures and the critical role of the family system. *Journal of Family Business Strategy, 10*(4), 100236.
Rubio-Mozos, E., García-Muiña, F. E., & Fuentes-Moraleda, L. (2019). Rethinking 21st-century businesses: An approach to fourth sector SMEs in their transition to a sustainable model committed to SDGs. *Sustainability, 11*(20), 5569.
Ryen, A. (2013). The qualitative interview: A privileged way to knowledge?. ISSBD bulletin. *Supplement to International Journal of Behavioral Development, 18.*

Santos, I. L. d., Vieira Marinho, S., & Santos, R. C. d. (2021). The systematization knowledge of articles adherent to familiness, entrepreneurial orientation and absorptive capacity. *Journal of Family Business Management, 11*(4), 512–537. https://doi.org/10.1108/JFBM-04-2020-0037

Stake, R. E. (2005). *Investigación con estudio de casos*. Morata.

Tato-Jiménez, J.-L., Carrillo-Durán, M.-V., & Garcia Garcia, M. (2021). The case of "La Hormiga Verde": Recycling electronic waste (e-waste) as a paradigmatic example of a new entrepreneurial trend in the fourth sector. In *Entrepreneurship in the fourth sector: Entrepreneurial ecosystems and sustainable business models* (pp. 193–218). Springer International Publishing. https://doi.org/10.1007/978-3-030-68390-0_10

United Nations. Transforming our world: The 2030 agenda for sustainable development, *United Nations*, United Nations (2015).

Welfens, P., Bleischwitz, R., & Geng, Y. (2017). Resource efficiency, circular economy and sustainability dynamics in China and OECD countries. *International Economics and Economic Policy, 14*(3), 377–382. https://doi.org/10.1007/s10368-017-0388-0

Yin, R. (1994). *Case Study Research: Design and Methods*. SAGE Publications.

Yin, R. K. (2009). *Case Study Reseach*. Sage.

Zehrer, A., & Leiß, G. (2019). Family entrepreneurial resilience – An intergenerational learning approach. Journal of Family Business Management, *ahead-of-print*(ahead-of-print). https://doi.org/10.1108/jfbm-09-2018-0037.

Zinko, R., Gentry, W., & Laird, M. (2016). A development of the dimensions of personal reputation in organizations. *International Journal of Organizational Analysis, 24*(4), 634–649. https://doi.org/10.1108/ijoa-04-2015-0854

GPSR Compliance

The European Union's (EU) General Product Safety Regulation (GPSR) is a set of rules that requires consumer products to be safe and our obligations to ensure this.

If you have any concerns about our products, you can contact us on

ProductSafety@springernature.com

In case Publisher is established outside the EU, the EU authorized representative is:

Springer Nature Customer Service Center GmbH
Europaplatz 3
69115 Heidelberg, Germany

www.ingramcontent.com/pod-product-compliance
Ingram Content Group UK Ltd.
Pitfield, Milton Keynes, MK11 3LW, UK
UKHW020240290926
15470UKWH00002B/650
* 9 7 8 3 0 3 1 4 8 0 7 8 2 *